T0199848

Methods of Non-α-Amino Acid Synthesis

Second Edition

Methods of Non-α-Amino Acid Synthesis

Second Edition

Michael Bryant Smith

University of Connectitcut

Storrs, USA

CRC Press
Taylor & Francis Group
Boca Raton London New York

CRC Press is an imprint of the
Taylor & Francis Group, an **informa** business

CRC Press
Taylor & Francis Group
6000 Broken Sound Parkway NW, Suite 300
Boca Raton, FL 33487-2742

First issued in paperback 2019

ISBN-13: 978-1-4665-7789-3 (hbk)
ISBN-13: 978-0-367-37934-6 (pbk)

Visit the Taylor & Francis Web site at
http://www.taylorandfrancis.com

and the CRC Press Web site at
http://www.crcpress.com

Contents

Preface to the Second Edition

The first edition of this book was published just under 20 years ago. During this time, many new publications have appeared that expand the synthetic work devoted to the preparation of non-α-amino acids. This new edition continues the work of the first edition to give a representative overview of the synthesis of non-α-amino acids beginning with the year 1962. New literature references expand this work to include literature after 1995 through 2012. As with the first edition, this edition will focus on acyclic amino acids of C_3–C_{10}, but also aminoalkanoic carboxylic acids, aminoalkenoic acids, and aminoalkynoic acids. The syntheses of amino carboxylic acids attached to or incorporated in rings of 3 to 10 carbons are also presented, including amino cycloalkanoic and amino cycloalkenoic acids. Saturated heterocyclic derivatives and aryl-substituted amino acids are discussed once again, but aromatic amino carboxylic acids and heteroaromatic amino carboxylic acids are not discussed, except where they are synthetic precursors to or related to aliphatic amino carboxylic acids. The first five chapters not only illustrate different synthetic strategies to prepare non-α-amino acids, but also serve as a useful review of a variety of synthetic methodology.

There are significant differences in this new edition. It has been extensively rewritten and reorganized. Apart from many updated references, there is a greater emphasis on the biological importance of non-α-amino acids, and a serious attempt to limit redundancies. The first edition was more or less a litany of synthetic methods, but there were few discussions that concerned why non-α-amino acids are important. Throughout the new edition there are many brief statements to illustrate the biological activity of important non-α-amino acids. Further, Chapter 6 in the new edition is greatly expanded to include several specific classes of non-α-amino acids with a focus on the biological importance as well as syntheses. The goal is to push this new edition away from a simple recitation of syntheses and methods toward a comprehensive discussion of both synthesis and the biological importance of this class of compounds.

As with the first edition, synthetic approaches to saturated amino acids of all types are discussed, as well as unsaturated, alkyl, and aryl-substituted derivatives, and amino acids that bear a heteroatom functional group. In general, four structural types of alkenyl amino acids are considered. In the first type, the double bond can be conjugated to the carbonyl group and the amine moiety can be attached directly to the double bond. Second, the amino group can be attached to the saturated carbon chain. The third and fourth types focus on the double bond, which can be in or out of conjugation with the carboxyl, and the amine group can be attached anywhere on the chain. Substituents can appear anywhere on the chain. For alkynyl amino acids, both conjugated and unconjugated are discussed.

As in the first edition, this new monograph is divided into seven chapters. However, the organization and content of these seven chapters are significantly different. The first chapter discusses synthetic methods that rely on substituent refunctionalization.

Chapter 2 discusses the conversion of cyclic precursors to acyclic amino acids. Chapter 3 is a relatively short chapter that places the focus on conjugate addition reactions. Chapter 4 discusses enolate anion reactions and condensation reactions that lead to non-α-amino acids. The stereoselectivity of many reactions is discussed in the Chapters 1–4. Chapter 5 discusses many of the synthetic methods presented in the first four chapters, but puts a major emphasis on reactions that proceed with good stereoselectivity. Those reactions and strategies that lead to good diastereoselectivity and enantioselectivity during synthesis are discussed. Chapter 6 discusses biologically important amino acids. There is an entirely new section on peptides and proteins that contain non-α-amino acids. Several biologically important amino acids are discussed in separate sections, including GABA, GABOB, carnitine, DAVA, statine, and others. Chapter 7 concludes the book and discusses amino acids for which the amino group is attached to the ring, and those in which the nitrogen is part of the ring (heterocyclic derivatives) are discussed.

It is hoped that this new edition will be particularly useful to those engaged in synthesis and synthetic manipulations involving amino acids. Its utility is not limited to that audience, however. Those engaged in a general study of amino acids and molecules containing amino acids, those involved in developing synthetic methods, and those pursuing the total synthesis of natural products should find the book useful because it contains a helpful review of strategies and reaction types found in nitrogen-containing molecules. It is my hope that students will find this collection useful and stimulating for new areas of research in this important area.

Where there are errors, I take complete responsibility. Every effort has been made to keep this book error-free, but if errors are found, please contact me and I will try to answer any questions that may be raised and correct errors if they exist.

This book was prepared on a MacBook Pro using Microsoft Word 2004. All drawings in this book were prepared using ChemBioDraw Ultra, v. 11.0.1 (350440), kindly provided as a gift from CambridgeSoft, located in Cambridge, Massachusetts.

I thank Lance Wobus and Fiona MacDonald of CRC Press, Taylor & Francis Group, who initially floated the idea of a second edition for this book. Their interest in this work made the new edition possible. I thank my editor, Hilary Rowe, and also David Fausel and Judith Simon of CRC Press, who were instrumental in making the new edition become a reality. Most of all, I thank my wife, Sarah, and my son, Steven, for their love and support.

Michael B. Smith
Professor of Chemistry
Storrs, Connecticut
April 2013
michael.smith@uconn.edu

Preface to the First Edition

Amino carboxylic acids (or just amino acids) are, as a class of organic molecules, among the most important and useful compounds known. The chemistry of amino acids has been studied for well over a century. The importance of α-amino acids [$RCH(NH_2)COOH$] in mammalian biology and as synthetic intermediates is well established and there have been many reviews, several monographs, and thousands of individual research papers in this area. The synthesis of α-amino acids has also been the subject of many books and reviews. In addition to α-amino acids, there are many other amino carboxylic acids where the amino group is not on the carbon immediately adjacent to the carboxyl group (the α position) but rather is attached to another carbon in the chain (e.g., the β, γ, δ carbon, etc.). These *non-α-amino acids* are also components of biologically important molecules, are important in the pharmaceutical industry, and are useful in materials science, particularly for polymer synthesis.

This monograph gives a representative overview of the synthesis of non-α-amino acids beginning with the year 1962. With a few exceptions, citations prior to that date are leading references taken from other citations. The work prior to 1962 is well represented, however, in the synthetic approaches incorporated into this book since the "early work" is the basis of most of those approaches. Every attempt was made to give the most common and useful methods for the synthesis of non-α-amino acids. This monograph will focus attention on acyclic amino acids of C_3–C_{10} and includes aminoalkanoic carboxylic acids, aminoalkenoic acids, and aminoalkynoic acids. The synthesis of amino carboxylic acids attached to or incorporated in rings of three to ten carbons is also presented, including amino cycloalkanoic and amino cycloalkenoic acids. Although saturated heterocyclic derivatives and aryl substituted amino acids are discussed, aromatic amino carboxylic acids and heteroaromatic amino carboxylic acids are not discussed, except where they are synthetic precursors to or related to aliphatic amino carboxylic acids.

The primary goal of this book is to summarize synthetic approaches to non-α-amino acids, particularly those amino acids that are key synthetic intermediates or important compounds in their own right. Achiral amino carboxylic acids are discussed throughout and special attention is focused on both chiral nonracemic and chiral racemic amino acids in chapter five, emphasizing the diastereoselectivity and/or enantioselectivity of each synthetic process. Having such synthetic information collected in one place will, it is hoped, facilitate current research and stimulate new research in this important area.

Saturated amino acids of all types are discussed, as well as unsaturated, alkyl and aryl substituted, and amino acids that bear a heteroatom functional group. In general, four structural types of alkenyl amino acids are considered. In the first type, the double bond can be conjugated to the carbonyl group and the amine moiety can be attached directly to the double bond. Secondly, the amino group can be attached to the saturated carbon chain. The third and fourth types focus on the double bond, which can be in or out of conjugation with the carboxyl, and the amine group can be attached anywhere on the chain. Substituents can appear anywhere on the chain. Less formalism is required for discussing structural variations in alkynyl amino acids. The triple bond is both conjugated and unconjugated and the amino group can appear at almost any point in the carbon. Only one derivative has the amino attached directly to the triple bond, however, and it is 3-amino 2-propynoic acid.

Non-α-amino acids have been important for over a century and their synthesis and chemical transformations constitute a rich and varied chapter in organic chemistry. Despite this, virtually no compilations exist that discuss these compounds. This stands in stark contrast to α-amino acids, for which thousands of articles have appeared, as well as numerous books, monographs, and reviews. Since part of my own research involves the used of non-α-amino acids in the development of synthetic methods, I was particularly interested in this type of reference book. This was the impetus for preparing this monograph, which is a compilation of synthetic methods that can be used to prepare non-α-amino acids.

This monograph is a comprehensive review of synthetic methods used to prepare non-α-amino acids. The source material surveyed in this book formally begins with literature after 1962, but there are many citations and examples taken from literature published prior to 1962. Virtually all of these early references were taken from literature citations in later articles. I am confident that the "older" literature is well represented, particularly in its survey of the synthesis of and the synthetic methods applied to non-α-amino acids.

This monograph is divided into seven chapters. In the first chapter, methods for synthesizing the fundamental structural types found in acyclic amino acids are presented. In this chapter, the synthetic methods rely on substituent refunctionalization. In chapter two, refunctionalization again plays a prominent role. The major difference is that cyclic precursors are used to prepare acyclic amino acids. Chapter three discusses those methods that rely on conjugate addition reactions. Chapter four brings together those synthetic strategies that rely on condensation reactions involving a nucleophilic species with a carbonyl compound, or a carbonyl surrogate. This technique is particularly useful for incorporating heteroatom substituents such as hydroxyl, oxo, or amino into an amino acid. Chapter five collects a variety of synthetic methods together under the umbrella of stereoselectivity. Those reactions and strategies that lead to good diastereoselectivity and/or enantioselectivity during the synthesis are discussed. Chapter six again collects several different strategic approaches for the synthesis of amino acids. The unifying theme of this chapter is the preparation of biologically important amino acids, with particular emphasis given to GABOB and carnitine as well as statine and its derivatives. Chapter seven concludes the book and discusses amino acids that have a carboxyl group attached to a ring. Those derivatives for which the amino group is attached to the ring and those in which the nitrogen is part of the ring (heterocyclic derivatives) are discussed.

This monograph will be particularly useful to those of us engaged in synthesis and synthetic manipulations involving amino acids. Its utility is not limited to that audience, however. Those involved in developing synthetic methods and those pursuing the total synthesis of natural products will also find the book useful because it contains a useful review of strategies and reaction types found in nitrogen-containing molecules. This monograph will be a useful reference to quickly find what synthetic methods have been used, and in what systems. With such a background, it is my hope that new areas of research in this important area will be discovered more quickly. I believe this work will be a valuable contribution to the ever-growing field of non-α-amino acids.

Where there are errors, I take complete responsibility. Every effort has been made to keep the manuscript error-free but if errors are found please contact me and I will try to answer any questions that may be raised.

This manuscript was prepared with a Macintosh IIci computer using Microsoft Word, version 5.1. All drawings in this book were prepared using CSC ChemDraw Plus, from Cambridge Scientific Computing, Inc., Cambridge, MA.

I want to thank Dr. Maurits Dekker, who made an intriguing suggestion several years ago concerning one of my papers. This suggestion for a book led to the preparation of this manuscript and I thank Dr. Dekker for inspiring this work. I want to thank Mr. Joe Stubenrauch and Ms. Anita Lekhwani of Marcel Dekker Inc., for editing this manuscript and providing the care and knowledge required to convert it into a book. I particularly want to thank the independent reviewers of this manuscript, with whom I must share credit for the final form of this monograph. Their reviews were insightful, very helpful, and I thank them very much. Most of all I want to thank my wife Sarah and my son Steven for their patience and understanding during the research for, and preparation of this manuscript.

Michael B. Smih

Professor of Chemistry
Storrs, Connecticut
February 6, 1995

Common Abbreviations

Other, less common abbreviations are given in the text when the term is used.

9-BBN	9-Borabicyclo[3.3.1]nonane	
Ac	Acetyl	–COMe
acac ligand	Acetylacetonate	
AIBN	*azo-bis*-Isobutyronitrile	
aq.	Aqueous	
ax	Axial	
BINAP	2*R*,3*S*-2,2'-*bis*-(Diphenylphosphino)-1,1'-binapthyl	
BINOL	1,1'-Bi(2-binaphthol)	
BMS	Borane methyl sulfide	
Bn	Benzyl	$-CH_2Ph$
Boc	*tert*-Butoxycarbonyl	$-CO_2t$-Bu
Bpy (Bipy)	2,2'-Bipyridyl	
Bs	Brosylate	*O*-(4-Bromophenyl)sulfonate
BSA	Bis(trimethylsilyl)acetamide	
Bu	*n*-Butyl	$-CH_2CH_2CH_2CH_3$
Bz	Benzoyl	
°C	Temperature in degrees Celsius	
CAN	Ceric ammonium nitrate	$(NH_4)_2Ce(NO_3)_6$
cat	Catalytic	
Cbz	Carbobenzyloxy	$-CO_2CH_2Ph$
Chap	Chapter(s)	
Chirald	(2*S*,3*R*)-(+)-4-Dimethylamino-1,2-diphenyl-3-Methylbutan-2-ol	
CIP	Cahn-Ingold-Prelog	
cod ligand	1,5-Cyclooctadienyl	
cot ligand	1,3,5-Cyclooctatrienyl	
Cp	Cyclopentadienyl	
Cy	Cyclohexyl	c-C_6H_{11}
%de	%diastereomeric excess	
DABCO	1,4-Diazabicyclo[2.2.2]octane	
dba ligand	Dibenzylidene acetone	
DBN	1,5-Diazabicyclo[4.3.0]non-5-ene	
DBU	1,8-Diazabicyclo[5.4.0]undec-7-ene	
DCC	1,3-Dicyclohexylcarbodiimide	c-C_6H_{11}–N=C=N–c-C_6H_{11}
DDQ	2,3-Dichloro-5,6-dicyano-1,4-benzoquinone	
DEA	Diethylamine	$HN(CH_2CH_3)_2$
DEAD	Diethylazodicarboxylate	$EtO_2C-N=NCO_2Et$
Dibal-H	Diisobutylaluminum hydride	$(Me_2CHCH_2)_2AlH$
DMAP	4-Dimethylaminopyridine	
DME	Dimethoxyethane	$MeOCH_2CH_2OMe$

DMF	*N,N'*-Dimethylformamide	
DMS	Dimethyl sulfide	
DMSO	Dimethyl sulfoxide	
dppb	1,4-Diphenylphosphinobutane	$Ph_2P(CH_2)_4PPh_2$
dppe	1,2-Diphenylphosphinoethane	$Ph_2PCH_2CH_2PPh_2$
dppf	*bis*-(Diphenylphosphino)ferrocene	
dppp	1,3-Diphenylphosphinopropane	$Ph_2P(CH_2)_3PPh_2$
%ee	% Enantiomeric excess	
e^-	Transfer of electrons	
Et	Ethyl	$-CH_2CH_3$
EDTA	Ethylenediaminetetraacetic acid	
Equiv	Equivalent(s)	
ESR	Electron spin resonance spectroscopy	
FMO	Frontier molecular orbital	
FVP	Flash vacuum pyrolysis	
GC	Gas chromatography	
gl	Glacial	
h	Hour (hours)	
$h\nu$	Irradiation with light	
HMPA	Hexamethylphosphoramide	$(Me_2N)_3P = O$
^{1}H NMR	Proton nuclear magnetic resonance spectroscopy	
HOMO	Highest occupied molecular orbital	
HPLC	High-performance liquid chromatography	
HSAB	Hard/soft acid/base	
i-Pr	Isopropyl	$-CH(Me)_2$
IR	Infrared spectroscopy	
LICA (LIPCA)	Lithium *N*-isopropyl-*N*-cyclohexylamide	
LDA	Lithium diisopropylamide	$LiN(i\text{-}Pr)_2$
LHMDS	Lithium hexamethyldisilazide	$LiN(SiMe_3)_2$
LTMP	Lithium 2,2,6,6-tetramethylpiperidide	
LUMO	Lowest unoccupied molecular orbital	
mcpba	*meta*-Chloroperoxybenzoic acid	
Me	Methyl	$-CH_3$
MEM	2-Methoxyethoxymethyl	$MeOCH_2CH_2OCH_2-$
Mes	Mesityl	2,4,6-tri-Me-C_6H_2
min	Minutes	
MOM	Methoxymethyl	$MeOCH_2-$
Ms	Methanesulfonyl	$MeSO_2-$
MS	Molecular sieves (3 or 4 Å)	
NBS	*N*-Bromosuccinimide	
NCS	*N*-Chlorosuccinimide	
Ni(R)	Raney nickel	
NIS	*N*-Iodosuccinimide	
NMO	*N*-Methylmorpholine *N*-oxide	
Nu (Nuc)	Nucleophile	
OBs	Brosylate = *O*-(4-bromophenyl)sulfonate	

Oxone®	$2\ KHSO_5 \bullet KHSO_4 \bullet K_2SO_4$	
●	Polymeric backbone	
PCC	Pyridinium chlorochromate	
PDC	Pyridinium dichromate	
PEG	Polyethylene glycol	
Ph	Phenyl	$C_6H_5^-$
PhH	Benzene	
PhMe	Toluene	
PPA	Polyphosphoric acid	
Pr	*n*-Propyl	$-CH_2CH_2CH_3$
Py	Pyridine	C_6H_5N
quant	Quantitative yield	
Red-Al	$[(MeOCH_2CH_2O)_2AlH_2]Na$	
rt	Room temperature	
s	seconds	
sBuLi	*sec*-Butyllithium	$CH_3CH_2CH(Li)CH_3$
Sec.	section(s)	
SET	Single electron transfer	
$(Sia)_2BH$	Disiamylborane (siamyl is *sec*-isoamyl)	
SOMO	Singly occupied molecular orbital	
TBAF	Tetrabutylammonium fluoride	$n\text{-}Bu_4N^+\ F^-$
TBDMS	*tert*-Butyldimethylsilyl	$t\text{-}BuMe_2Si$
TBDPS	*tert*-Butyldiphenylsilyl	$t\text{-}BuPh_2Si$
TBHP (*t*-BuOOH)	*t*-Butylhydroperoxide	Me_3COOH
t-Bu	tert-Butyl	$-CMe_3$
TEBA	Triethylbenzylammonium	$Bn(Et_3)_3N^+$
TEMPO	2,2,6,6-Tetramethylpiperidine-1-oxyl free radical	
TFA	Trifluoroacetic acid	CF_3COOH
tfa ligand	Trifluoroacetyl as a ligand	CF_3COO-
Tf (OTf)	Triflate	$-SO_2CF_3\ (-OSO_2CF_3)$
$ThexBH_2$	Thexylborane (*tert*-hexylborane)	
THF	Tetrahydrofuran	
THP	Tetrahydropyran	
TMEDA	Tetramethylethylenediamine	$Me_2NCH_2CH_2NMe_2$
TMS	Trimethylsilyl	$-Si(CH_3)_3$
TMS	Tetramethylsilane	$-Si(CH_3)_4$
Tol	Tolyl	$4\text{-}(Me)C_6H_4$
TOSMIC	Toluenesulfonylmethyl	
TPAP	Tetrapropylammoniumperruthenate	
TPP and tpp ligand	Triphenylphosphine	$P(PPh_3)_3$
Tr	Trityl	$-CPh_3$
Ts(Tos)	Tosyl = *p*-toluenesulfonyl	$4\text{-}(Me)C_6H_4SO_2$
UV	Ultraviolet spectroscopy	

Introduction

The chemistry of amino carboxylic acids, or just amino acids, has been studied for well over a century. The importance of α-amino acids [$RCH(NH_2)COOH$] in mammalian biology and as synthetic intermediates is well established, and there have been many reviews, several monographs, and thousands of individual research papers in this area. Indeed, amino acids are among the most important and useful compounds known, particularly α-amino acids. However, there are many other amino carboxylic acids with a structure that has the amino group on a carbon that is not immediately adjacent to the carboxyl group (the α position), but rather attached to another carbon in the chain (the β, γ, δ carbon, etc.). These *non-α-amino acids* are components of biologically important molecules, are important in the pharmaceutical industry, and are useful starting materials for many areas of organic chemistry.

This book focuses attention primarily on acyclic non-α-amino acids of C_3–C_{10} and will include aminoalkanoic carboxylic acids, aminoalkenoic acids, and aminoalkynoic acids. The synthesis of amino carboxylic acids attached to or incorporated in rings of 3 to 10 carbons is also presented, including amino cycloalkanoic and amino cycloalkenoic acids. Although saturated heterocyclic derivatives and aryl-substituted amino acids are discussed in a limited manner, aromatic amino carboxylic acids and heteroaromatic amino carboxylic acids are not discussed, except where they are synthetic precursors to or related to aliphatic amino carboxylic acids.

The goal of this book is to illustrate the synthetic approaches to and the importance of non-α-amino acids, particularly those amino acids that are key synthetic intermediates or important compounds in their own right. Achiral as well as chiral amino carboxylic acids will be discussed throughout, and special attention is focused on both chiral nonracemic and chiral amino acids (see Chapter 5), emphasizing the diastereoselectivity and enantioselectivity of synthetic processes. Having such synthetic information collected in one place will hopefully facilitate further research and stimulate new research in this important area.

$H_2N(CH_2)_n\,CO_2H$

1

2

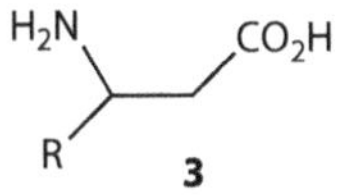

3

Glycine (H_2N-CH_2-COOH) is arguably the prototype α-amino acid, and the formal name is 2-aminoethanoic acid. An ω-aminoalkanoic acid is defined as an amino acid in which the amino group is on the carbon farthest removed from (distal to) the carboxyl group (see ***1***). 3-Aminopropanoic acid (***2***) is the simplest ω-amino carboxylic acid, and it has the common name of β-alanine (β-Ala). There are many non-α-amino acids that have the amine moiety at positions other than the α-carbon proximal to the carboxyl group. Derivatives of ***3*** that have alkyl substituents at C3 and also an amino group at C3 (see ***3***) are referred to as β-amino acids.* As with ***2*** and ***3***, it is often useful to view an ω-amino acid as the parent structure for non-α-amino acids, particularly functionalized and substituted amino acids.

Apart from ***2***, important amino acids include ***1*** (n = 3–9), and they appear quite often in synthesis and in biologically important systems. These amino acids ***1*** include 4-aminobutanoic acid (n = 3), also known as γ-aminobutyric acid and given the abbreviation GABA (γ-Abu).† 5-Aminopentanoic acid (n = 4) is known as δ-aminovaleric acid and given the abbreviation DAVA, but the three-letter code is 5-Ava. 6-Aminohexanoic acid (n = 5) is known as Amicar, ε-aminocaproic acid, and has the three-letter code ε-Ahx. 7-Aminoheptanoic acid (n = 5) is known as 7-aminoenanthic acid, ζ-aminoenanthic acid, or ω-aminoenanthic acid. 8-Aminooctanoic acid (n = 7), 9-aminononanoic acid (n = 8), and 10-aminodecanoic acid (n = 9), which is also known as aminocapric acid, complete the straight-chain non-α-amino carboxylic acids in this series.

There are several synthetic routes to these structurally simple yet important classes of compounds, including functional group transformations that generate either the amino group or the carboxyl group. In some cases, both groups are incorporated in a single synthetic step. Biological aspects of these amino acids will be presented when available, but in the context of the specific compound that is synthesized, although Chapter 6 contains more detailed discussions of biological importance.

* For a review on stereoselective syntheses of β-amino acids, see Liu, M.; Sibi, M.P. *Tetrahedron* ***2002***, *58*, 7991. Also see Juaristi, E.; Soloshonok, V.A., eds. *Enantioselective Synthesis of* β-*Amino Acids*, 2nd ed. Wiley Interscience, Hoboken, NJ, ***2005***. For a discussion of chiral stationary phases in the high-performance liquid chromatographic enantioseparation of unusual β-amino acids, see Ilisz, I.; Berkecz, R.; Forró, E.; Fülöp, F.; Armstrong, D.W.; Antal Péter, A. *Chirality* ***2009***, *21*, 339.

† For a review on stereoselective syntheses of γ-amino acids, see Ordóñez, M.; Cativiela, C. *Tetrahedron Asym.* ***2007***, *18*, 3.

1 Functional Group Exchanges

1.1 REACTIONS WITH AMMONIA, AMINES, OR AMINE SURROGATES

Incorporation of an amine moiety into a molecule by reaction with an alkyl halide or an alkylsulfonate ester is a common synthetic method. This approach includes reactions that produce amino acids, often using ammonia or amines as nucleophiles. There are problems associated with such reactions, however, and several amine surrogates have been developed that allow incorporation of nitrogen into the molecule, and the amine moiety is unmasked later.

1.1.1 Reaction with Ammonia

Amino acids can be prepared by the reaction of α-halo-carboxylic acid or a halo-carboxyl with ammonia or ammonium hydroxide. Ammonium hydroxide was reacted with 9-bromononanoic acid (*1*) in one example to give 9-aminononanoic acid (*2*) in 79% yield by this procedure.[1] Similarly, reaction of bromo-ester ***3*** with aqueous ammonia for 9 days gave ethyl 2,2-diethyl-5-aminopentanoate (*4*).[2] The second example is shown to illustrate that the yield of direct displacement reactions of a halide by ammonia is sometimes rather low.

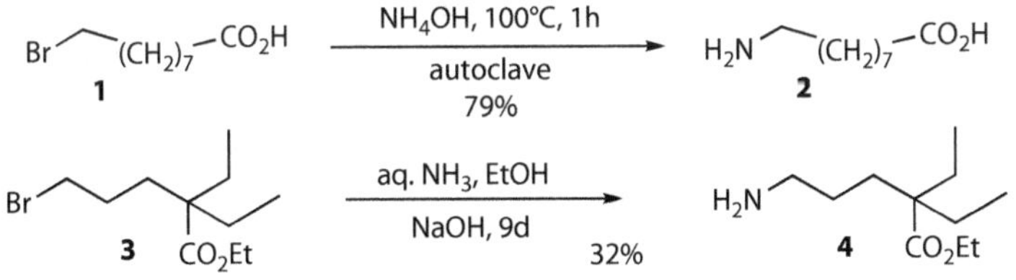

Methods for incorporation of a halide leaving group into a carboxylic acid vary, and it is often possible to incorporate other functionalities. The reaction of methyl acrylate with bromine and aqueous silver nitrate gave methyl 3-bromo-2-hydroxypropanoic acid.[3] Subsequent reaction with ammonia led to isoserine. Alkenes with an allylic position react with *N*-bromosuccinimide to give an allylic bromide.[4]

---

[1] Temple, C., Jr.; Elliott, R.D.; Montgomery, J.A. *J. Med. Chem.* ***1988***, *31*, 6976.
[2] Baler, J.A.; Harper, J.F. *J. Chem. Soc. C* ***1967***, 2148.
[3] Leibman, K.C.; Fellner, S.K. *J. Org. Chem.* ***1962***, *27*, 438.
[4] Pinza, M.; Pifferi, G. *J. Pharm. Sci.* ***1978***, *67*, 120.

Subsequent reaction with aqueous ammonia gives 4-aminobut-2-enoic acid derivatives. The reaction of *E*-pent-2-enoic acid with *N*-bromosuccinimide, for example, led to ***5***,[5] and subsequent treatment with ammonia gave 4-aminopent-2-enoic acid (***6***). An alternative method reacted ***8*** with sodium azide and then reduced the azide with zinc and acetic acid.[5]

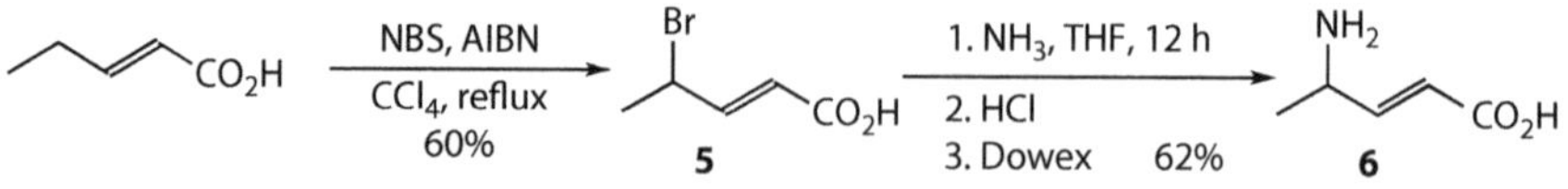

An example that generates an alkynyl amino acid is shown by the mono-chlorination of 1,4-but-2-yne diol (***7***)[6] followed by oxidation of the remaining alcohol moiety to give ***8***.[7] Subsequent reaction with ammonia gave aminotetrolic acid (***9***, 4-aminobut-2-ynoic acid).

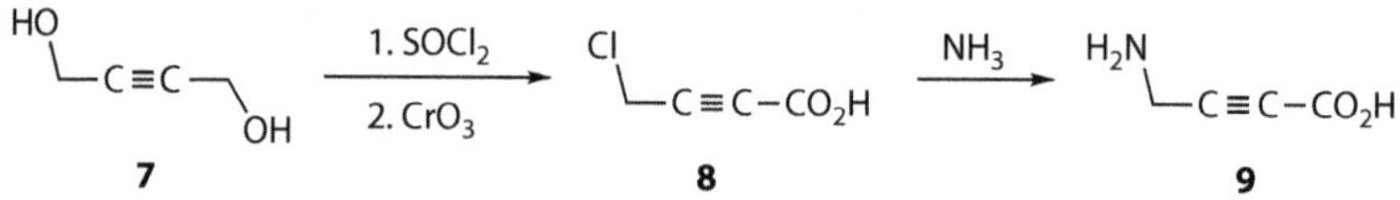

Ammonia reacts with functional groups other than alkyl halides, including epoxides. The reaction of ammonia with the β-carbon of an α,β-epoxy ester is clearly related to the Michael reaction.[8] In one synthesis that exploited the reaction of an epoxide with ammonia, conjugated ester ***10*** was prepared by the reaction of 3-methyl-butanal reacted with an appropriate ylid. Subsequent epoxidation with trifluoroperoxoyacetic (buffered with disodium phosphate to trap the trifluoroacetic acid by-product) gave ***11***. Treatment with ammonium hydroxide led to a ring opening reaction with ammonia to give 3-amino-2-hydroxy-5-methylhexanoic acid, ***12***.[9]

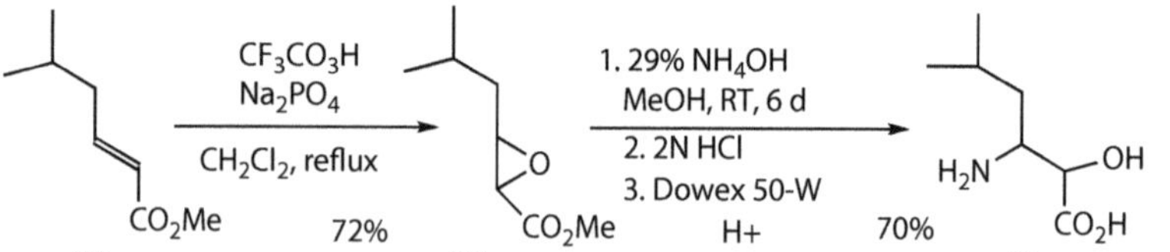

[5] Allan, R.D. *Aust. J. Chem.* ***1979***, *32*, 2507.

[6] Bailey, W.J.; Fujiwara, E. *J. Am. Chem. Soc.* ***1955***, *77*, 165.

[7] Beart, P.M.; Johnston, G.A.R. *Aust. J. Chem.* ***1972***, *25*, 1359. Also see Allan, R.D.; Johnston, G.A.R.; Twitchin, B. *Aust. J. Chem.* ***1980***, *33*, 1115.

[8] (a) Michael, A. *J. Prakt. Chem.* ***1887***, *35*, 379; (b) Bergmann, E.D.; Gingberg, D.; Pappo, R. *Org. React.* ***1959***, *10*, 179; (c) Perlmutter, P. *Conjugative Addition Reactions in Organic Synthesis*. Pergamon Press, Oxford, ***1992***; (d) see Smith, M.B. *Organic Synthesis*, 3rd ed. Wavefunction, Inc./Elsevier, Irvine, CA/London, England, ***2010***, pp. 877–888; (e) Smith, M.B. *March's Advanced Organic Chemistry*, 7th ed. John Wiley & Sons, Hoboken, NJ, ***2013***, pp. 943–949.

[9] Kato, K.; Saino, T.; Nishizawa, R.; Takita, T.; Umezawa, H. *J. Chem. Soc. Perkin Trans. I* ***1980***, 1618. Also see Igarashi, M.; Tamura, M.; Yanagi, M. *Bull. Chem. Soc. Jpn.* ***1971***, *44*, 3468.

1.1.2 Reaction with Amines

Amines react with halo-acids in an identical manner to the ammonia reactions shown above. Both *N*-alkyl and *N,N*-dialkylamino acids can be produced by this approach. A simple example is the reaction of 2-chloroethanoic acid (***13a***), 3-chloropropanoic carboxylic acid (***13b***), or 6-chlorohexanoic acid (***13c***) with a 25% aqueous solution of dimethylamine. The products of these reactions are 69% of ***14a***, 54% of ***45b***, and 53% of ***14c***.[10] In general, the yields in nucleophilic displacement reactions with amines are better than the analogous reaction with ammonia, but in many cases they remain low. Allylic halides can also be used, as in the reaction of ***15*** with diethylamine to give ethyl 4-(*N,N*-diethylamino)but-2-enoate, ***16***.[11]

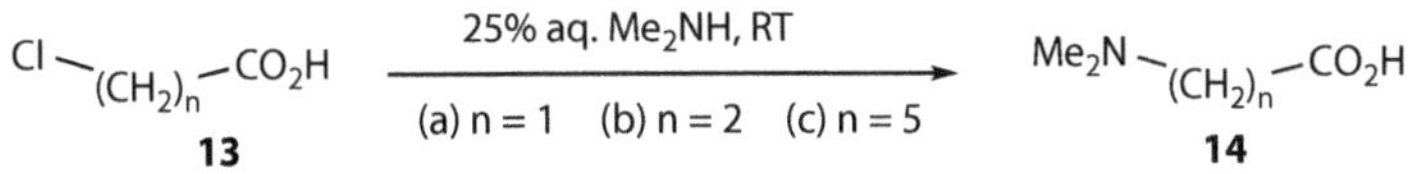

13 **14**

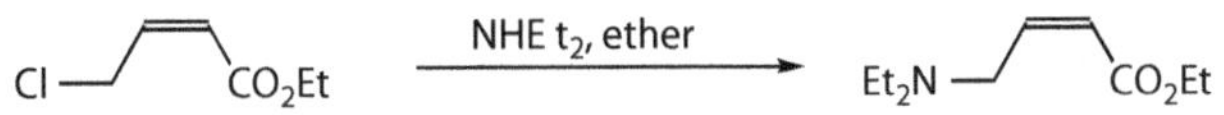

15 **16**

1.1.3 Reaction with Amine Surrogates

The reaction of ammonia or amines with alkyl halides, even allylic halides, often gives poor yields of the amine product. Such poor yields are usually improved by a two-step sequence that uses an amine surrogate (a reagent that incorporates a nitrogen functional group into a molecule and is later converted to an amino group). In most cases, the first step is the displacement reaction (with halides or sulfonate esters) that introduces the functional group, and the second step converts the functional group (the surrogate) into an amine. Common surrogates are phthalimide, nitro (NO_2), cyanide (C≡N), or azide (N_3), and hydrolysis or reduction leads to either an aminomethyl moiety[12] (–CH_2NH_2) or an amino group (–NH_2).

The reaction of halides with phthalimide to generate an *N*-alkyl phthalimide is well known (the Gabriel synthesis).[13,14] One of the more important methods for converting an *N*-alkyl phthalimide to the targeted amine is by treatment with hydrazine (the Ing-Manske procedure).[15] This classical reaction sequence is an excellent method for the synthesis of amino carboxylic acids, as illustrated by the conversion of ***18***

---

[10] Olomucki, M. *Bull. Soc. Chim. Fr.* ***1963***, 2067.

[11] Vessiere, R. *Ann. Fac. Sci. Univ. Clermont Chim.* ***1960***, 1 [*Chem. Abstr.* ***1963***, 59: 13811g]. Also see Ziv, D.; Olomucki, M. *Bull. Soc. Chim. Fr.* ***1970***, 1025.

[12] (a) Soffer, L.M.; Katz, M. *J. Am. Chem. Soc.* ***1956***, 78, 1705; (b) Gribble, G.W.; Switzer, F.L.; Soll, R.M. *J. Org. Chem.* ***1988***, *53*, 3164; (c) Rylander, P.N. *Catalytic Hydrogenation in Organic Synthesis*. Academic Press, New York, ***1979***, pp. 140–141.

[13] Gibson, M.S.; Bradshaw, R.W. *Angew. Chem. Int. Ed. Engl.* ***1968***, *7*, 919.

[14] See (a) Smith, M.B. *Organic Synthesis*, 3rd ed. Wavefunction, Inc./Elsevier, Irvine, CA/London, England, ***2010***, p. 125; (b) Smith, M.B. *March's Advanced Organic Chemistry*, 7th ed. John Wiley & Sons, Hoboken, NJ, ***2013***, p. 494.

[15] Ing, H.R.; Manske, R.H.F. *J. Chem. Soc.* ***1926***, 2348.

to 4-aminobutanoic acid (GABA).[16] In this particular report, Ganem and coworkers introduced a mild method for the deprotection of phthalimides using $NaBH_4$ followed by treatment with acid and then passage through Dowex-50 resin. In a different example, conversion of the carboxyl unit of 2-alkylated derivatives of ***17*** to an (*R*)-pantolactone ester derivative allowed resolution to give chiral amino acids.[17] Other functional groups can be tolerated in this phthalimide protocol.[18,19] The reaction of propargylic halides with phthalimide leads to alkynyl amino acids, and subsequent reaction with HCl leads to the corresponding vinyl chloride.[20,21]

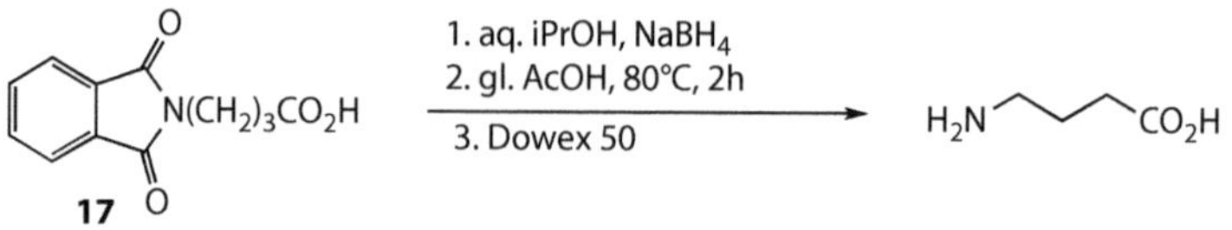

17

This amino acid synthesis can exploit known chemistry to prepare more substituted or functionalized derivatives. Phthalimide reacted with the epoxide moiety in ***18***, for example, in this case catalyzed by a Pd catalyst, to give *N*-phthalimido methyl 4-amino-5-hydroxyhex-2-enoate, ***19***.[22] The allylic position in ***18*** was clearly the more reactive for nucleophilic attack.

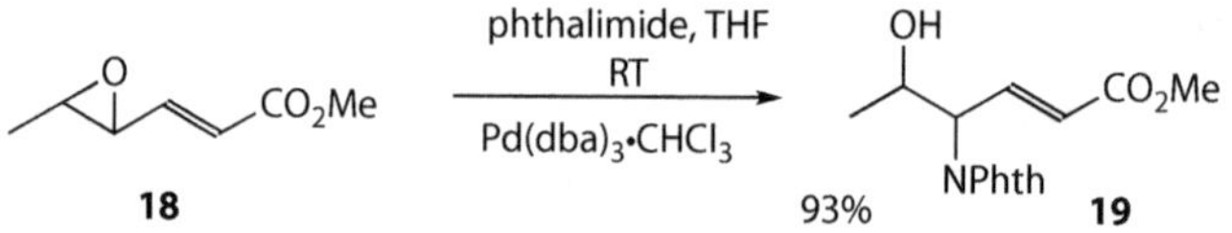

18 **19**

Cyanide is an excellent amine surrogate because it is a good nucleophile in various reactions, and subsequent reduction of a nitrile will generate an aminomethyl moiety. The fundamental reaction involves S_N2 displacement of a halide by cyanide (:NC:$^-$), as in the reaction of ethyl 2-bromopropanoate and sodium cyanide to give ***20***.[23] Catalytic hydrogenation, in this case using a Raney nickel catalyst, converted the cyano group into an aminomethyl group ($-CH_2NH_2$) and acid hydrolysis of the ester to give 2-methyl-3-aminopropanoic acid, ***21***. Note that an ion exchange resin was required for isolation of the neutral amino acid, which exists as a zwitterion. When a zwitterionic amino acid is the final product, ion exchange is often necessary

---

[16] Osby, J.O.; Martin, M.G.; Ganem, B. *Tetrahedron Lett.* ***1984***, *25*, 2093.

[17] Duke, R.K.; Chebib, M.; Hibbs, D.E.; Mewett, K.N.; Johnston, G.A.R. *Tetrahedron Asym.* ***2004***, *15*, 1745.

[18] (a) Mitsui Toatsu Chemicals, Inc. *Jpn. Kokai Tokkyo* 81 20,556 [*Chem. Abstr.* ***1981***, *95*: P62707f]; (b) Mitsui Toatsu Chemicals, Inc. *Jpn. Kokai Tokkyo Koho* 80 147,246 [*Chem. Abstr.* ***1981***, *94*: P157263b].

[19] Uchimaru, F.; Sato, M.; Kosasayama, E.; Shimizu, M.; Takashi, H. *Jpn.* 70 16,682 [*Chem. Abstr.* ***1970***, *73*: P77617w].

[20] (a) Bowden, K.; Heilbron, I.M.; Jones, E.R.H.; Weedon, B.C.L. *J. Chem. Soc.* ***1946***, 39; (b) Heilbron, I.M.; Jones, E.R.H.; Sondheimer, F. *J. Chem. Soc.* ***1949***, 604; (c) Haynes, L.J.; Heilbron, I.M.; Jones, E.R.H.; Sondheimer, F. *J. Chem. Soc.* ***1947***, 1583; (d) Helbron, I.M.; Jones, E.R.H.; Sondheimer, J. *J. Chem. Soc.* ***1947***, 1586.

[21] Allan, R.D.; Johnston, G.A.R.; Twitchin, B. *Aust. J. Chem.* ***1980***, *33*, 1115.

[22] Tsuda, T.; Horii, Y.; Nakagawa, Y.; Ishida, T.; Sawgusa, T. *J. Org. Chem.* ***1989***, *54*, 977.

[23] Takano, S.; Ogasawara, K.; Seikiguchi, Y.; Kasai, N.; Sakaguchi, K. *Jpn. Kokai Tokkyo Koho* JP 63 174,957 [*Chem. Abstr.* ***1989***, *110*: P24308x].

for its purification. In another example, methyl 6-amino-3-methoxyaminohexanoate was prepared in an identical manner from methyl 5-cyano-2-methoxypentanoate.[24]

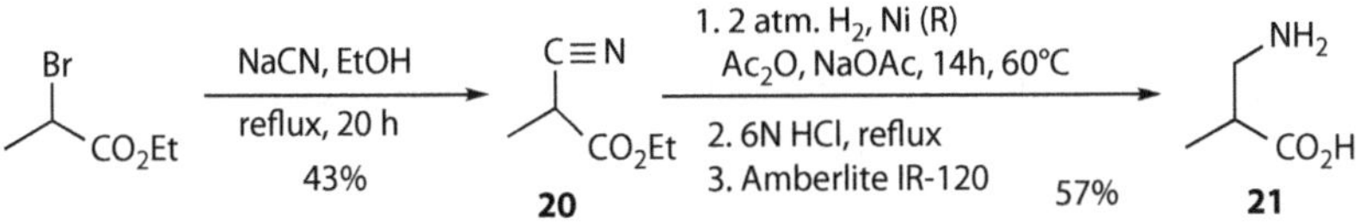

20 21

The reaction of an alkyl nitrile with a nucleophilic organometallic such as a Grignard reagent generated an iminium salt (*22*).[25] Aqueous acid hydrolysis liberated imine ***23***, which may be converted to a carbonyl under the reaction conditions by loss of ammonia. If a reducing agent is used rather than aqueous acid hydrolysis, however, the imine is reduced to the corresponding amine, ***24***.[26] Amino acids may be prepared if the cyano group also has a carboxyl group elsewhere in the molecule. Ethyl cyanoacetate and similar compounds are examples of such molecules, and they react with Grignard reagents such as ethylmagnesium bromide to give the imine. Upon hydrolysis the imine isomerizes to the enamine form (an enamine ester), ethyl 3-aminopent-2-enoate, ***25***.[27] Ethyl 3-aminohex-2-enoate (13% yield) and ethyl 3-aminohept-2-enoate (16% yield) were also prepared from the appropriate cyano-ester. As seen in these examples, the yields were often poor. The reaction of ethyl α-cyanoacetate with phenylmagnesium bromide, which gave ethyl 3-amino-3-phenylprop-2-enoate in 83% yield is an exception.[27]

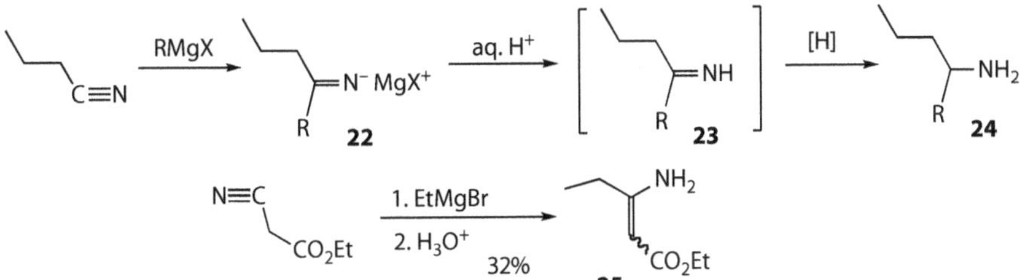

22 23 24 25

For the most part, the reaction of halides and nitrite (NO_2^-) does not provide a good general synthesis of nitroalkanes, although silver nitrate has been used to prepare nitro compounds. The most common method for introducing a nitro group and a carboxyl group into the same molecule is via reactions that utilize nitroalkanes as a starting material. Conversion to the corresponding enolate anion via reaction with a suitable base (lithium diisopropylamide (LDA), hexamethyldisilazide, etc.) can be followed by reaction with a variety of electrophilic reagents, including carboxyl-containing molecules or carboxyl surrogates (see Chapter 4, Section 4.3). To illustrate the method, the reaction of the enolate anion of nitromethane with methyl

[24] Garritsen, J.W.; Penders, J.M. *Neth. Appl.* 72 16,661 [*Chem. Abstr.* ***1974***, *81*: P151570j].

[25] For reactions of this type, see Larock, R.C. *Comprehensive Organic Transformations*, 2nd ed. Wiley-VCH, New York, ***1999***, pp. 1420–1422.

[26] See Smith, R.G.; Lucas, R.A.; Wasley, J.W.F. *J. Med. Chem.* ***1980***, *23*, 952.

[27] Lukes R.; Kloubek, J. *Coll. Czech. Chem. Commun.* ***1960***, *25*, 607.

3-chloropropanoate gave nitroester ***26***.[28] In this particular case, reduction of the nitro group was accomplished with ammonium formate, to give methyl 4-aminobutanoate (***27***). This method is limited only by the availability of the nitroalkane precursor and the relative reactivity and availability of the ω-halo-ester. It is noted that $LiAlH_4$ will also reduce a nitro compound to an amine,[29] but in the sequence shown, the ester (or acid) moiety would also be reduced to an alcohol. For this reason, nitro groups in multifunctional molecules are more commonly reduced by catalytic hydrogenation or a more selective reagent such as ammonium formate.[30]

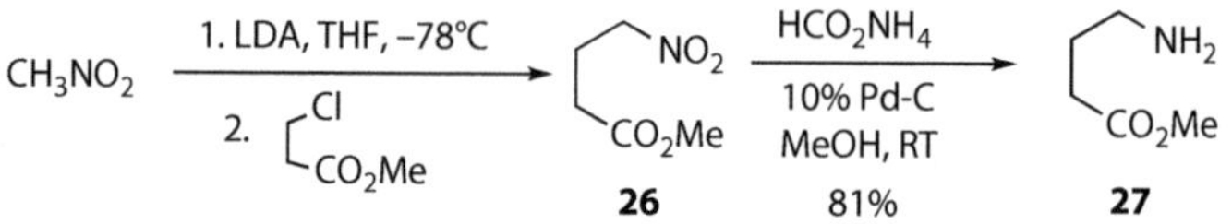

26 **27**

Azide is a useful amine surrogate, although the alkyl azide products may be unstable and sometimes explosive. Therefore, care should be exercised when using this surrogate, and experimental conditions carefully reviewed. Nonetheless, the reaction of azide ion is a good way to introduce nitrogen into a molecule, particularly on a laboratory scale of milligrams to a few grams. In one example, the carboxylic acid moiety in *S*-***28*** was converted to an acid chloride, allowing reduction of the COOH moiety to a hydroxymethyl moiety in ***29***.[31] Reaction with tosyl chloride and pyridine gave the tosylate, which was displaced by the nucleophilic azide ion to give the alkyl azide. In this case, $NaBH_4$ was used to reduce the azide moiety to an amine. It is known that $NaBH_4$ will reduce esters,[32] but in most cases that reaction is slow relative to more labile functional groups such as azide. Indeed, examples are known where this reagent will not reduce certain esters at all.[33] The final product in the example shown was ethyl 3-amino-2-fluoro-2-methylpropanoate, ***30***. The literature contains several synthetic methods for preparing fluorinated amino acids,[34] including β-trifluoromethyl and β-fluoromethyl GABA derivatives.[35]

---

[28] Ram, S.; Ehrenkaufer, R.E. *Synthesis*, ***1986***, 133.

[29] For example, see Ohta, H.; Kobayashi, N.; Ozaki, K. *J. Org. Chem.* ***1989***, *54*, 1802.

[30] Ram, S.; Ehrenkaufer, R.E. *Tetrahedron Lett.* ***1984***, *25*, 3415.

[31] Kitatsume, T. *Jpn. Kokai Tokkyo Koho* JP 63 60,954 [*Chem. Abstr.* ***1988***, *109*: P190870y].

[32] (a) Brown, H.C.; Krishnamurthy, S. *Aldrichimica Acta*, ***1979***, *12*, 3; (b) Schenker, R. *Angew. Chem.* ***1961***, *73*, 81; (c) Fieser, L.F.; Fieser, M. *Reagents for Organic Synthesis*, vol. 1. Wiley, New York, ***1967***, p. 1049.

[33] For an example, see Keusenkothen, P.F.; Smith, M.B. *Synth. Commun.* ***1989***, *19*, 2859.

[34] See Qiu, X.-L.; Meng, W.-D.; Qing, F.-L. *Tetrahedron* ***2004***, *60*, 6711.

[35] Gerus, I.I.; Mironets, R.V.; Shaitanova, E.N.; Kukhar, V.P. *J. Fluorine Chem.* ***2010***, *131*, 224.

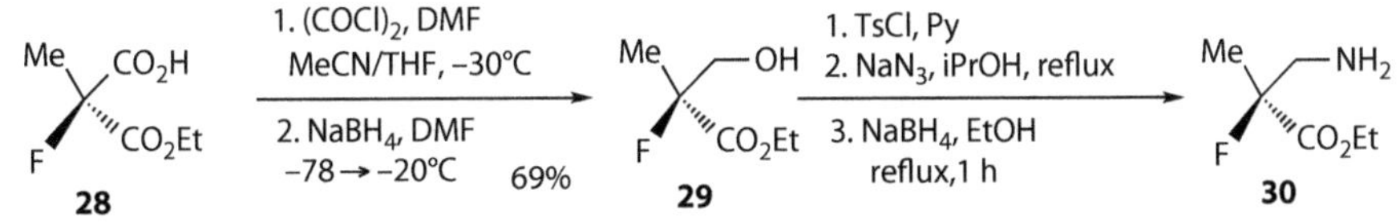

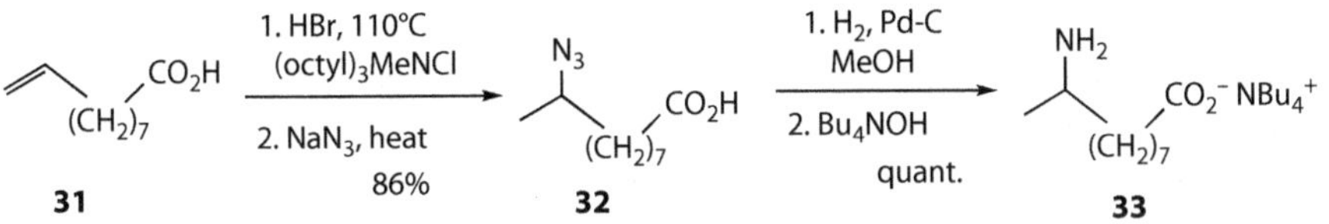

A different example that uses an azide surrogate is the reaction of oct-9-enoic acid (***31***) with HBr to generate the secondary bromide, which reacted with sodium azide to give ***32***.[36] Catalytic hydrogenation converted the azido moiety to an amine ($-N_3 \rightarrow -NH_2$), and 9-aminodecanoic acid was isolated as its tetrabutylammonium salt, ***33***. Other reagents have been used in conjunction with hydrogen gas, as in the conversion of azide ***34*** (n = 1, $R^1 = R^2 = H$; $R^3 = Me$) to ***35*** using triphenylphosphine and hydrogen.[37] Reduction of ***34*** produced 7-amino-2-methyl-hept-2-enoate (***35a***), 7-amino-4-methylhept-2-enoate (***35b***), and 7-aminooct-2-enoate (***35c***).

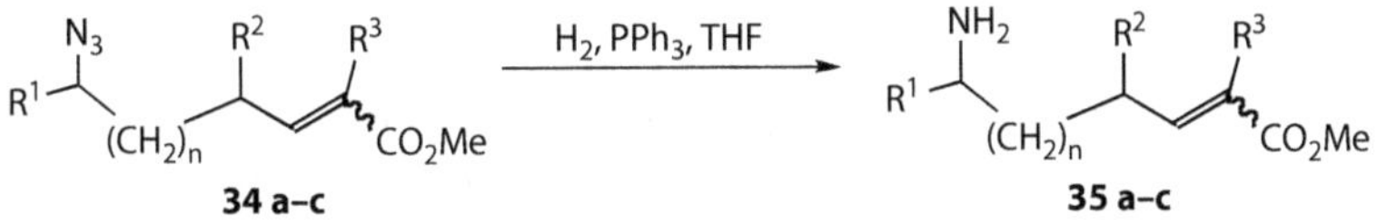

35	n	R¹	R²	R³	% Yield
a	2	H	H	Me	86
b	2	H	Me	H	83
c	2	Me	H	H	82

Azide is a potent nucleophile that reacts with electrophilic substrates other than halides and sulfonate esters, which is useful for the preparation of functionalized amino acids. Reaction of azide with the epoxide moiety in ***36***, for example, was followed by catalytic hydrogenation and hydrolysis of the ester to give isoserine (3-amino-2-hydroxypropanoic acid), ***37***.[38] This synthetic approach to isoserine dates back to 1879.[39]

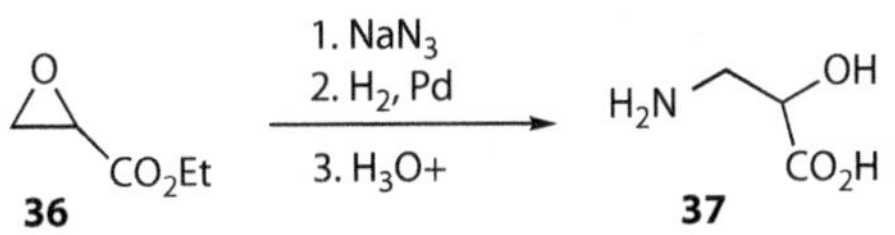

[36] Bartra, M.; Vilarrasa, J. *J. Org. Chem.* ***1991***, *56*, 5132.
[37] Knouzi, N.; Vaultier, M.; Toupet, L.; Carrie, R. *Tetrahedron Lett.* ***1987***, *28*, 1757.
[38] Williams, T.M.; Crumbie, R.; Mosher, H.S. *J. Org. Chem.* ***1985***, *50*, 91 and references cited therein.
[39] (a) Melikoff, P. *Ber.* ***1879***, *12*, 2227; (b) Erlenmeyer, E. *Ber.* ***1880***, *13*, 1077; (c) Fisher, E.; Leuchs, H. *Ber.* ***1902***, *35*, 3787; (d) Nakajima, Y.; Kinishi, R.; Oda, J.; Inouye, Y. *Bull. Chem. Soc. Jpn.* ***1977***, *50*, 2025.

2-Hydroxy-3-amino acids have been prepared in water, utilizing the nucleophilic reaction of azide ion to open an epoxide ring. Kynostatin ***38*** (KNI-272) is a human immunodeficiency virus (HIV)-1 protease inhibitor.[40] The hydroxyamino acid moiety (a norstatine; see Chapter 6, Section 6.1.6) is "boxed" in ***38***. The synthesis of this amino acid moiety used the copper nitrate-catalyzed reaction of ***39*** with sodium azide at 65°C to give azido alcohol ***40***, and reduction with two molar equivalents of sodium borohydride in the same reaction flask (a one-pot reaction) gave ***41*** in 72% yield.[41] It is noted that 2-hydroxy-3-amino acids have also been prepared by hydroxylation of perhydropyrimidin-4-ones, prepared from a parent amino acid (also see Section 1.4).[42]

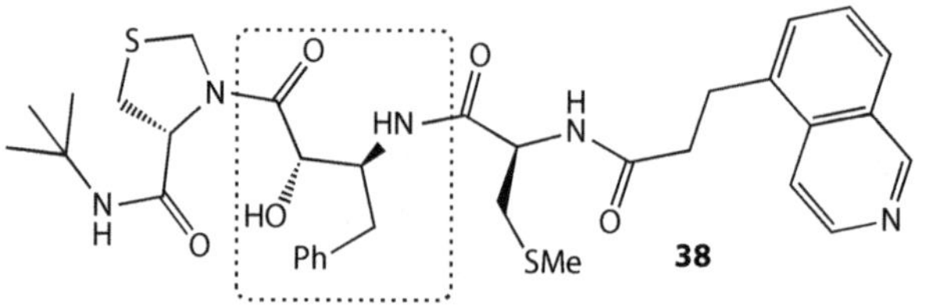

38

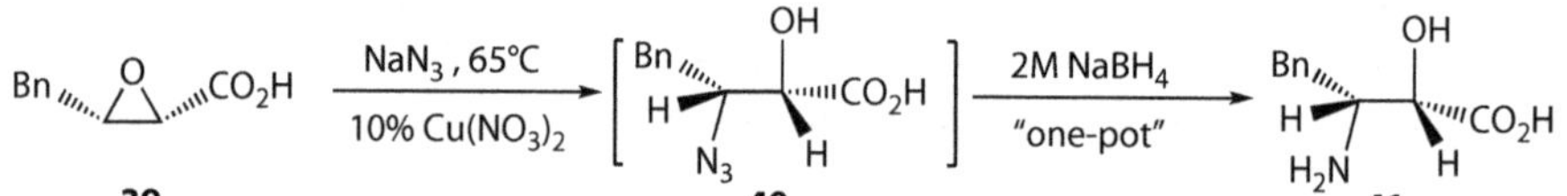

39 **40** **41**

Other azide reagents have been developed that avoid the use of sodium or potassium azide. Reaction of the acid moiety in ***42*** with diphenylphosphoryl azide (DPPA), at 90°C, led to a rearrangement that converted the $–CO_2H$ moiety into a $–NHCO_2H$ moiety. This intermediate was trapped as the 4-methylbenzyloxy carbamate (***43***), as shown.[43] The half-esters of dicarboxylic acids were utilized in this sequence for the synthesis of amino acids. Several methods are known to prepare the half-ester of a dicarboxylic acid, including selective enzymatic hydrolysis. Where the half-ester is available or can be easily prepared, this sequence is a good synthetic approach.

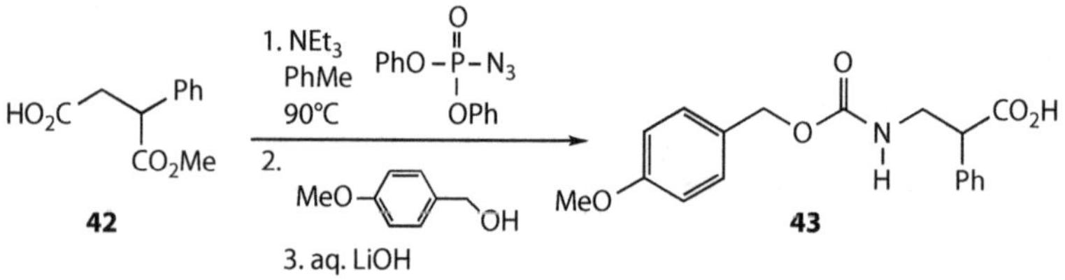

42 **43**

Other amine surrogates may be used to prepare ω-amino acids in reactions with halides or other electrophilic species. Most are rather specific, but one shows some generality. Effenberger and coworkers showed that reaction of methyl 5-bromohexanoate

[40] See Mimoto, T.; Hattori, N.; Takaku, H.; Kisanuki, S.; Fukazawa, T.; Terashima, K.; Kato, R.; Nojima, S.; Misawa, S.; Ueno, T.; Imai, J.; Enomoto, H.; Tanaka, S.; Sakikawa, H.; Shintani, M.; Hayashi, H.; Kiso, Y. *Chem. Pharm. Bull.* ***2000***, *48*, 1310.

[41] Fringuelli, F.; Pizzo, F.; Rucci, M.; Vaccaro, L. *J. Org. Chem.* ***2003***, *68*, 7041.

[42] Escalante, J.; Juaristi, E. *Tetrahedron Lett.* ***1995***, *36*, 4397.

[43] Talley, J.J. *Eur. Pat. Appl.* EP 396,526 [*Chem. Abstr.* ***1991***, *114*: 229382w].

with potassium isocyanate gave ***44***. In this case, the isocyanate is the amine surrogate, and hydrolysis with concentrated HCl at 100°C gave 6-aminohexanoic acid, ***45***.[44] One use of 6-aminohexanoic acid (also, aminocaproic acid, known as Amicar) is to treat excessive postoperative bleeding.[45] 6-Aminohexanoic acid, which has antifibrinolytic properties (breakdown of fibrin), has been used empirically as a treatment for exercise-induced pulmonary hemorrhage (EIPH), but studies have shown that it was not effective in preventing or reducing the severity of EIPH.[46] It is also a competitive inhibitor of plasminogen binding.[47] 4-Aminobutanoic acid, 3-aminopropanoic acid, and 5-aminopentanoic acid were prepared by the same route shown for the preparation of ***45***.[44] Branched-chain and cyclic 6-aminohexanoic acid esters have been used as the hydrophobic part of skin permeation enhancers.[48]

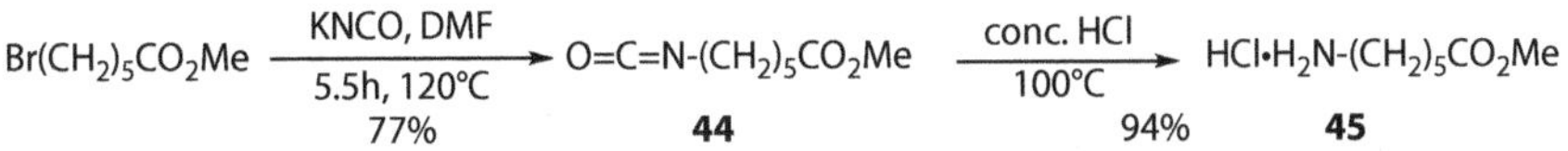

44 **45**

Another amine surrogate is illustrated by the direct α-amination reaction of 2-keto-esters with azodicarboxylates such as diethyl azodicarboxylate. In the presence of a chiral bis(oxazoline)–copper(II) complex as the catalyst, derivatives such as ***46*** were formed with good enantioselectivity.[49] Reduction and reaction with trimethylsilyldiazomethane led to cyclization to oxazolidinone ***47***, and these steps were followed by deprotection and reduction, and reprotection with Boc to give syn-β-amino-α-hydroxy ester ***48*** in 33% overall yield, again with good overall enantioselectivity.

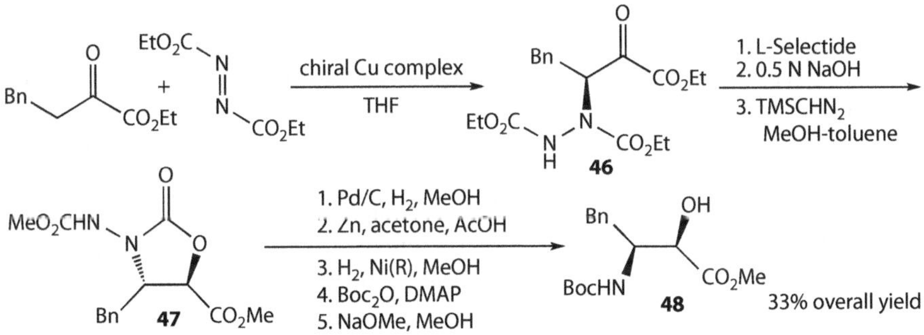

46 **47** **48**

57% overall yield (89% ee)

Previous examples in this section discuss reactions of amines or amine surrogates. There are other methods that will introduce an amine moiety into a molecule such that an amino acid is produced. Reduction of amides to amines provides another route to

---

[44] (a) Effenberger, F.; Drauz, K.; Förster, S. Müller, W. *Chem. Ber.* ***1981***, *114*, 173; (b) Effenberger, F.; Drauz, K. *Angew. Chem.* ***1979***, *91*, 504.
[45] Willoughby, J.S.; *Anaesth. Intensive Care* ***1984***, *12*, 35860.
[46] Buchholz, B.M.; Murdock, A.; Bayly, W.M.; Sides, R.H. *Equine Vet. J.* ***2010***, *42*, 256.
[47] Slaughter, T.F.; Greenberg, C.S. *Am. J. Hematol.* ***1997***, *56*, 32.
[48] Hrabálek, A.; Vávrová, K.; Doležal P., Macháček, M. *J. Pharm. Sci.* ***2005***, *94*, 1494.
[49] Juhl, K.; Jørgensen, K.A. *J. Am. Chem. Soc.* ***2002***, *124*, 2420.

amino acids. One sequence reacted α,ω-di-carboxylic acids such as ***49*** with an amine to give the corresponding ammonium salt, ***50***. Subsequent heating to 180°C generated the amide (***51***), which was electrolytically reduced to the *N,N*-dimethylamino acid (***52***).[50] Several straight-chain amino acids were prepared by this route.

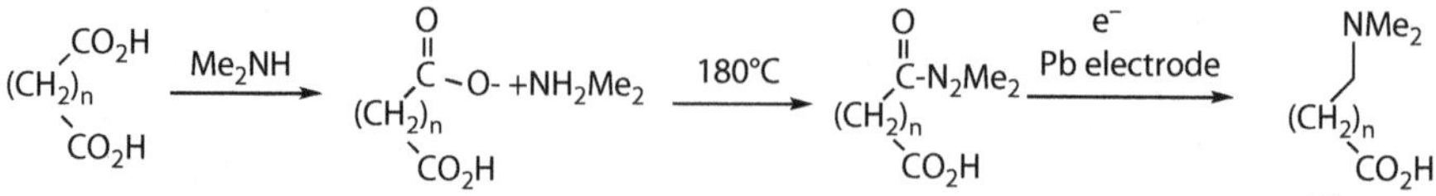

49 50 51 52

1.2 CONVERSION OF CARBONYL COMPOUNDS TO AMINES

1.2.1 From β-Keto-Acids or β-Diketones

Primary amines are known to react with aldehydes or ketones to form imines.[51] Amines or ammonia react similarly with 1,3-diketones (β-diketones), but the initially formed imine tautomerizes to the more stable enamine form, which in this case is an enamino-ketone. The reaction of 2,4-pentanedione with ammonia, for example, gave ***53***.[52] As a class of compounds enamines were first used in syntheses by Stork,[53] and are well known.

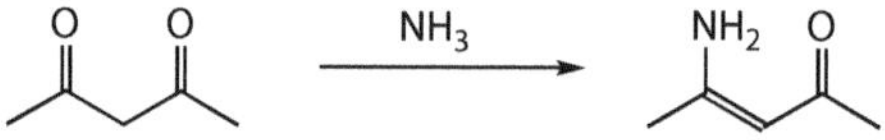

53

If a β-keto-acid derivative reacts with ammonia, a similar reaction occurs to generate an enamino acid derivative. The reaction of ***54*** with ammonia was reported in 1887, for example, and gave ***55***.[54] Amines react in a same manner to ammonia, as illustrated by the reaction of dimethylamine and ethyl acetoacetate to give ethyl 3-(*N,N*-dimethylamino)but-2-enoate, ***56***.[55] The reaction of ammonia with ethyl acetoacetate gave methyl 3-aminobut-2-enoate in 90% yield.[56] In the presence of an acid catalyst, $Ph_3P{=}N{-}SiMe_3$[triphenyl(trimethylsilylamino)phosphorane][57]

[50] Prelog, V. Coll. *Czech. Chem. Commun.* **1930**, 2, 712.

[51] See Dayagi, S.; Degani, Y., in Patai, S. *The Chemistry of the Carbon-Nitrogen Double Bond.* Wiley, New York, ***1970***, pp. 64–83; Reeves, R.L., in Patai, S. *The Chemistry of the Carbonyl Group*, pt. 1. Wiley, New York, ***1966***, pp. 600–614. Also see Guzen, K.P.; Guarezemini, A.S.; Órfão, A.T.G.; Cella, R.; Pereira, C.M.P.; Stefani, H.A. *Tetrahedron Lett.* ***2007***, *48*, 1845.

[52] Combes, A.; Combes, C. *Bull. Soc. Chim. Fr.* ***1892***, *7*, 778.

[53] (a) Stork, G.; Terrell, R.; Szmuszkovicz, J. *J. Am. Chem. Soc.* ***1954***, *76*, 2029; (b) Mundy, B.F.; Ellerd, M.G. *Name Reactions and Reagents in Organic Synthesis.* Wiley, New York, ***1988***, pp. 206–207.

[54] Conrad, M.; Epstein, W. *Deutsh. Geschel. Ber.* ***1887***, *20*, 3052.

[55] (a) Glickman, S.A.; Cope, A.C. *J. Am. Chem. Soc.* ***1945***, *67*, 1019; (b) Michaelis, A. *Ann.* ***1909***, *366*, 324 (see p. 337); (c) Brühl, J.W. *Z. Physik. Chem.* ***1895***, *16*, 216.

[56] (a) Gasteiger, J.; Strauß, U. *Chem. Ber.* ***1981***, *114*, 2336; (b) also see Züblin, J. *Ber.* ***1878***, *11*, 1417; (c) Bülow, C.; Neber, P. *Ber.* ***1912***, *45*, 3732; (d) Parmerter, S.M. *Org. React.* ***1959***, *10*, 1; (e) Prager, B. *Ber.* ***1901***, 34, 3600.

[57] Birkofer, L.; Ritter, A.; Richter, P. *Chem. Ber.* ***1963***, *96*, 2750.

reacted with ethyl acetoacetate to give ***56*** in 73% yield.[58] One modification used a mixture of ammonia and *p*-toluenesulfonic acid to induce the reaction.[59] Such enamino derivatives are actually 3-aminobutenoic acid derivatives and can be used as such or reduced to 3-aminobutenoic acid derivatives.

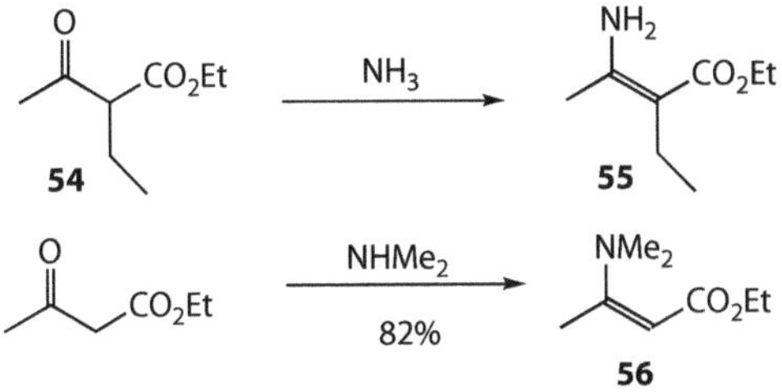

54 **55** **56**

This approach can be used to prepare many substituted alkene-amino acid derivatives from an appropriate β-keto-ester.[60] The reaction of ethyl 4-methyl-3-oxopentanoate with ammonia in ethanol gave ethyl 3-amino-4-methylpent-2-enoate in 75% yield.[61] Diketo-ester ***57*** reacted with ammonia to give exclusively ethyl 4-amino-6-oxohept-4-enoate, ***58***.[62] β-Keto-esters can also be condensed with nitriles to give the corresponding alkenyl amino acid.[63]

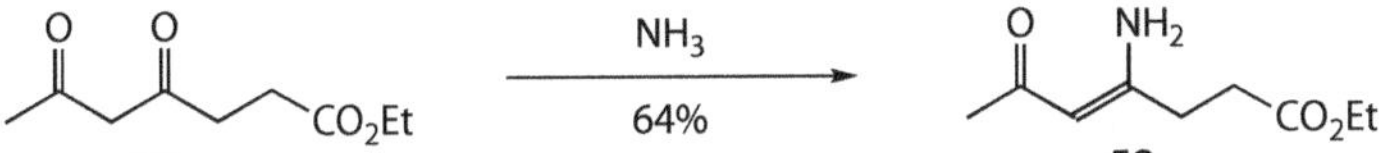

57 **58**

The aminoalkenoic acids generated from diketones or keto-esters and easily reduced to give the corresponding aminoalkanoic acid. The platinum-catalyzed hydrogenation of ***59*** gave methyl 4-methyl-3-(*N*-acetylamino)pentanoate (***60***), for example.[64] The *N*-acetyl methyl esters of several 3-aminoalkanoic acids were prepared by hydrogenation of the alkenylamino acid precursor. The yield of saturated products can be low using catalytic hydrogenation, as seen for ***60***, and other methods for reducing the alkenyl moiety have been developed, including sodium cyanoborohydride.[65]

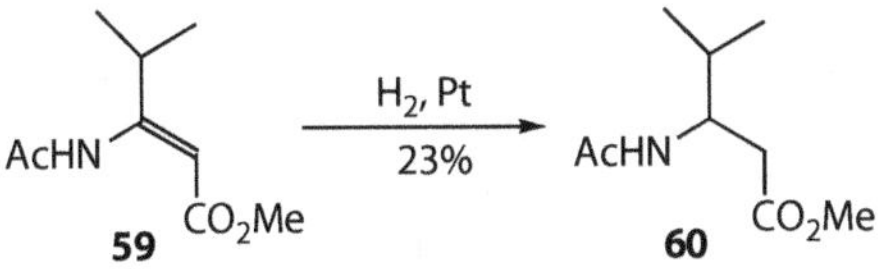

59 **60**

---

[58] Kloek, J.A.; Leschinsky, K.L. *J. Org. Chem.* ***1978***, *43*, 1460.

[59] Buckler, R.T.; Hartzler, H.E. *J. Med. Chem.* ***1975***, *18*, 509.

[60] (a) Buckler, R.T.; Hartzler, H.E. *J. Med. Chem.* ***1975***, *18*, 509; (b) Potesil, T.; Potesilová H. *J. Chromatogr.* ***1982***, *249*, 131; (c) Furukawa, M.; Okawara, T.; Noguchi, Y.; Terawaki, Y. *Chem. Pharm. Bull.* ***1979***, *27*, 2223.

[61] Aberhart, D.J.; Lin, H.-J. *J. Org. Chem.* ***1981***, *46*, 3749.

[62] Gelin, S.; Rouet, J. *Bull. Soc. Chim. Fr.* ***1971***, 2179.

[63] Veronese, A.C.; Gandolfi, V.; Basato, M.; Corain, B. *J. Chem. Res. Synop.* ***1988***, 246.

[64] Slopianka, M.; Gossauer, A. *Liebigs Ann. Chem.* ***1981***, 2258.

[65] Furukawa, M.; Okawara, T.; Noguchi, Y.; Terawaki, Y. *Chem. Pharm. Bull.* ***1979***, *27*, 2223.

Another variation generated an enamino-ester derivative using a different synthetic approach. Acetonitrile reacted with HCl and then with Meldrum's acid[66] to give ***61*** via a Pinner synthesis.[67] Hydrolysis was followed by decarboxylation (in situ) to give ethyl 3-aminoprop-2-enoate, ***62***.[68] Other amino acid derivatives were prepared using this method.

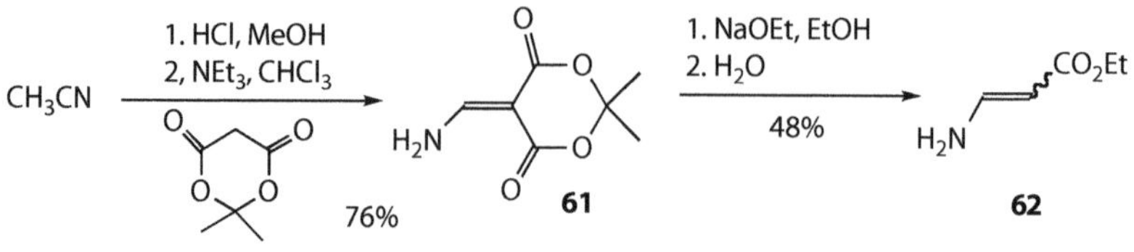

1.2.2 From Other Keto-Acids or Aldehyde Acids

Mono-aldehydes and mono-ketones are converted to amines by several methods. A common method to accomplish this transformation is reductive amination (also see Section 1.4.3). Amino acids that are produced from α-keto- or β-keto-acid derivatives are discussed in Section 1.2.1. Reductive amination of a keto-acid or a keto-ester, when the carbonyl is not proximal to the carbonyl unit, leads to an amino acid or an amino-ester, respectively.

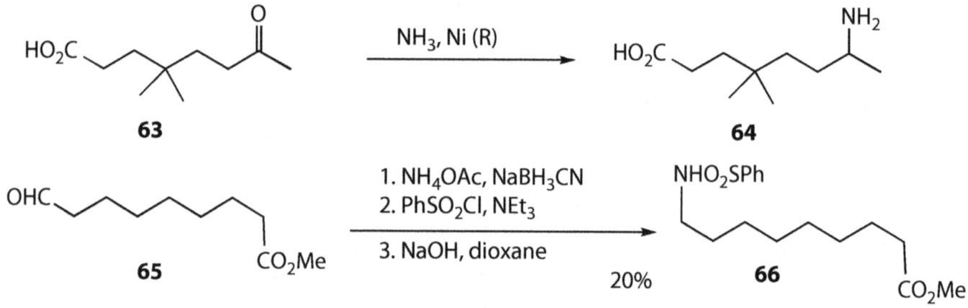

Both ketones and aldehydes may be used. An imine is generally an intermediate in reductive amination reactions of this type, but the intermediate iminium salt need not be isolated, as illustrated in the conversion of keto-acid ***63*** to amino acid ***64***.[69] In another example, the amine resulting from reaction of aldehyde ***65*** with $NaBH_3CN$/NH_4OAc was converted to the benzenesulfonate, giving methyl *N*-benzenesulfonyl 9-aminononanoate, ***66*** (20% overall yield).[70]

---

[66] (a) Meldrum, A.N. *J. Chem. Soc.* ***1908***, *93*, 598; (b) Hanford, W.E.; Sauer, J.P. *Org. React.* ***1946***, *3*,124.

[67] (a) Pinner, A.; Klein, F. *Ber.* ***1889***, *10*, 1889; (b) Pinner, A.; Klein, F. *Ber.* ***1878***, *11*, 4, 1475; (c) Pinner, A.; Klein, F. *Ber.* ***1883***, *16*, 352, 1643; (d) Roger, R.; Neilson, D. *Chem. Rev.* ***1961***, *61*, 179; (e) Gray, A.P.; Dipinto, V.M.; Solomon, I.J. *J. Org. Chem.* ***1977***, *41*, 2428.

[68] Célérier, J.P.; Deloisy, E.; Kapron, P.; Lhommet, G. *Synthesis*, ***1981***, 130.

[69] Merger, F.; Rischer, R.; Harder, W.; Priester, C.U.; Vagt, U. *Ger. Offen.* DE 3,843,792 [*Chem. Abstr.* ***1990***, *113*: P230789v].

[70] Ladouceur, G.; Mais, D.E.; Jakubowski, J.A.; Utterback, B.G.; Robertson, D.W. *Bioorg. Med. Chem. Lett.* ***1991***, *1*, 173.

In some cases, it is more convenient to isolate the imine prior to the reduction step. Aldehydes and ketones are useful precursors to amino acids when they are converted to an imine. Subsequent reduction of the imino C = N moiety generates the amine. When ***67*** (m = 1, 2) was treated with ammonia, for example, the resulting iminium salt (***68***) was hydrogenated using a Rh catalyst to give either 8-aminooctanoic acid (***69***) or 12-aminododecanoic acid (***70***).[71] A common form of the reaction combines ammonia with a hydrogenation catalyst such as Raney nickel.[72] Reduction of the imine intermediate can also be done using zinc and acetic acid.[73]

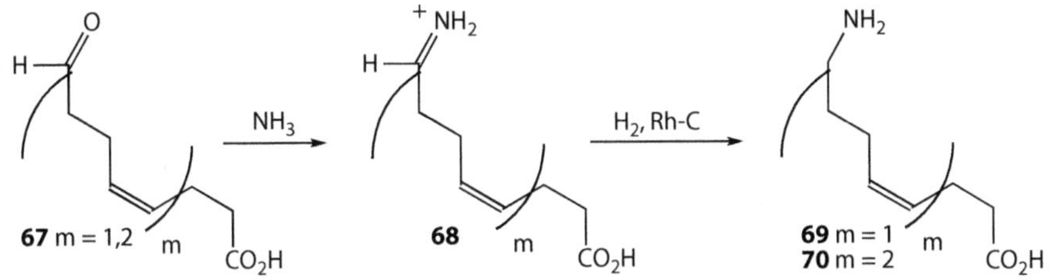

In some cases the direct reaction of an aldehyde or ketone with an amine followed by reduction, or even with direct reductive amination, gives poor yields of products or sometimes unwanted by-products. There are at least two important variations of the reductive amination process that can improve the yield of amine; one involves conversion of the aldehyde or ketone to an oxime, and the other involves conversion to a hydrazone.

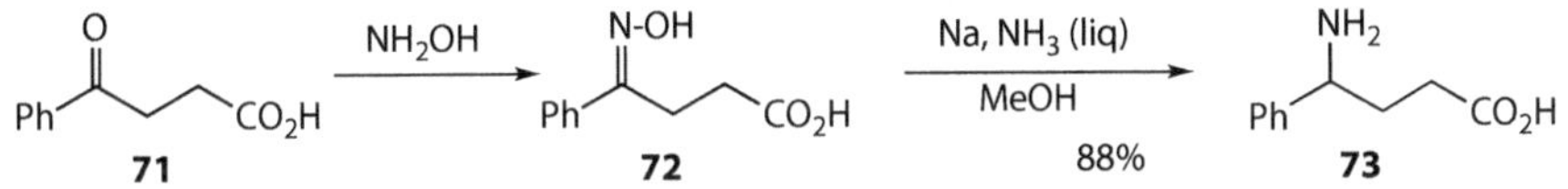

The reaction of a ketone or aldehyde moiety with hydroxylamine gives an oxime. This reaction can be applied to the preparation of amino acids if a carboxyl moiety is present elsewhere in the molecule. The reaction of 4-oxo-4-phenylbutanoic acid ***71*** with hydroxylamine leads to oxime ***72***. Subsequent dissolving metal reduction of the oxime gave 4-phenyl-4-aminobutanoic acid, ***73***, in 88% yield.[74] A related procedure converted 5-phenyl-4-oxopentanoic acid to 5-phenyl-4-aminopentanoic acid in 77% yield.[75] Indeed, catalytic hydrogenation can also be used for the reduction of an oxime to an amine.[76] Catalytic hydrogenation is perhaps the most common procedure used with this approach.

---

[71] Siclari, F.; Rossi, P.P.; De Gaetano, M. *Ger. Offen.* 2,608,203 [*Chem. Abstr.* ***1977***, *77*: 4923b].
[72] See Mileo, J.-C.; Sillion, B.; de Gaudemaris, G. *C.R. Acad. Sci. Ser. C* ***1968***, *267*, 93.
[73] Vaccher, C.; Berthelot, P.; Flouquet, N.; Debaert, M. *Synth. Commun.* ***1989***, *19*, 2049.
[74] Howe, R.K.; Gruner, T.A; Carter, L.G.; Franz, J.E. *J. Heterocyclic Chem.* ***1978***, *15*, 1001.
[75] Tomoeda, M.; Tani, Y.; Okada, H. *Yakugaku Zasshi* ***1966***, *86*, 1213 [*Chem. Abstr.* ***1967***, *67*: 2960v].
[76] Charles, J.; Descotes, G. *Bull. Soc. Chim. Fr.* ***1966***, 102.

Oximes derived from aldehydes may be used as well, and in the conversion of 4,4-dimethyl-glutaraldehyde (***74***) to 5-amino-4,4-dimethylpentanoic acid (***76***).[77,78] The initial product of this reaction was an oxime, and reduction led to lactam ***75***. Subsequent acid hydrolysis was required to generate ***76***. In some cases, reduction in this type of system can lead to a mono-amide as an intermediate product.[79]

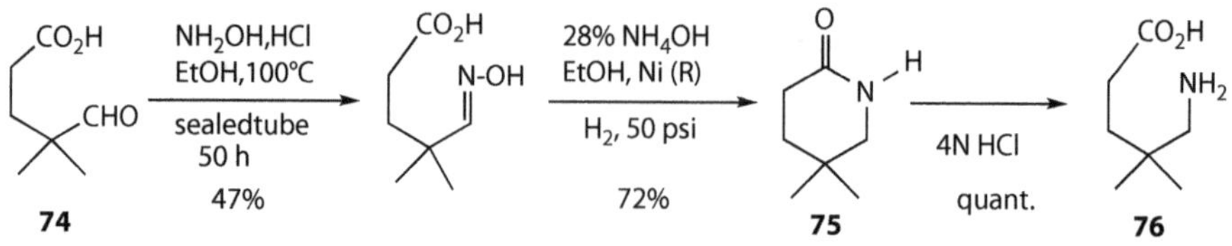

If the carbonyl unit of an aldehyde or ketone is sufficiently close to an ester/acid moiety, reaction with hydroxylamine can lead to an isoxazolidine derivative rather than a lactam. The reaction of the proximal ester moiety in ***77*** (prepared by literature methods)[80] with hydroxylamine, for example, gave isoxazolidone ***78***[81] via reaction of the initially formed oxime. This heterocyclic ring was subsequently cleaved with sodium in isopropanol and the carboxylic acid product was esterified to give methyl 3-amino-2,2-dimethylhept-6-enoate, ***79***.

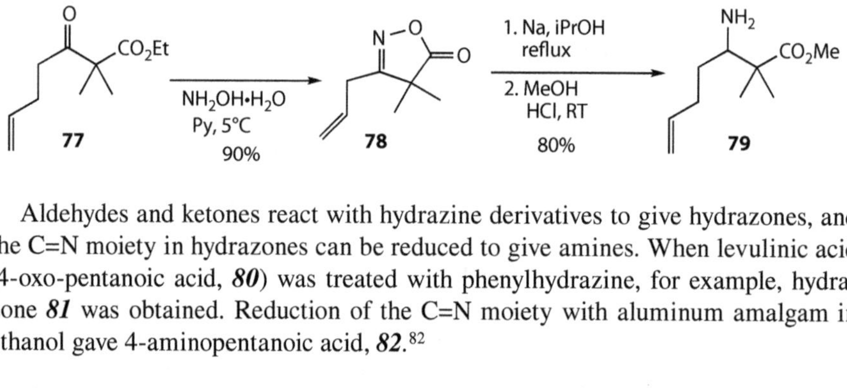

Aldehydes and ketones react with hydrazine derivatives to give hydrazones, and the C=N moiety in hydrazones can be reduced to give amines. When levulinic acid (4-oxo-pentanoic acid, ***80***) was treated with phenylhydrazine, for example, hydrazone ***81*** was obtained. Reduction of the C=N moiety with aluminum amalgam in ethanol gave 4-aminopentanoic acid, ***82***.[82]

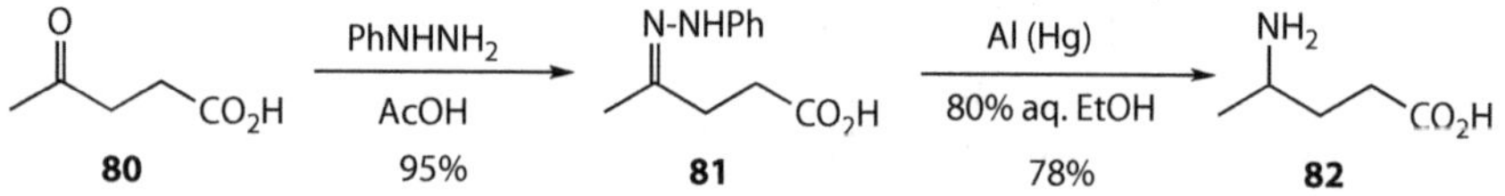

---

[77] Brannock, K.C.; Bell, A.; Burpitt, R.D.; Kelly, C.A. *J. Org. Chem.* ***1964***, *29*, 801.

[78] Weintraub, L.; Wilson, A.; Goldhamer, D.L.; Hollis, D.P. *J. Am. Chem. Soc.* ***1964***, *86*, 4880.

[79] Arata, Y.; Wada, H. *Kanazawa Daigaku Yakugakuba Kenyu Nempo* ***1963***, *13*, 72 [*Chem. Abstr.* ***1964***, *60*: 7907b]. Also see Arata, Y.; Ohashi, T.; Okumura, Z.; Wada, Y.; Ishiwata, M. *Yakugku Zasshi* ***1962***, *82*, 782 [*Chem. Abstr.* ***1963***, *58*: 2432h].

[80] Weiler, L. *J. Am. Chem. Soc.* ***1970***, *92*, 6702.

[81] (a) Shibuya, M.; Kuretani, M.; Kuboto, S. *Tetrahedron Lett.* ***1981***, *22*, 4453; (b) Shibuya, M.; Kubota, S. *Heterocycles* ***1980***, *14*, 601.

[82] Cane, D.E.; Shiao, M.-S. *J. Am. Chem. Soc.* ***1978***, *100*, 3203. Also see Fles, D.; Seke, V.; Dadic, M. *J. Polym. Sci. Part C*, ***1969***, *22*, 971.

1.3 REFUNCTIONALIZATION TO GENERATE CARBOXYL GROUPS

Just as there are several methods for incorporating an amino group into an amino acid, there are several strategies for incorporating a carboxyl moiety (–COOH) into a molecule that already contains an amine unit or an amine surrogate. Another way to say this is that there are several carboxyl surrogates.

1.3.1 Carboxyl Equivalents

A well-known approach uses the reaction of an organometallic with the electrophilic carbonyl of carbon dioxide, giving a carboxyl group directly.[83] Poor yields of acid in the reaction of carbon dioxide and Grignard reagents are common, except when aryl Grignard reagents are used.[84] It is possible to generate an amino acid from halo-amines via Grignard reagents. Formation of the Grignard reagent of ***83*** and subsequent reaction with carbon dioxide, for example, gave ethyl 4-(*N*,*N*-diethylamino)-but-2-enoate (***84***), but the yield was poor.[85]

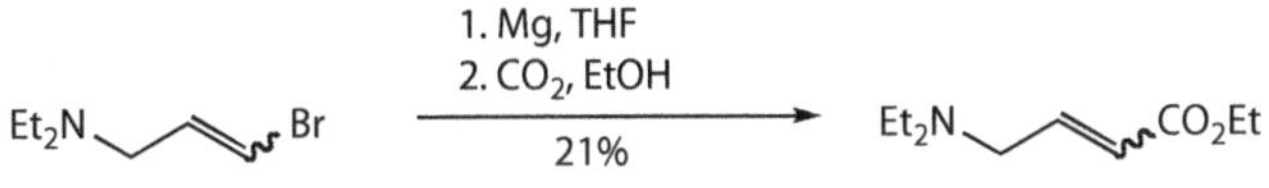

83 **84**

A variation used the alkyne anion of an amino-alkyne rather than a Grignard reagent. Treatment of ***85*** with sodium amide and then with carbon dioxide gave 4-(*N*,*N*-dimethylamino)but-2-ynoic acid, ***86***.[86] Controlled catalytic hydrogenation of the triple bond gave 4-(*N*,*N*-dimethylamino)but-2-enoic acid, ***87***.[86] In addition to ***87***, 5-(*N*,*N*-dimethylamino)pent-2-enoic acid and 8-(*N*,*N*-dimethylamino)oct-2-enoic acid were prepared by

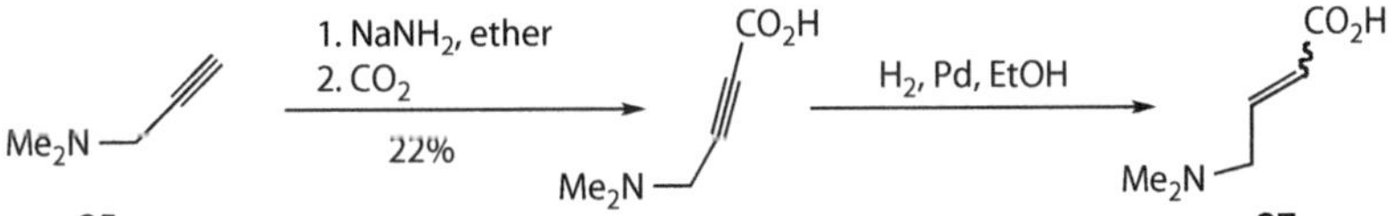

85 **86** **87**

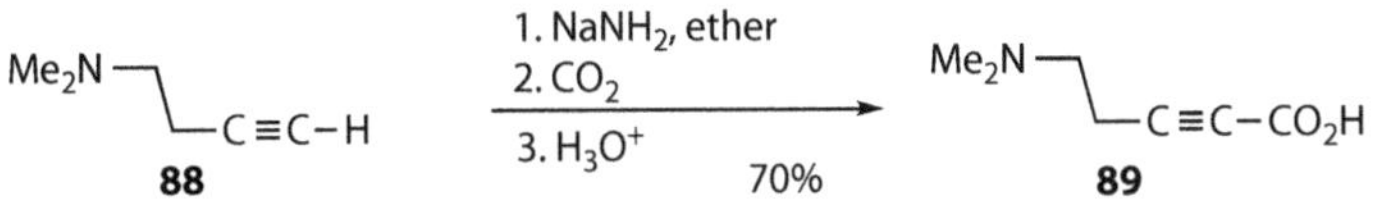

88 **89**

this method using the appropriate amino-alkyne.[86] The yield of the alkyne anion reaction with carbon dioxide was very poor, but extending the carbon chain between

---

[83] See Oppolzer, W.; Kündig, E.P.; Bishop, P.M.; Perret, C. *Tetrahedron Lett.* ***1982***, *23*, 3901.

[84] (a) Sneeden, R.P.A., in *The Chemistry of Carboxylic Acids and Esters*, Patai, S., ed. Interscience, New York, ***1969***, pp. 137–175; (b) Kharasch, M.S.; Reinmuth, O. *Grignard Reactions of Nonmetallic Substances*. Prentice-Hall, New York, ***1954***, pp. 913–948; (c) Eberson, L. *Acta Chem. Scand.* ***1962***, *16*, 781.

[85] Caubère, P. *Bull. Soc. Chim. Fr.* ***1964***, 148.

[86] Marszak, I.; Olomucki, M. *Bull. Soc. Chim. Fr.* ***1959***, 182.

the triple bond and the dimethylamino group, as in the reaction of ***88*** with sodium amide and then carbon dioxide,[87] led to a greatly improved yield of acid. Amino acid ***89*** [5-(*N*,*N*-dimethylamino)pent-2-ynoic acid] was isolated in 70% yield.[87] It should be noted that other bases can be used to generate the alkyne anion, including *n*-butyllithium.[88]

The reaction of organometallics with a chloroformate is an important direct method for introducing a carboxyl group, but as an ester. There are also several reagents such as cyanide and carbon dioxide that can be used to insert a carboxyl into an amino acid. In one example,

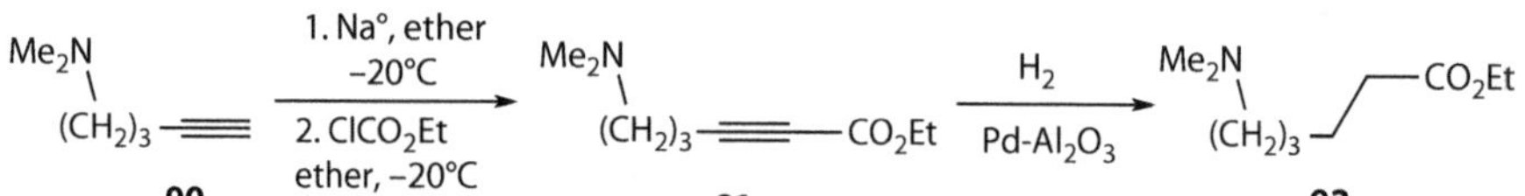

90 **91** **92**

amino-alkyne ***90*** was converted to an amino-ester via reaction with sodium metal to give the alkyne anion, and subsequent reaction with ethyl chloroformate gave ethyl 6-(*N*,*N*-dimethylamino)hex-2-ynoate, ***91***.[89] Hydrogenation of the triple bond in ***91*** gave ethyl 7-(*N*,*N*-dimethylamino)heptanoate, ***92***.[89] Amino-alkynes such as ***90*** are usually prepared by reaction of a chlorinated alkyne, such as 6-chloro-1-hexyne, and dimethylamine.[90] It is also possible to do the amination step *after* formation of, or incorporation of, the acid moiety, giving aminotetrolic acids (4-aminobut-2-ynoic acids).[91]

A common method for incorporating an acid moiety into a molecule involves the reaction of cyanide (usually with NaCN or KCN) with a molecule having a suitable leaving group to give a nitrile. In Section 1.3, cyanide was used as an aminomethyl surrogate, but hydrolysis of the cyano group gives a COOH unit. Subsequent hydrolysis of that cyano group leads to an acid moiety. Therefore, cyanide functions as both a carboxyl surrogate and an aminomethyl surrogate. The synthesis of an ω-amino acid begins with reaction of potassium phthalimide and 4-methyl-1,7-dibromoheptane to give ***93***. A subsequent reaction with KCN led to 5-methyl-8-aminooctanoic acid (***94***) after acid hydrolysis.[92] In this case, the phthalimide group functions as an amine surrogate (see Section 1.1.2) and cyanide as a carboxyl surrogate.

---

[87] (a) Olomucki, M.; Marszak, I. *Compt. Rend.* ***1956***, *242*, 1338; (b) Olomucki, M.; Ziv, D. *Bull. Soc. Chim. Fr.* ***1968***, 4923; (c) Marszak, I.; Olomucki, M. *Bull. Soc. Chim. Fr.* ***1959***, 182.

[88] MacInnes, I.; Walton, J.C. *J. Chem. Soc. Perkin Trans. I* ***1987***, 1077.

[89] Ziv, D.; Olomucki, M.; Marszak, I. *Bull. Soc. Chim. Fr.* ***1970***, 150.

[90] (a) Olomucki, M.; Marszak, I. *C.R. Acad. Sci.* ***1961***, *253*, 2239; (b) Lövgren, K.; Nilsson, J.L.G. *Acta Pharm. Suec.* ***1975***, *12*, 379; (c) also see Corrium R.J.P.; Huynh, V.; Moreau, J.J.E. *Tetrahedron Lett.* ***1984***, *25*, 1887 and Bestmann, H.J.; Wolfel, G. *Chem. Ber.* ***1984***, *117*, 1250.

[91] (a) Olomucki, M.; Marszak, I. *Bull. Soc. Chim. Fr.* ***1964***, 1767; (b) Olomucki, M.; Marszak, I. *Bull. Soc. Chim. Fr.* ***1959***, 315.

[92] Delpierre, G.R.; Eastwood, F.W.; Gream, G.E.; Kingston, D.G.I.; Sarin, P.S.; Todd, L.; Williams, D.H. *J. Chem. Soc. C* ***1966***, 1653.

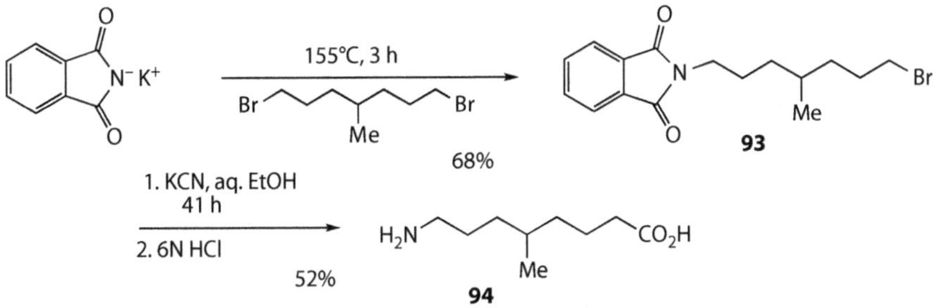

Nitriles can be prepared via acyl addition reactions and subsequently modified to generate amino acids. The Boc-protected aldehyde derived from alanine (known as *N*-Boc alanol, ***95***) was condensed with potassium cyanide to give a cyanohydrin. Hydrolysis of the cyano group gave an acid and removal of the *N*-Boc group gave 3-amino-2-hydroxybutanoic acid, ***96***.[93]

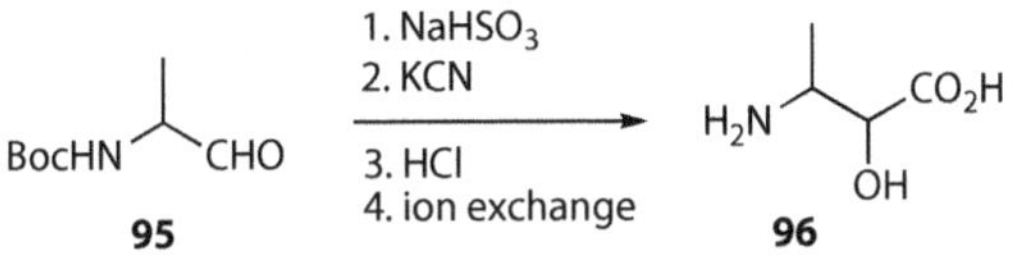

By use of the Passerini reaction with aldehydes or ketones,[94,95] isonitriles can be used as carboxyl surrogates in a three-component coupling strategy. The paclitaxel side chain (***97***) and bestatin (***98***) both contain 2-hydroxy-3-amino carboxylic acid moieties, the so-called norstatine unit (see the "boxed" fragments). In a synthesis that used the Passerini reaction, the reaction of Boc-protected amino aldehyde ***99*** with *tert*-butylisonitrile, in the presence of trifluoroacetic acid, gave a 78% yield of ***100***.[96] This approach may be used for the synthesis of norstatines, but the selectivity was only 1:1 to 1:3 in the reaction shown.

---

[93] Peet, N.P.; Burkhart, J.P.; Angelastro, M.R.; Giroux, E.L.; Mehdi, S.; Bey, P.; Kolb, M.; Neises, B.; Schirlin, D. *J. Med. Chem.* ***1992***, *33*, 394.

[94] See (a) Armstrong, R.W.; Combs, A.P.; Tenpest, P.A.; Brown, S.D.; Keating, T.A. *Acc. Chem. Res.* ***1996***, *29*, 123; (b) Lumma, W. *J. Org. Chem.* ***1981***, *46*, 3668.

[95] See (a) Smith, M.B. *Organic Synthesis*, 3rd ed. Wavefunction, Inc./Elsevier, Irvine, CA/London, England, ***2010***, p. 627; (b) Smith, M.B. *March's Advanced Organic Chemistry*, 7th ed. John Wiley & Sons, Hoboken, NJ, ***2013***, p. 1247.

[96] Semple, J.E.; Owens, T.D.; Nguyen, K.; Odile, E.; Levy, O.E. *Org. Lett.* ***2000***, *2*, 2769.

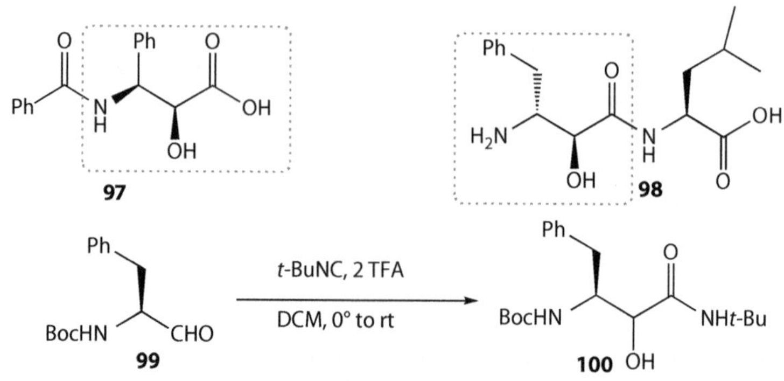

β-Ketosulfoxides serve as carboxyl precursors[97] thanks to the Pummerer rearrangement, followed by selective acyl migration. Sulfoxide ***101*** was prepared in several steps from phenylalanine, and reaction with acetic anhydride with pyridine and 4-dimethylaminopyridine (DMAP) led to a Pummerer rearrangement[98] to give ***102***.[99] Subsequent treatment with 1,8-diazabicyclo[5.4.0]undec-7-ene (DBU) gave thioester ***103***, with good selectivity, and subsequent hydrolysis led to the β-amino-α-hydroxyamino acid.

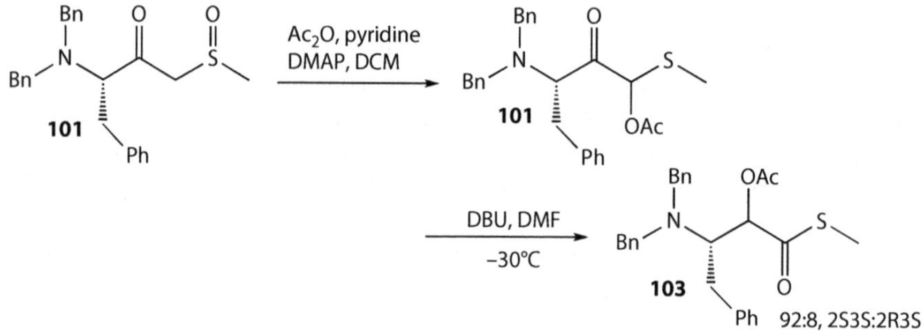

allo-Threonine (via amide ***106***) was prepared from 3*S*-aminobutanoic acid (***104***), which was prepared in enantiopure form by the enzymatic hydrolysis with penicillin G acylase. Reaction with benzyl chloride was followed by conversion to oxazoline ***105***. Hydrolysis gave *N*-benzamide ***106***, which was later hydrolyzed to *allo*-threonine.[100] This transformation could be discussed in Section 1.4.2, which shows the transformation of one amino acid to another. However, the reaction is placed here because the oxazolidine moiety functions as a carboxyl surrogate, in this case, as a protected carboxylic acid.

[97] (a) Sharma, A.K.; Swern, D. *Tetrahedron Lett.* ***1974***, 1503; (b) Sharma, A.K.; Ku, T.; Dawson, A.D.; Swern, D. *J. Org. Chem.* ***1975***, *40*, 2758; (c) Pummerer, R. *Berichte* ***1910***, *43*, 1401.

[98] See (a) Smith, M.B. *Organic Synthesis*, 3rd ed. Wavefunction, Inc./Elsevier, Irvine, CA/London, England, ***2010***, p. 241; (b) Smith, M.B. *March's Advanced Organic Chemistry*, 7th ed. John Wiley & Sons, Hoboken, NJ, ***2013***, pp. 1565–1566.

[99] See Suzuki, T.; Honda, Y.; Izawa, K.; Williams, R.M. *J. Org. Chem.* ***2005***, *70*, 7317.

[100] Cardillo, G.; Tolomelli, A.; Tomasini, C. *Eur. J. Org. Chem.* ***1999***, 155.

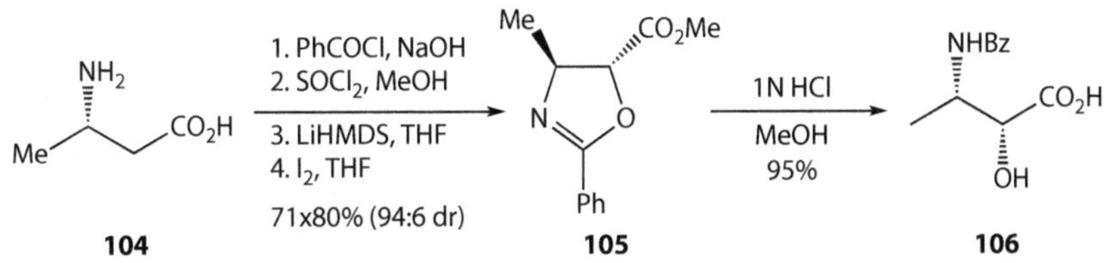

1.3.2 Carboxyl-Bearing Carbanions and Ylids

A useful method that incorporates a carboxyl group or a carboxyl surrogate into a molecule relies on carbanion or ylid derivatives, in reactions with aldehydes, ketones, or alkyl halides. Ester and carboxylic acid enolate anions, for example, react with a variety of electrophilic reagents to produce new ester or acid derivatives. If the electrophilic agent contains an amine or an amine surrogate, or functionality that will allow the introduction of such species at a later time, amino acids may be synthesized. Ylids bearing an acid or ester moiety react with aldehydes or ketones bearing an amine moiety via a Wittig reaction[101] or a Horner-Wadsworth-Emmons reaction[102] to give alkenyl amino acids.

"Simple" ester enolates (those derived from mono-esters) can be used to prepare amino acids. The lithium enolate of *tert*-butyl acetate displaced the methoxy group in ***107***, for example, to give the methyl carbamate of *tert*-butyl 3-aminobutanoate, ***109***.[103] An acid surrogate was used with ***107*** to produce an amino acid. Meyers developed an oxazolidine derivative (***108***)[104] that was readily converted to the carbanion by reaction with lithium diisopropylamide. In the presence of Ti(IV), this carbanion condensed with ***109***, and hydrolysis liberated the methyl carbamate of methyl 3-aminobutanoate, ***110***, in good yield and with high asymmetric induction (see Chapter 5).[104]

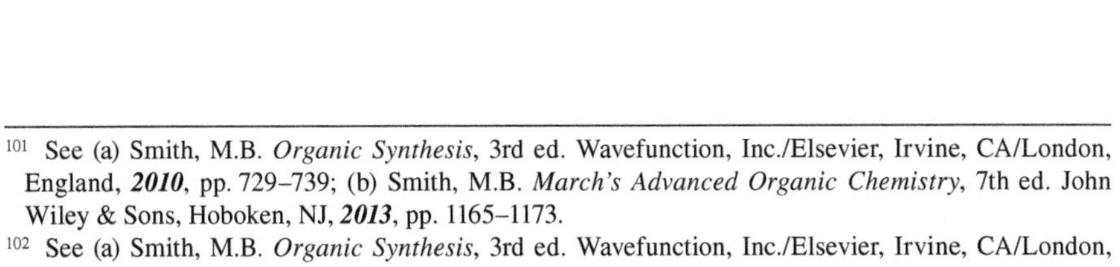

---

[101] See (a) Smith, M.B. *Organic Synthesis*, 3rd ed. Wavefunction, Inc./Elsevier, Irvine, CA/London, England, ***2010***, pp. 729–739; (b) Smith, M.B. *March's Advanced Organic Chemistry*, 7th ed. John Wiley & Sons, Hoboken, NJ, ***2013***, pp. 1165–1173.

[102] See (a) Smith, M.B. *Organic Synthesis*, 3rd ed. Wavefunction, Inc./Elsevier, Irvine, CA/London, England, ***2010***, pp. 739–744; (b) Smith, M.B. *March's Advanced Organic Chemistry*, 7th ed. John Wiley & Sons, Hoboken, NJ, ***2013***, pp. 1169–1170.

[103] Shono, T.; Kise, N.; Sanda, F.; Ohi, S.; Tsubata, K. *Tetrahedron Lett.* ***1988***, *29*, 231.

[104] (a) Meyers, A.I.; Nabeya, A.; Adickes, H.W.; Politzer, I.R. *J. Am. Chem. Soc.* ***1969***, *91*, 763; (b) Meyers, A.I.; Nabeya, A.; Adickes, H.W.; Politzer, I.R.; Kovelesky, A.C.; Nolen, R.L.; Portnoy, R.C. *J. Org. Chem.* ***1973***, *38*, 36.

Malonate anions are convenient sources of ester enolate anions and carbanion equivalents, and reaction with suitable substrate leads to an amino acid. An example used phthalic anhydride as a starting material in a reaction with 2-aminoethanol to give ***111***.[105] Conversion of the alcohol moiety in ***111*** to the *O*-benzenesulfonate ester allowed displacement by sodium diethyl malonate to give ***112***. Treatment with aqueous sulfuric acid and neutralization with barium carbonate hydrolyzed the ester, cleaved the phthalimide group, hydrolyzed the ester moieties, and induced decarboxylation to give 4-aminobutanoic acid.[105]

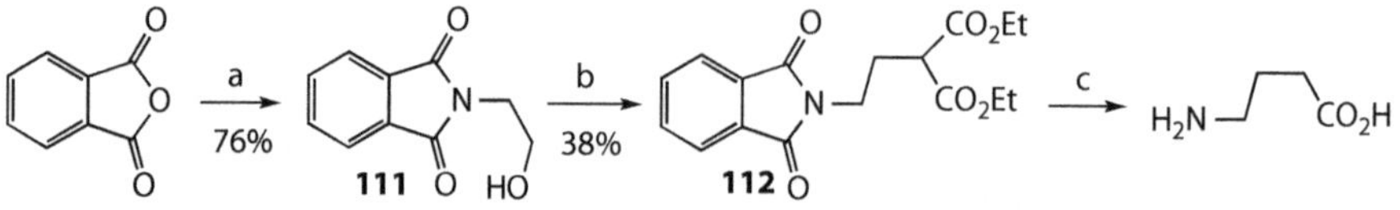

(a) $HOCH_2CH_2NH_2$, PhMe (b) 1. $PhSO_2Cl$, Py 2. $CH_2(CO_2Et)_2$, PhMe, Na°, reflux (c) 1. aq. H_2SO_4, reflux 2. aq. $BaCO_3$

A variation in this approach used the lithium enolate of diethyl malonate (generated in situ from diethyl malonate and lithium hexamethyldisilazide) in a Pd-catalyzed π-allyl displacement reaction with the allylic acetate moiety in ***113***. In the presence of palladium(0), formed in situ from the reaction of (tetrakis)triphenylphosphino palladium and triphenylphosphine,[106] the malonate anion reacted in an addition-elimination sequence that gave ethyl 6-(*N*-Boc amino)-7-(4-benzyloxyphenyl)-2-carboethoxyhept-4-enoate, ***114***.[107]

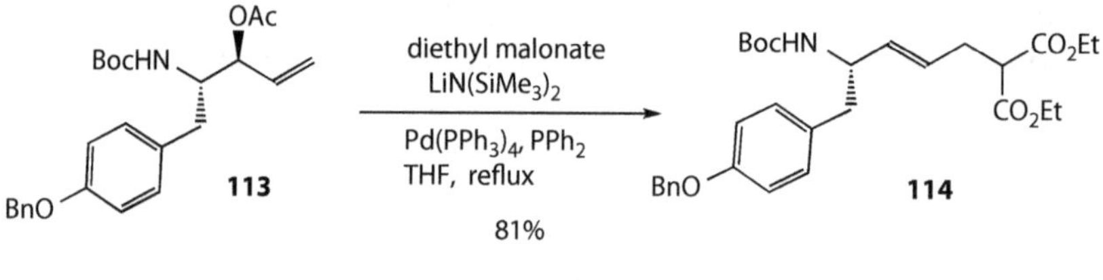

Carboxy-substituted ylids react with aldehydes and ketones to give alkenyl esters. With a suitable reactant, alkenyl amino acids can be prepared using this approach. One example used a phosphonate ester ylid in a Horner-Wadsworth-Emmons

[105] Jakobiec, T. *Acta Polon. Pharm.* ***1966***, *23*, 111 [*Chem. Abstr.* ***1966***, *65*: 10492b]. Also see Edwards, D.; Hamer, D.; Stewart, W.H. *J. Pharm. Pharmacol.* ***1964***, *16*, 618.

[106] (a) Trost, B.M.; Weber, L.; Strege, P.E.; Fullerton, T.J.; Dietsche, T.J. *J. Am. Chem. Soc.* ***1978***, *100*, 3416; (b) Trost, B.M.; Weber, L.; Strege, P.E.; Fullerton, T.J.; Dietsche, T.J. *J. Am. Chem. Soc.* ***1978***, *100*, 342; (c) Trost, B.M.; Verhoeven, T.R. *J. Am. Chem. Soc.* ***1976***, *98*, 630; (d) Trost, B.M.; Verhoeven, T.R. *J. Am. Chem. Soc.* ***1978***, *100*, 3435; (e) Takahashi, K.; Miyake, A.; Hata, G. *Bull Chem. Soc. Jpn.* ***1970***, *45*, 230,1183; (f) Trost, B.M.; Verhoeven, T.R. *J. Org. Chem.* ***1976***, *41*, 3215; (g) Trost, B.M.; Verhoeven, T.R. *J. Am. Chem. Soc.* ***1980***, *102*, 4730.

[107] Thompson, W.J.; Tucker, T.J.; Schwering, J.E.; Barnes, J.L. *Tetrahedron Lett.* ***1990***, *31*, 6819.

reaction[108,109] with Cbz-protected amino aldehyde ***115***. The product after hydrolysis was conjugated acid ***116***.[110] Cleavage of the *N*-Cbz group with trifluoroacetic acid gave 3-(1-cyclopropyl-1-amino)prop-2-enoic acid, ***117***.

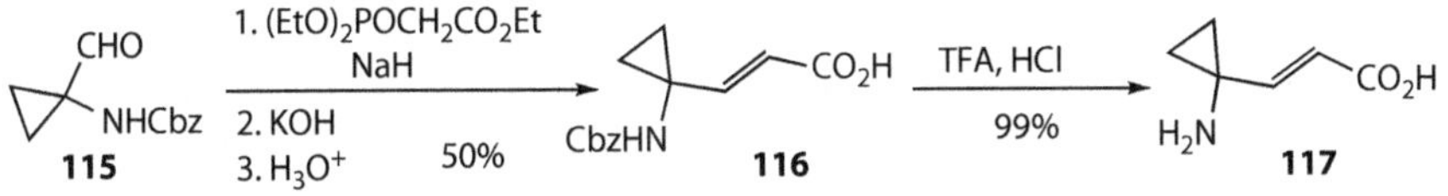

A Wittig olefination[111] reaction was used to prepare an enamino-ester. The requisite starting material (***119***) was prepared by reaction of *n*-butylmagnesium bromide and isothiocyanate ***118***.[112] Ethyl *N*-thioacyl urethane ***119*** reacted with the Wittig reagent shown to give ***120***.[113] This sequence was also used with methylmagnesium bromide to give the 3-aminobut-2-enoate derivative, with ethylmagnesium bromide to give the 3-aminopent-2-enoate derivatives, and with *n*-propylmagnesium bromide to give 3-aminohex-2-enoate.[112]

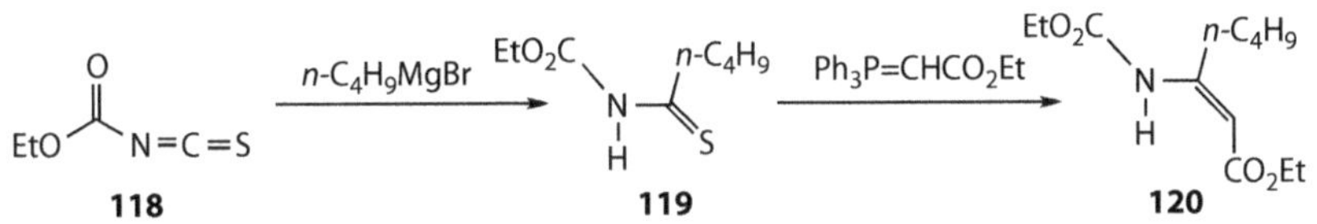

1.3.3 Oxidative Methods

There are several oxidation reactions that will "unmask" a suitable precursor to give a carboxylic acid. Perhaps the most obvious is oxidation of a primary alcohol to a carboxylic acid using chromium(VI) reagents,[114] illustrated by oxidation of the primary alcohol moiety in phthalimidoyl alcohol ***121*** (prepared by literature methods).[115] Subsequent catalytic hydrogenation of the triple bond gave 4-phthalimidoylbut-2*Z*-enoic acid, ***122***.[116] Treatment with diethylamine unmasked the phthalimide group to give 4-aminobut-2*Z*-enoic acid, ***123***. Although protected amines or amine surrogates

[108] (a) Horner, L.; Hoffmann, H.; Wippel, J.H.; Klahre, G. *Ber.* ***1959***, *92*, 2499; (b) Wadsworth, W.S., Jr.; Emmons, W.D. *J. Am. Chem. Soc.* ***1961***, *83*, 1733; (c) Boutagy, J.; Thomas, R. *Chem. Rev.* ***1974***, *74*, 87.

[109] See (a) Smith, M.B. *Organic Synthesis*, 3rd ed. Wavefunction, Inc./Elsevier, Irvine, CA/London, England, ***2010***, pp. 739–744; (b) Smith, M.B. *March's Advanced Organic Chemistry*, 7th ed. Wiley & Sons, Hoboken, NJ, ***2013***, pp. 1169–1170.

[110] Silverman, R.B.; Invergo, B.J.; Mathew, J. *J. Med. Chem.* ***1986***, *29*, 1840. Also see Liu, S.; Hanzlik, R.P. *J. Med. Chem.* ***1992***, *35*, 1067.

[111] See (a) Smith, M.B. *Organic Synthesis*, 3rd ed. Wavefunction, Inc./Elsevier, Irvine, CA/London, England, ***2010***, pp. 729–739; (b) Smith, M.B. *March's Advanced Organic Chemistry*, 7th ed. John Wiley & Sons, Hoboken, NJ, ***2013***, pp. 1165–1173.

[112] Gossauer, A.; Roessler, F.; Zilch, H. *Liebigs Ann. Chem.* ***1979***, 1309.

[113] Slopianka, M.; Gossauer, A. *Synth. Commun.* ***1981***, *11*, 95.

[114] Wiberg, K.B., in *Oxidation in Organic Chemistry, Part A*, Wiberg, K.B., ed. Academic Press, New York, ***1965***, p. 174.

[115] Allan, R.D.; Johnston, G.A.R.; Twitchin, B. *Aust. J. Chem.* ***1980***, *33*, 1115.

[116] Allan, R.D.; Johnston, G.A.R.; Kazlauskas, R. *Aust. J. Chem.* ***1985***, *38*, 1647.

are commonly used, as in ***121***, it is possible to oxidize a primary alcohol in the presence of an amine unit. The reaction of amino-alkyne ***124*** with ethylmagnesium bromide give the alkyne anion, and a subsequent standard by condensation with formaldehyde gave ***125***. Subsequent oxidation of the alcohol moiety to the acid moiety gave 4-(*N*,*N*-diethylamino)hept-2-ynoic acid, ***126***.[117] A number of other alkynyl amino acids were prepared by this method, including 4-(*N*,*N*-diethylamino)pent-2-ynoic acid, 4-(*N*,*N*-diethylamino)but-2-ynoic acid, and 4-(*N*,*N*-diethylamino)-4-(2-furyl)but-2-ynoic acid.[117]

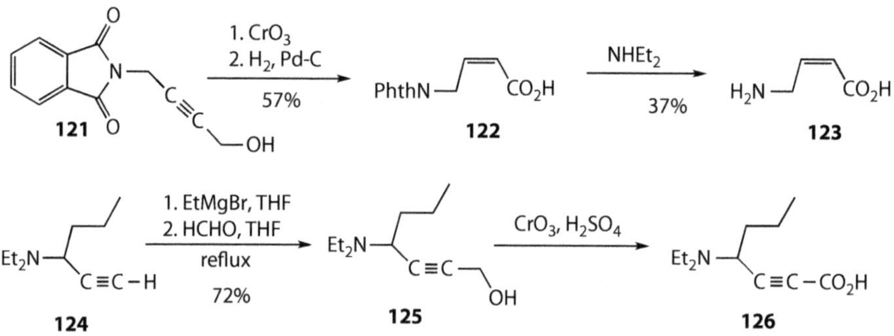

Another synthetic option that uses this strategy oxidizes an alcohol to an acid earlier in the sequence, with the amine moiety incorporated later, as seen in the conversion of ***127*** to ***128*** in 29% yield.[118] Conversion of alkyne ***127*** to the corresponding alkyne anion allowed reaction with the functionalized alkyl halide shown. Hydrogenation of the alkyne moiety allowed hydrolysis of the protecting group on oxygen in ***127*** to the alcohol. Oxidation of the primary alcohol to the corresponding acid was followed by conversion to the methyl ester. Deprotection of the acetal and conversion to the corresponding amine via reductive amination was followed by protection as the benzenesulfonamide and saponification of the methyl ester to give ***128***.[118]

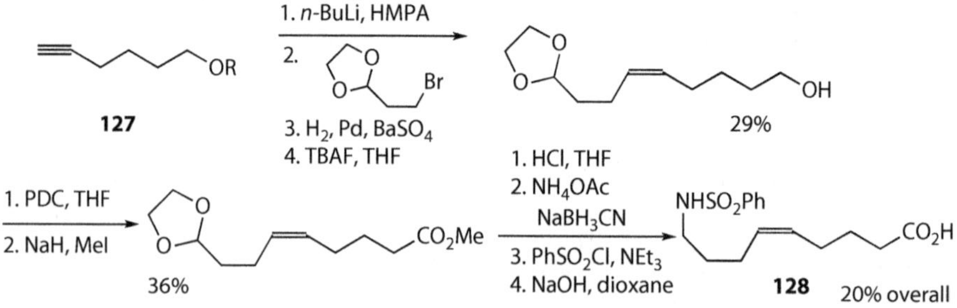

ω-Amino acids ranging from 3-aminopropanoic acid to 13-aminotridecanoic acid have been functionalized to produce new bioactive derivatives. Used in combination

[117] Crika, A.; Kupetis, G.; Mozolis, V. *Liet. TSR Mokslu Akad. Darb. Ser. B* ***1969***, 101 [*Chem. Abstr.* ***1970***, *72*: 89715u].

[118] Ladouceur, G.; Mais, D.E.; Jakubowski, J.A.; Utterback, B.; Robertson, D.W. *Bioorg. Med. Chem. Lett.* ***1991***, *1*, 173.

with heparin, the amino acid derivatives led to an increase in both plasma heparin concentrations (anti-factor Xa) and clotting times, APTT (activated partial thromboplatin time).[119] An example is the conversion of undec-10-en-1-amine (***129***) to acyl derivative ***130***. Subsequent oxidation of the alkenyl moiety gave the amino acid derivative, ***131***.

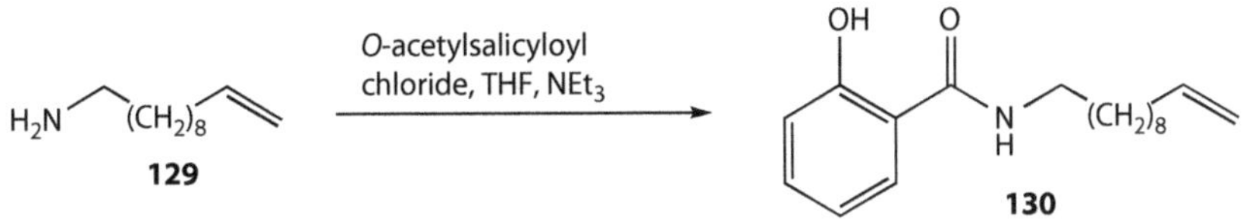

129

130

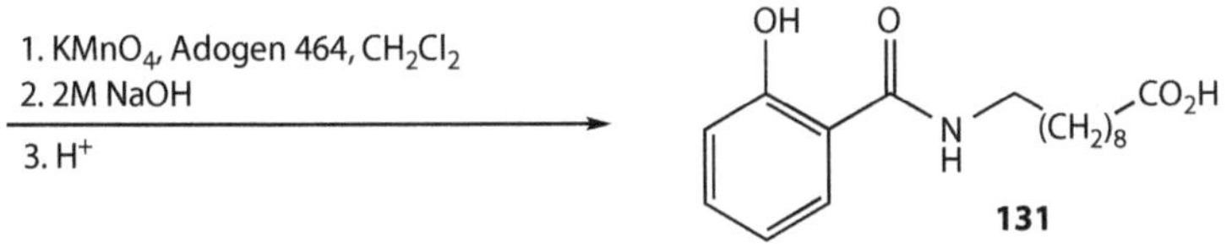

131

1.4 REFUNCTIONALIZATION OF AMINO ACIDS

Rather than prepare amino acids via the de novo syntheses described in previous sections, an alternative approach simply modifies another amino acid. Many transformations involve the conversion of chiral, nonracemic α-amino acids to chiral, nonracemic non-α-amino acids, and these approaches will be discussed further in Chapter 5. There are, however, several procedures that generate functionality in an amino acid product in which the stereogenic center is retained, racemized, or lost altogether. There are many sequences that use chiral, racemic amino acid precursors to prepare substituted non-α-amino acids. All 20 α-amino acids have been converted to the corresponding *N*-methyl-β-amino acid, for example, by conversion to the corresponding 1,3-oxazolidin-5-one or 1,3-oxazinan-6-one.[120] Chain extension was accomplished by homologation using the Arndt-Eistert procedure.[121]

1.4.1 From α-Amino Acids

Glycine is perhaps the simplest α-amino acid, and glycine derivatives are useful templates to prepare other amino acids. In one example, ethyl *N*-benzoyl glycine was converted to an enolate anion by reaction with LDA. Subsequent reaction with the mixed anhydride shown led to displacement of the acetate moiety to give ***132***.[122] Acid hydrolysis generated a β-keto-amino acid, which decarboxylated under the reaction conditions to give 4-oxo-5-aminopentanoic acid (***133***, also known as

[119] O'Toole, D.G., Wang, E.; Harris, E.; Rosado, C.; Rivera, T.; DeVincent, A.; Tai, M.; Mercogliano, F.; Agarwal, R.; Leipold, H.; Baughman, R.A. *J. Med. Chem.* ***1998***, *41*, 1163.

[120] Hughes, A.B.; Sleebs, B.E. *Helv. Chim. Acta* ***2006***, *89*, 2611.

[121] Meier, H.; Zeller, K. *Angew. Chem. Int. Ed.* ***1975***, *14*, 32; Kirmse, W. *Carbene Chemistry*, 2nd ed. Academic Press, New York, ***1971***, pp. 475–493; Whittaker, D., in Patai, S. *The Chemistry of Diazonium and Diazo Compounds*, pt. 2. Wiley, New York, ***1978***, pp. 593–644.

[122] Evans, D.A.; Sidebottom, P.J. *J. Chem. Soc. Chem. Commun.* ***1978***, 753.

5-aminolevulinic acid). Amino acid ***133*** is important in the porphyrin synthesis pathway and has been proposed for use as a biodegradable herbicide, an insecticide, and a plant growth regulator.[123]

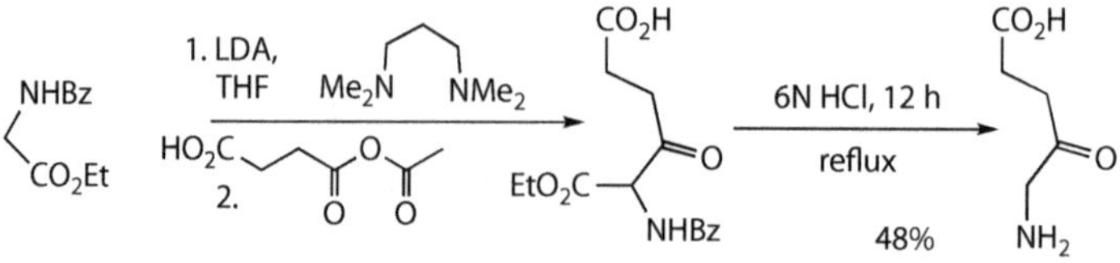

132 **133**

Glycine surrogates can be used in this strategy. The phthalimido derivative of glycine (***134***, prepared from the reaction of glycine and phthalic anhydride)[124] was condensed with diethyl 2-fluorosuccinate and then hydrolyzed (which proceeded with decarboxylation) to give 2-fluoro-4-phthalimidobut-2-enoic acid, ***135***, but in very poor yield.[125] Despite the poor conversion, this is a fast route to these compounds.

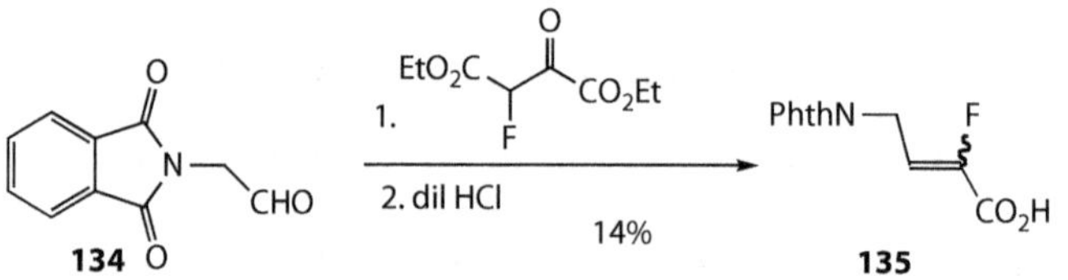

134 **135**

The carboxyl unit of an α-amino acid can be refunctionalized and then used to prepare other amino acids. *N*-Boc alanine, for example, was converted to the acid chloride and then condensed with the Mg enolate shown, derived from diethyl malonate, to give ***136***.[126] Catalytic hydrogenation of the ketone moiety gave the alcohol group in ***137***. Subsequent conversion to the alkyl chloride and base-induced dehydrohalogenation gave ethyl 4-(*N*-Boc-amino)pent-2-enoate (***138***).[127]

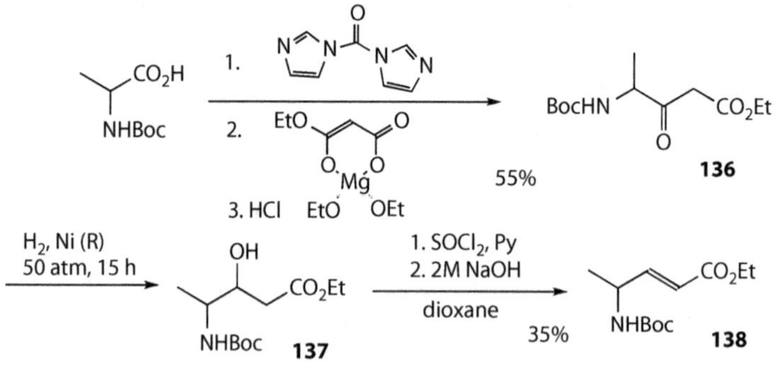

[123] Fukuda, H.; Casas, A.; Baatile, A. *Int. J. Biochem. Cell Biol.* ***2005***, *37*, 272.
[124] Siedel, W.; Nahm, H. *Ger. Pat.* 928,711 [*Chem. Abstr.* ***1958***, *52*: P5471].
[125] Bergmann, E.D.; Cohen. A. *Tetrahedron Lett.* ***1965***, 2085.
[126] Grzonka, Z.; Lammek, B. *Synthesis*, ***1974***, 661.
[127] Honoré, T.; Hjeds, H.; Krogsgaard-Larsen, P.; Christiansen, T.R. *Eur. J. Med. Chem. Chim. Ther.* ***1978***, *13*, 429.

The acid unit of an amino acid can be reduced to an aldehyde, allowing refunctionalization. Reduction of phenylalanine to the aldehyde (phenylalanal, ***139***) was followed by Horner-Wadsworth-Emmons olefination[109] to give the alkene, and removal of the *N*-Boc group with trifluoroacetic acid gave methyl 4-amino-5-phenylpent-2-enoate, ***140***.[128] In a related method, amino aldehydes can also be condensed with triphenylphosphonium ylids in a Wittig olefination reaction.[129,130] These compounds were prepared as protease inhibitors.[128]

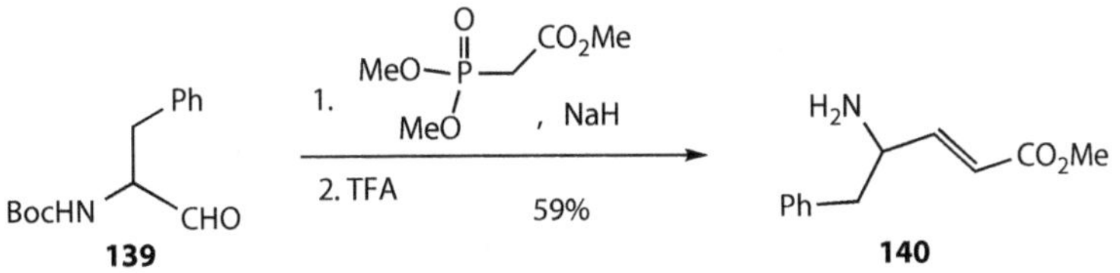

Another route used *N*-Boc phenylalanine as well as other α-amino acids in a reaction with Meldrum's acid to give a keto-diester that cyclized to keto-lactam ***141***.[131] Alternatively, heating in toluene led to the analogous *N*-Boc 5-benzyl-2-pyrrolidinone rather than the keto-lactam. Reduction of the ketone moiety in ***141*** gave the hydroxy-lactam ***142***, and base-induced hydrolysis gave *N*-Boc-3-hydroxy-4-amino-5-phenylpentanoic acid, ***143***. Similarly, *N*-Boc-5-benzyl-2-pyrrolidinone was converted to *N*-Boc-4-amino-5-phenylpentanoic acid.[131]

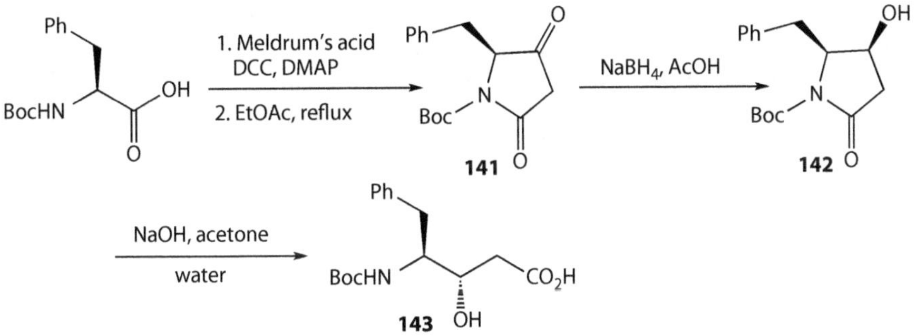

There are, of course, many α-amino acids that may be used to prepare new amino acids. One is arginine derivative ***144***, which reacted with HONO (generated in situ by reaction with sodium nitrite and acid) to convert one amino group to an alcohol

[128] Thompson, S.A.; Andrews, P.R.; Hanzlik, R.P. *J. Med. Chem.* ***1986***, *29*, 104.

[129] (a) Wittig, G.; Rieber, M. *Ann.* ***1949***, *562*, 187; (b) Wittig, G.; Geissler, G. *Ann.* ***1953***, *580*, 44; (c) Wittig, G.; Schöllkopf, U. *Chem. Ber.* ***1954***, *87*, 1318; (d) Gensler, W.J. *Chem. Rev.* ***1957***, *57*, 191 (see p. 218).

[130] See (a) Smith, M.B. *Organic Synthesis*, 3rd ed. Wavefunction, Inc./Elsevier, Irvine, CA/London, England, ***2010***, pp. 729-739; (b) Smith, M.B. *March's Advanced Organic Chemistry*, 7th ed. John Wiley & Sons, Hoboken, NJ, ***2013***, pp. 1165-1173.

[131] Smrcina, M.; Majer, P.; Majerová, E.; Guerassina, T.A.; Eissenstat, M.A. *Tetrahedron* ***1997***, *53*, 12867.

moiety. The final product was 2-hydroxy-4-aminobutanoic acid (***145***).[132] It is noted that the aminoglycoside antibiotic pyrankacin contains the 4-amino-2-hydroxybutanoic acid unit (***145***).[133]

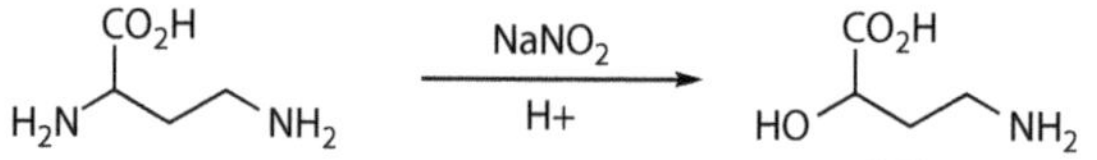

144 **145**

3-Amino-4-oxoamino acids have been prepared by refunctionalization of *L*-homoserine. Conversion of homoserine to the *N*-benzenesulfonamide allowed a nickel-catalyzed reaction with various alkyl Grignard reagents to give the amido-ketone.[134] Jones oxidation[135] of the primary alcohol moiety was followed by deprotection to give the targeted ***146***. This amino acid was used to prepare a β-peptide trimer.[134]

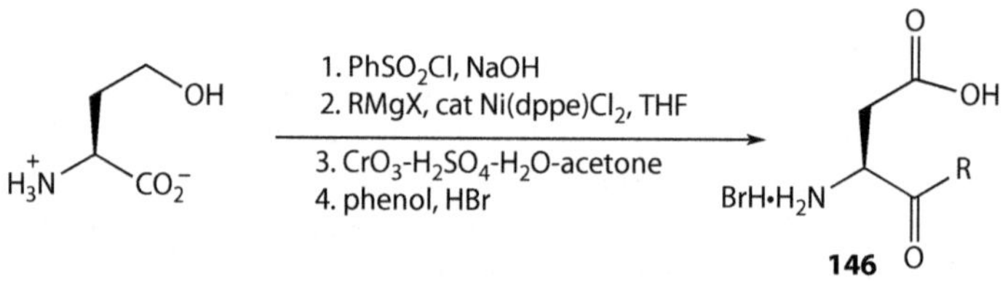

146

Both aspartic acid[136] and glutamic acids were converted to β-amino acid derivatives via the corresponding benzotriazole (Bt) derivative. The methyl ester of trifluoroacetic-protected aspartic acid was converted to benzotriazole derivative ***147***, and subsequent Friedel-Crafts acylation[137] with indole gave ***148*** in 70% yield.[138] Reduction with sodium borohydride gave ***149*** in 87% yield. Katritsky et al. noted in this work that both β- and γ-amino acids are constituents of many biologically important compounds.[138]

---

[132] Horiuchi, Y.; Akita, E.; Ito, T. *Agric. Biol. Chem.* ***1976***, *40*, 1649. Also see Nomoto, S.; Shiba, T. *Chem. Lett.* ***1978***, 589.

[133] Rai, R.; Chen, H.N.; Czyrca, P.G.; Li, J.; Chang, C.-W.T. *Org. Lett.* ***2006***, *8*, 887.

[134] Sharma, A.K.; Hergenrother, P.J. *Org. Lett.* ***2003***, *5*, 2107.

[135] See (a) Smith, M.B. *Organic Synthesis*, 3rd ed. Wavefunction, Inc./Elsevier, Irvine, CA/London, England, ***2010***, pp. 233–234; (b) Smith, M.B. *March's Advanced Organic Chemistry*, 7th ed. John Wiley & Sons, Hoboken, NJ, ***2013***, pp. 1142–1145.

[136] Also see Jefford, C.W.; McNulty, J.; Lu, Z.-H.; Wang, J.B. *Helv. Chim. Acta* ***1996***, *79*, 1203.

[137] See (a) Smith, M.B. *Organic Synthesis*, 3rd ed. Wavefunction, Inc./Elsevier, Irvine, CA/London, England, ***2010***, pp. 1203–1210; (b) Smith, M.B. *March's Advanced Organic Chemistry*, 7th ed. John Wiley & Sons, Hoboken, NJ, ***2013***, pp. 621–625.

[138] Katritzky, A.R.; Tao, H.; Jiang, R.; Suzuki, K.; Kirichenko, K. *J. Org. Chem.* ***2007***, *72*, 407.

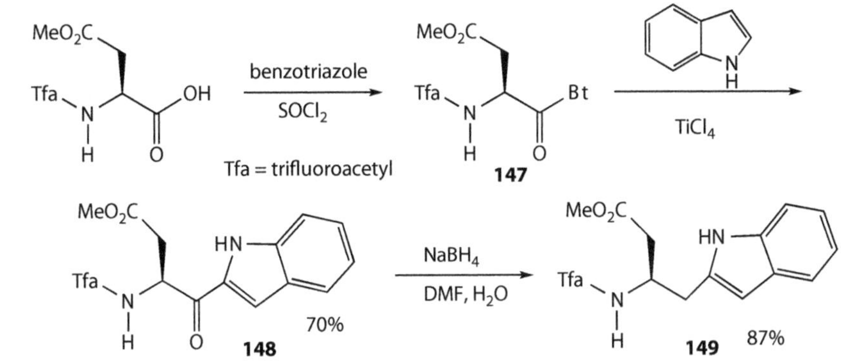

Aspartic acid was used as a precursor to prepare substituted amino acids by cyclization to an anhydride, followed by selective reduction of one carbonyl group to give lactone ***150***.[139] Ring opening gave the acyclic iodide, and reaction with several organocuprates gave ***151***. Deprotection and hydrolysis gave the β-amino acid, ***152***. In this case, the carboxyl group of the α-amino acid was selectively reduced, allowing addition of a targeted alkyl group. Diversification of this scheme, including the use of chiral auxiliaries, led to other derivatives, including components of bestatin (see Chapter 4, Section 4.4 and Chapter 6, Section 6.4.2) and microginin.[139]

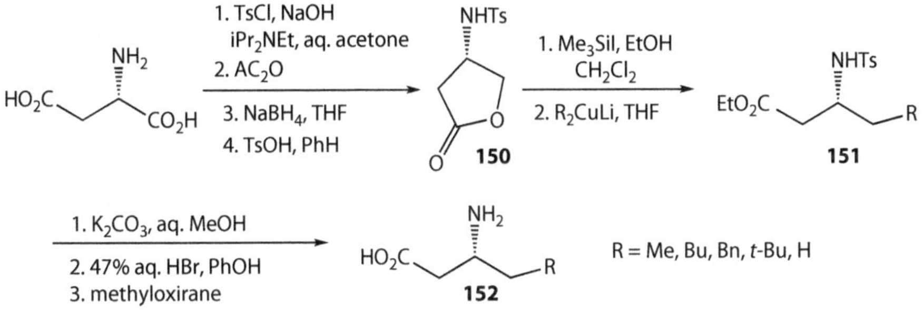

The amino acid balinol (***156***) is a fungal metabolite that has been shown to have inhibitory properties toward the protein kinase C, which plays a role in cell growth, signal transduction, and differentiation.[140] The synthesis of α-amino acid ***153*** provided the starting material for this synthesis. Subsequent conversion to cyclic amine ***154*** and deprotection gave diamino-alcohol ***155***. This synthesis constitutes a formal synthesis of ***156***.[141]

[139] Jefford, C.W.; McNulty, J.; Lu, Z.-H.; Wang, J.B. *Helv. Chim. Acta* ***1996***, *79*, 1203.

[140] See (a) Kulanthaivel, P.; Hallock, Y.F.; Boros, C.; Hamilton, S.M.; Janzen, W.P.; Ballas, L.M.; Loomis, C.R.; Jiang, J.B.; Katz, B.; Steiner, J.R.; Clardy, J. *J. Am. Chem. Soc.* ***1993***, *115*, 6452; (b) Nishizuka, Y. *Nature* ***1984***, *308*, 693; (c) Nishizuka, Y. *Science* ***1986***, *233*, 305.

[141] Phansavath, P.; Duprat de Paule, S.; Ratovelomanana-Vidal, V.; Genêt, J.-P. *Eur. J. Org. Chem.* ***2000***, *23*, 3903.

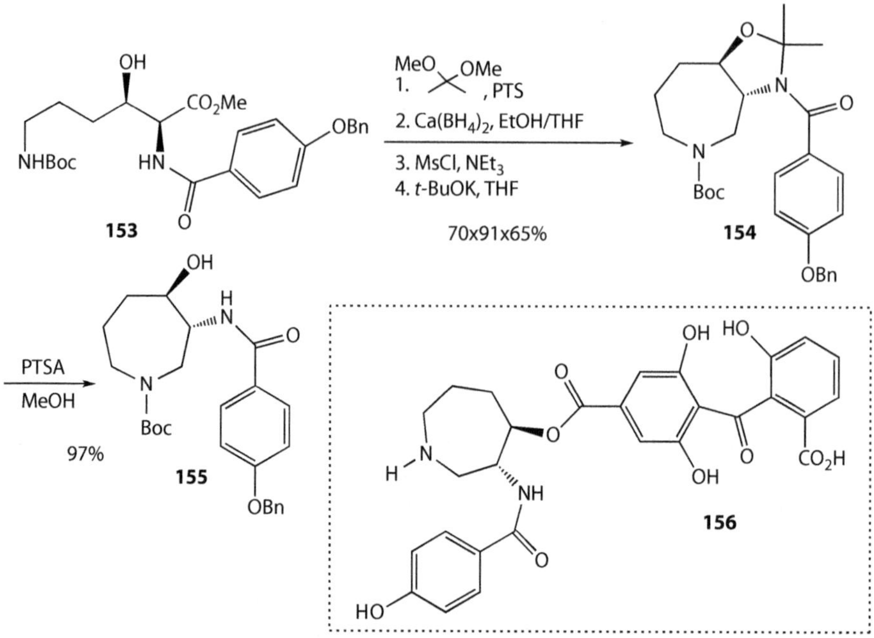

An amino acid homologation sequence was reported using an α-amino acid such as *L*-valine, which was converted to nitrile ***189***. A Blaise reaction[142,143] converted ***189*** to ***190***, and subsequent reduction of the enamino-ester gives the protected amino acid ***191***.[144]

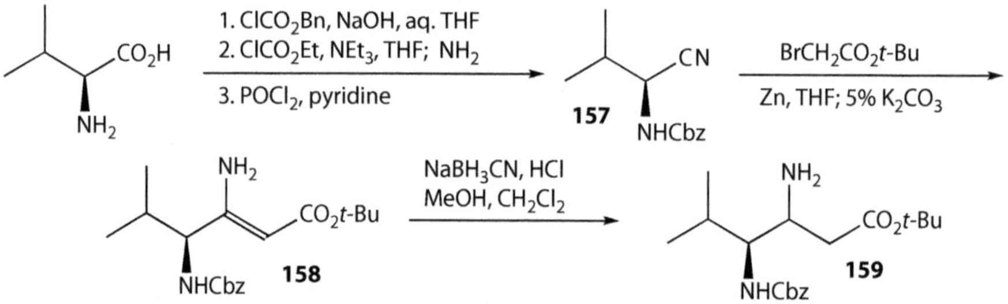

[142] (a) Blaise, E.E. *C.R. Hebd. Sci. Acad. Sci.* ***1901***, *132*, 478; (b) Cason, J.; Rinehart, K.L.; Thornton, S.D. *J. Org. Chem.* ***1953***, *18*, 1594.

[143] See Hannick, S.M.; Kishi, Y. *J. Org. Chem.* ***1983***, *48*, 3833.

[144] Hoang, C.T.; Bouillère, F.; Johannesen, S.; Zulauf, A.; Panel, C.; Pouilhès, A.; Gori, D.; Alezra, V.; Kouklovsky, C. *J. Org. Chem.* ***2009***, *74*, 4177.

1.4.2 From Other Non-α-Amino Acids

An abundant and widely used amino acid is γ-aminobutyric acid (GABA, 4-aminobutanoic acid), along with several synthetic derivatives of GABA.[145] This stems from the fact that GABA is a mammalian neural transmitter (see Chapter 6, Section 6.3), and preparation of GABA and the synthesis of analogs have received quite a bit of attention. GABA exhibits other important biological activity (see Chapter 6, Section 6.3).

Several methods are available to prepare functionalized or substituted GABA derivatives. A simple example reacted *N*-phthaloyl-4-aminobutanoic acid with ethanolic bromine and red phosphorus to give ethyl 2-bromo-4-phthalimido butanoate, ***160***.[146] The labile α-bromo moiety can be converted to other functionality, making ***160*** a useful synthetic intermediate. Chlorination is also possible, and the preparation of 4-amino-2-chlorobutanoic acid (***161***) was followed by conversion to the 2-fluoro derivative, ***162***.[147] It is noted that a synthetic approach to polyfluoro GABA derivatives has been reported.[148]

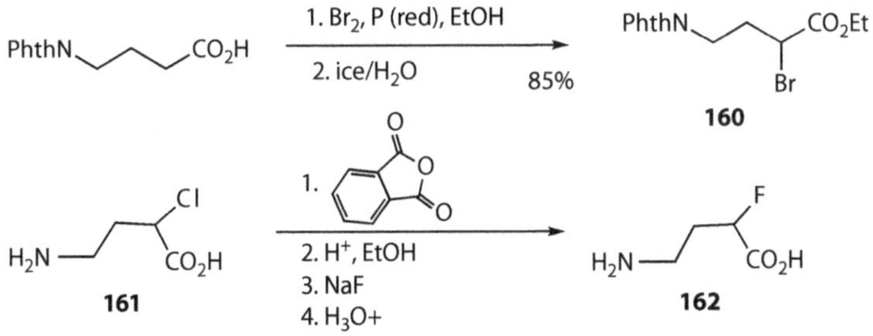

A synthetic derivative of GABA, tosylate ***163*** (prepared by literature methods),[149] was converted to the 3-methoxy derivative (***164***) via initial reaction with sodium methoxide in methanol.[150] Acid hydrolysis gave 3-methoxy-4-aminobutanoic acid, ***165***.

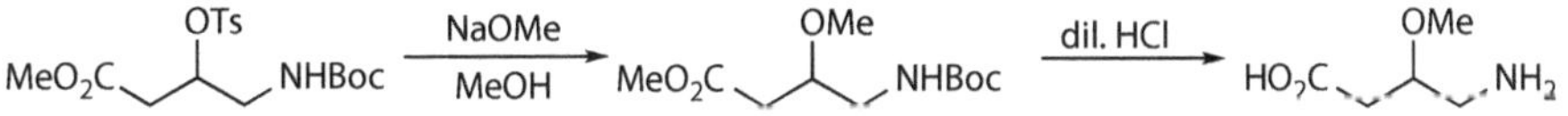

---

[145] For a discussion of GABAA and GABAB receptor agonists, antagonists, and modulators, see Krogsgaard-Larsen, P.; Frølund, B.; Kristiansen, U.; Frydenvang, K.; Ebert, B. *Eur. J. Pharm. Sci.* ***1997***, *5*, 355.

[146] Sakai, S.; Miyaji,Y.; Furutani, H.; Kobata, M.; Hachisuka, T.; Nakayama, A.; Takada, S.; Hayashi, T. *Jpn.* 12,264 ('62) [*Chem. Abstr.* ***1963***, *59*: P9805e].

[147] (a) Suzuki, Y.; Ogura, Y.; Takahashi, H. *Jpn. Kokai* 73 52,721 [*Chem. Abstr.* ***1974***, *80*: P47465x]. Also see (b) Bergmann, E.D.; Cohen, A. *Isr. J. Chem.* ***1967***, *5*, 15; (c) Nefkens, G.H.L.; Tesser, G.I.; Nivard, R.J.F. *Recuil. Trav. Chim.* ***1960***, *79*, 688.

[148] Shaitanova, E.N.; Gerus, I.I.; Belik, M.Yu.; Kukhar, V.P. *Tetrahedron Asym.* ***2007***, *18*, 192.

[149] Krogsgaard-Larsen, P.; Larsen, A.L.N.; Thyssen, K. *Acta Chem. Scand.* ***1978***, *B32*, 469.

[150] Falch, E.; Hedegaard, A.; Nielsen, L.; Jensen, B.R.; Hjeds, H.; Krogsgaard- Larsen, P. *J. Neurochem.* ***1986***, *47*, 898.

Peptidomimetics such as ***170*** were prepared from either 4-aminobutanoic acid (***166***, n = 1) or 6-aminohexanoic acid (***166***, n = 3) derivatives.[151] Initial coupling with *L*-isoleucine gave dipeptide ***167*** after basic hydrolysis. Subsequent coupling of ***167*** with the previously prepared ***168*** gave ***169***, allowing cyclization using *L*-proline and CuI to give ***170*** (n = 1 or n = 3).[151]

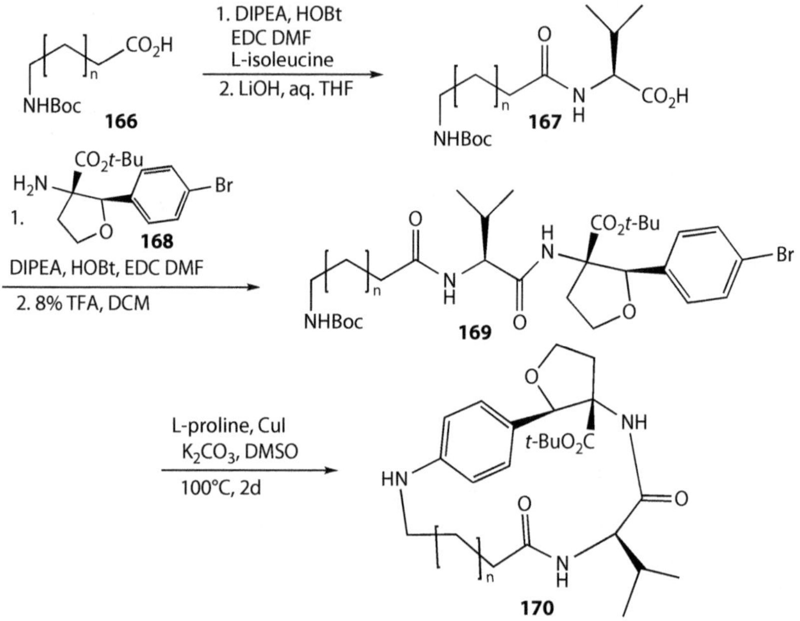

Many non-α-amino acids can be synthetic precursors. Methyl nipecotate (***171***), for example, was protected to give ***172***, and then converted to 5-amino-2-mercapto-methylpentanoic acid (***174***) in about eight steps, via ***173***.[152]

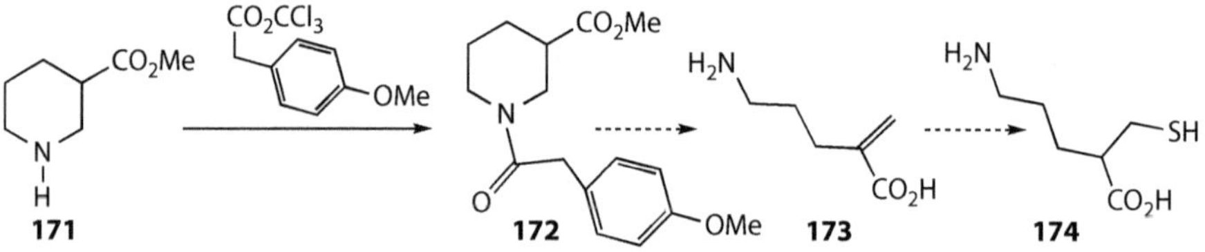

Conjugated amino acids (see ***175***) are available for synthetic manipulation and can be further functionalized. 4-Aminobut-2-enoic acid (***176***), for example, reacted with thiophenol to give 4-amino-3-phenylthiobutanoic acid, ***177***. This phenyl sulfide was then oxidized with hydrogen peroxide to give 4-amino-3-phenylsulfonylbutanoic acid, ***178***.[153] Alkenyl amino acids can also be converted to alkynes. Bromination of

[151] Grauer, A.; Späth, A.; Ma, D.; König, B. *Chem. Asian J.* ***2009***, *4*, 1134.
[152] Ondetti, M.A.; Condon, M.E. *Ger. Offen.* 2,801,911 [*Chem. Abstr.* ***1978***, *89*: P180372s].
[153] Hayashi, T.; Kondo, M.; Tanaka, M.; Ishiyama, N. *Jpn.* 72 24,007 [*Chem. Abstr.* ***1972***, *77*: P151679u].

178, for example, gave *179*, and subsequent treatment with potassium *tert*-butoxide gave *180* (methyl 4-(*N*,*N*-dimethylamino)prop-2-ynoate).[154]

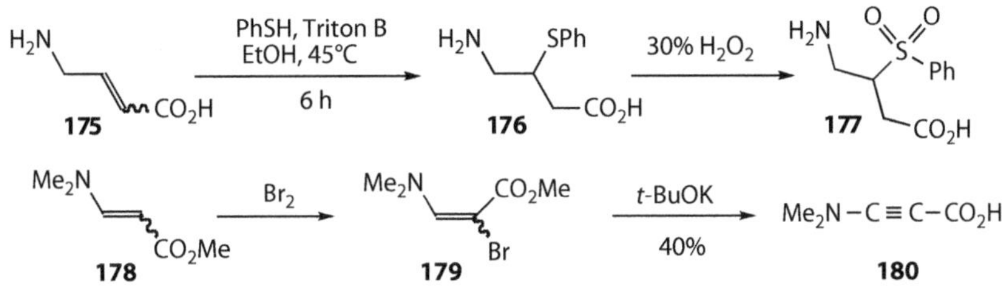

Aziridines are formed from 2-amino alcohols, and function as amino acid precursors via formation of an aziridine, using cyanide as a carboxyl surrogate. Reaction of chiral 2-amino 1-propanol with nosyl chloride (nosyl = nitrobenzenesulfonyl) followed by basification led to the *N*-nosyl aziridine, *181*.[155] Subsequent ring opening with NaCN and hydrolysis give 3*S*-aminobutanoic acid in 75% overall yield. The alkyl group in the β-amino acid is derived from the starting amino alcohol, and several derivatives were prepared. Fluorinated β-amino acids have been prepared from *trans*-*N*-benzyl-3-trifluoromethylaziridine-2-carboxylates.[156]

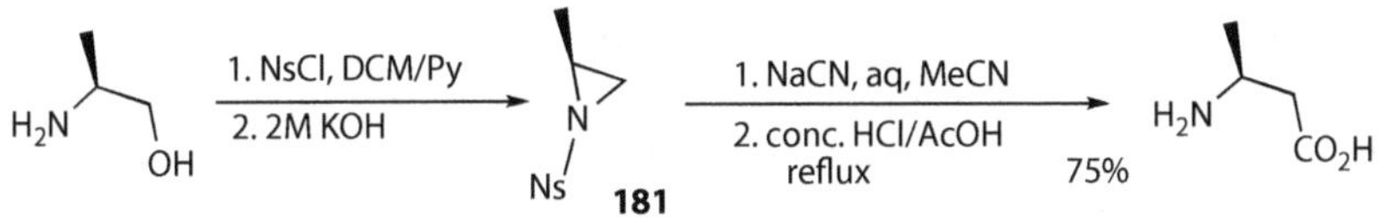

Substituted and functionalized amino acids such as *183* are prepared from β-amino acid derivatives such as *181*. Cyclization to aziridine derivative *182* was followed by hydroxylation and deprotection to give *183*.[157]

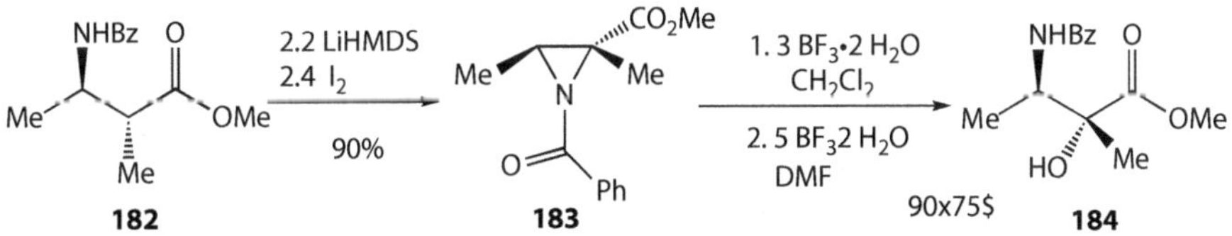

1.4.3 N-Alkylation or N-Arylation

Another, but somewhat limited, method to prepare amino acid derivatives from other amino acids is to alkylate the nitrogen atom of the amino acid. Several methods are available, however, ranging from direct reaction with alkyl halides to reductive

---

[154] Ostroumov, I.G.; Tsil'ko, A.E.; Maretina, I.A. *Zh. Org. Khim.* ***1988***, *24*, 2321 (Engl. 2092).

[155] Farràs, J.; Ginesta, X.; Sutton, P.W.; Taltavull, J.; Egeler, F.; Romea, P.; Urpí, F.; Vilarras, J. *Tetrahedron* ***2001***, *57*, 7665.

[156] Davoli, P.; Forni, A.; Franciosi, C.; Moretti, I.; Prati, F. *Tetrahedron Asym.* ***1999***, *10*, 2361.

[157] Papa, C.; Tomasini, C. *Eur. J. Org. Chem.* ***2000***, *8*, 1569.

alkylation with carbonyl compounds. In some cases, the *N*-alkylation protocol will be illustrated with α-amino acids, but the general approach is usually appropriate for non-α-amino acids as well. Indeed, with non-α-amino acids, the amine moiety often behaves more or less like a simple amine.[158]

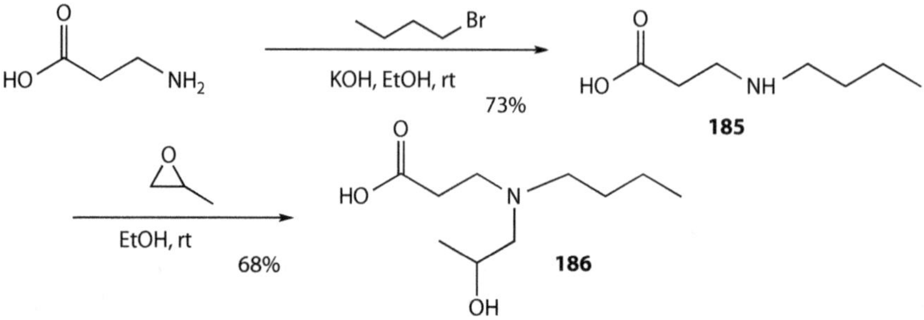

185

186

The amine moiety of non-α-amino acids can react directly with alkyl halides. The direct, base-mediated reaction of an amino acid such as β-alanine with 1-bromobutane is an example of this direct approach, and β-alanine gave ***185*** in 73% yield.[159] Interestingly, ***185*** was subsequently reacted with 2-methyloxirane to give ***186***.

The mono-*N*-alkylation of amino acids used cesium bases to prepare the secondary amine, exclusively.[160] It was shown that the cesium base suppressed overalkylation. In one example, the reaction of isoleucine with benzyl bromide and cesium carbonate, for example, gave a 55% yield of ***187***.[160] Lithium hydroxide (LiOH) has also been used to mediate the *N*-alkylation of amino acids with alkyl halides, particularly with benzylic halides.[161]

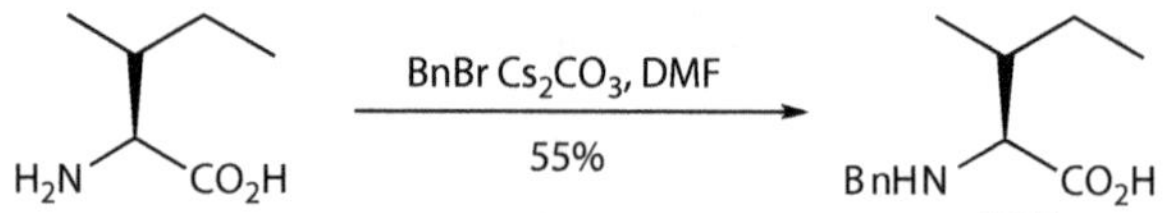

187

N-Alkylation is possible via the *N*-carbamate derivative of amino acids. The *N*-Boc derivative of valine was treated with sodium hydride and iodomethane to give the *N*-methyl derivative ***188***.[162]

---

[158] See Aurelio, L.; Brownlee, R.T.C.; Hughes, A.B. *Chem. Rev.* ***2004***, *104*, 5823.

[159] Santra, S.; Perez, J.M. *Biomacromolecules* ***2011***, *12*, 3917. For a related method, see Kawabata, T.; Kawakami, S.; Majumdar, S. *J. Am. Chem. Soc.* ***2003***, *125*, 13012.

[160] Salvatore, R.N.; Nagle, A.S.; Jung, K.W. *J. Org. Chem.* ***2002***, *67*, 674.

[161] Cho, J.H.; Kim, B.M. *Tetrahedron Lett.* ***2002***, *43*, 1273.

[162] Malkov, A.V.; Vranková, K.; Cerný, M.; Kocovský, P. *J. Org. Chem.* ***2009***, *74*, 8425.

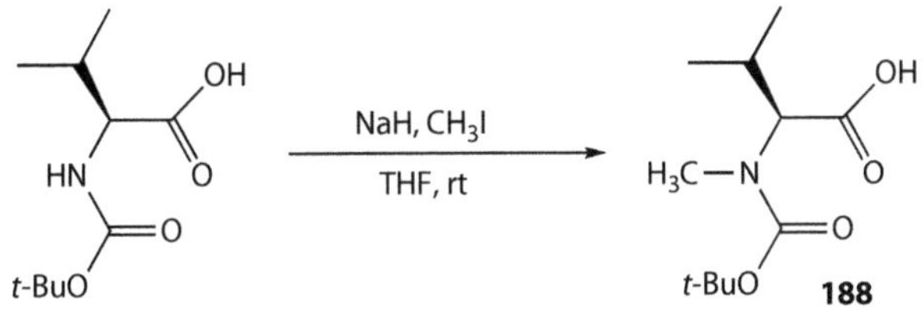

188

Reductive amination is an important method for the alkylation of amines. In this reaction, an aldehyde or ketone reacts with an amine in the presence of a reducing agent such as hydrogenation with a metal catalyst[163] or sodium borohydride to give the corresponding *N*-alkyl amine.[164] Alkylation of amino acids has been reported using this method, as in the conversion of valine to the *N*-cyclohexyl derivative ***189***.[165] Alcohols have been used as the alkylating agent in related hydrogenation-mediated reduction amination reactions of amino acids.[166] Sodium cyanoborohydride has been used as the reducing agent, and dimethylamino derivatives can be prepared from the amine by treatment with formaldehyde and cyanoborohydride.[167] *N*-alkylation of amino acid esters by reductive amination using sodium cyanoborohydride may often be problematic, and in such cases, sodium triacetoxyborohydride was found to be a superior reagent, and it has been used to alkylate various amino acid methyl esters.[168]

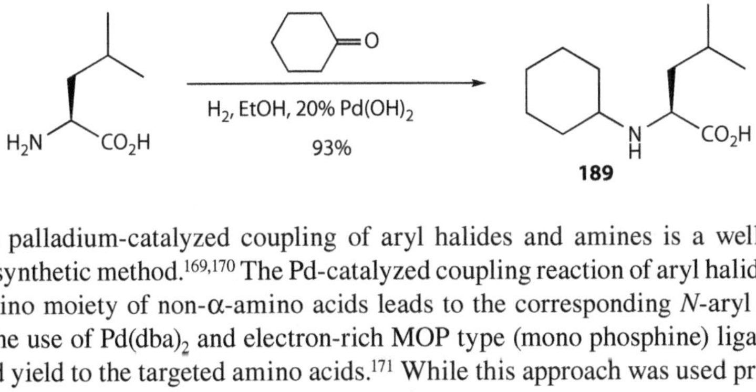

189

The palladium-catalyzed coupling of aryl halides and amines is a well-established synthetic method.[169,170] The Pd-catalyzed coupling reaction of aryl halides with the amino moiety of non-α-amino acids leads to the corresponding *N*-aryl derivative. The use of $Pd(dba)_2$ and electron-rich MOP type (mono phosphine) ligands led to good yield to the targeted amino acids.[171] While this approach was used primarily

[163] See Rylander, P.N. *Hydrogenation Methods*. Academic Press, New York, ***1985***, pp. 82–93; Klyuev, M.V.; Khidekel, M.L. *Russ. Chem. Rev.* ***1980***, *49*, 14; Rylander, P.N. *Catalytic Hydrogenation over Platinum Metals*. Academic Press, New York, ***1967***, pp. 291–303.

[164] For examples, see (a) Bru, C.; Thal, C.; Guillou, C. *Org. Lett.* ***2003***, *5*, 1845; (b) Bhattacharyya, S. *Synth. Commun.* ***2000***, *30*, 2001.

[165] Song, Y.; Sercel, A.D.; Johnson, D.R.; Colbry, N.L.; Sun, K.-L.; Roth, B.D. *Tetrahedron Lett.* ***2000***, *41*, 8225.

[166] Xu, C.-P.; Xiao, Z.-H.; Zhuo, B.-Q.; Wang, Y.-H.; Huang, P.-Q. *Chem. Commun.* ***2010***, *46*, 7834.

[167] McEvoy, F.J.; Allen, G.R., Jr. *J. Med. Chem.* ***1974***, *17*, 281.

[168] Ramanjulu, J.M.; Joullié, M.M. *Synth. Commun.* ***1996***, *26*, 1379.

[169] See Anderson, K.W.; Tundel, R.E.; Ikawa, T.; Altman, R.A.; Buchwald, S.L. *Angew. Chem. Int. Ed.* ***2006***, *45*, 6523. Also see Harris, M.C.; Huang, X.; Buchwald, S.L. *Org. Lett.* ***2003***, *4*, 2885; Coldham, I.; Leonori, D. *Org. Lett.* ***2008***, *10*, 3923; Shen, Q.; Hartwig, J.F. *J. Am. Chem. Soc.* ***2006***, *128*, 10028.

[170] Smith, M.B. *March's Advanced Organic Chemistry*, 7th ed. John Wiley & Sons, Hoboken, NJ, ***2013***, pp. 751–755.

[171] Ma, F.; Xie, X.; Ding, L.; Gao, J.; Zhang, Z. *Tetrahedron* ***2011***, *67*, 9405.

for the arylation of α-amino acids, several 3-, 4-, and 5-amino acids were arylated, including the arylation of 5-aminopentanoic acid, to give ***190*** in 66% yield. Arylation was also reported for substituted and functionalized aryl halides.[171]

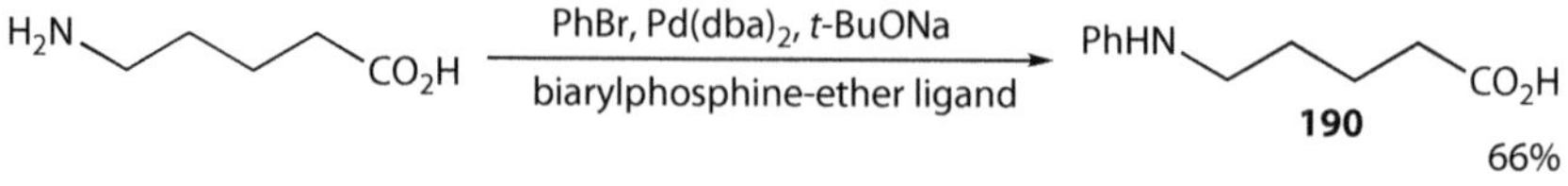

The preparation of ***190*** is but one example of aryl coupling with amines. In another example, the CuI-catalyzed coupling reaction of aryl halides with β-amino acids or β-amino-esters gave the *N*-aryl derivative. The reaction of ***192*** with iodobenzene in the presence of CuI and potassium carbonate gave a 62% yield of ***193***.[172] This route to amino acids was used to prepare SB-214857 (lotrafiban, ***194***), a GPIIb/IIIa receptor antagonist used for the protection of heart attack and stroke.[173]

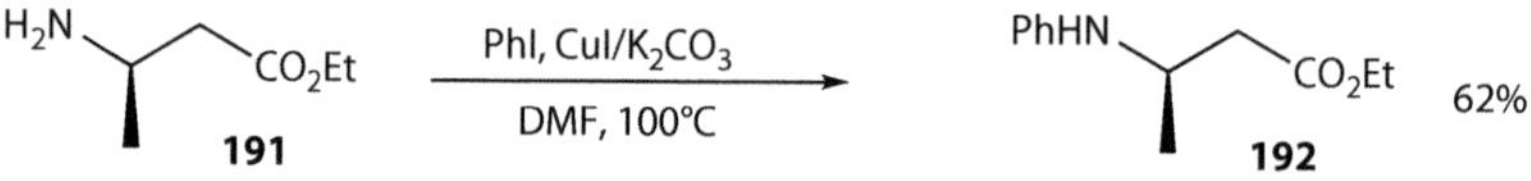

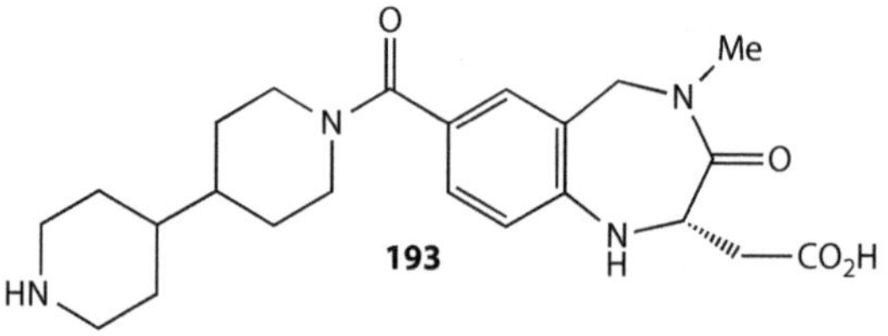

A similar coupling reaction was used to prepare *N*-aryl non-α-amino acids that were incorporated into cyclic peptides. Cyclic peptides occur in many naturally occurring bioactive compounds, many of which are viable drug candidates.[174] Dipeptide ***194*** was prepared from valine and 4-aminobutanoic acid in order to prepare a 15-membered ring cyclic peptide, and the 6-aminohexanoic acid dipeptide was also prepared in order to prepare the 17-membered ring analog.[175] In the case of ***194***, subsequent amide formation by reaction with ***195*** led to ***196***, after deprotection of the resulting tripeptide. Final copper-mediated cyclization (a coupling reaction) led to cyclic peptide ***197***, but in only 7% yield. In this work, cyclic peptide mimics were prepared in which the *N*-aryl amide moiety was replaced with an aryl ether moiety.

---

172 Ma, D.; Xia, C. *Org. Lett.* ***2001***, *3*, 2583.

173 See Scarborough, R.M.; Gretler, D.D. *J. Med. Chem.* ***2000***, *43*, 3453.

174 See, for example, (a) Lambert, J.N.; Mitchell, J.P.; Roberts, K.D. *J. Chem. Soc. Perkin Trans. I* ***2001***, 471; (b) Hamada, Y.; Shioiri, T. *Chem. Rev.* ***2005***, *105*, 4441; (c) Tan, N.-H.; Zhou, J. *Chem. Rev.* ***2006***, *106*, 840. For activity as enzyme inhibitors, see (d) Reid, R.C.; Pattenden, L.K.; Tyndall, J.D.A.; Walsh, M.T.; Fairlie, D.P. *J. Med. Chem.* ***2004***, *47*, 1641; (e) Mak, C.C.; Brik, A.; Lewrner, D.L.; Elder, J.H.; Morris, G.M.; Olson, A.J.; Wong, C.-H. *Bioorg. Med. Chem.* ***2003***, *28*, 247.

175 Grauer, A.; Späth, A.; Ma, D.; König, B. *Chem. Asian J.* ***2009***, *4*, 1134.

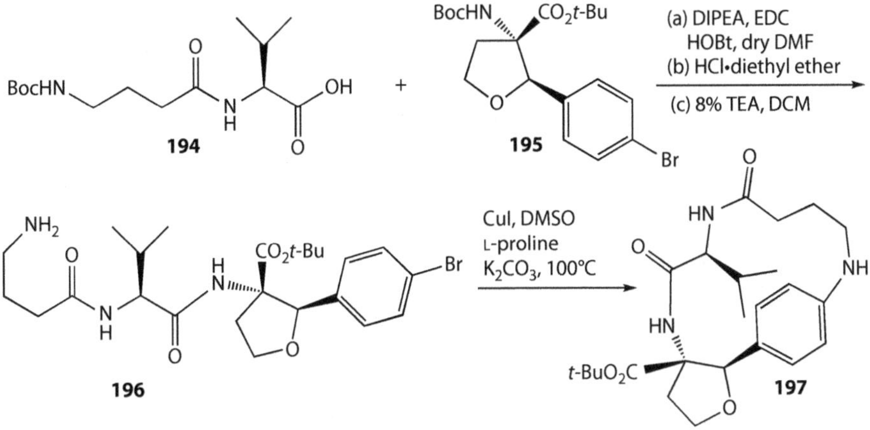

1.5 SPECIALIZED METHODS

There are, of course, methods for the preparation of amino acids other than those presented previously in Sections 1.1 to 1.4. Often, these methods are specific to one system. Some of these methods may be general, but are difficult to categorize.

1.5.1 Oxidation and Reduction

Nonconjugated alkene-amines can be used to prepare amino acids via oxidative carboxylation. Formic acid adds to alkenes in the presence of oxygen and hydrogen to give the corresponding acid. If an amine group is present, amino acids can be prepared, but conditions must be used that do not lead to oxidation of the amine moiety. Allyl amine reacted with formic acid and oxygen to give a 38% yield of 2-methyl-3-aminopropanoic acid,[176] and homoallylic amine ***198*** gave 2-methyl-4-aminobutanoic acid (***199***) under the same conditions. The value of this procedure is the direct conversion of an alkenyl amine to an amino acid. The yields are often poor to moderate, however, as shown in the example. In another example using this method, 4-amino-2-phenylbutanoic acid was prepared in 16% yield from 3-phenyl-1-amino-2-propene.[177]

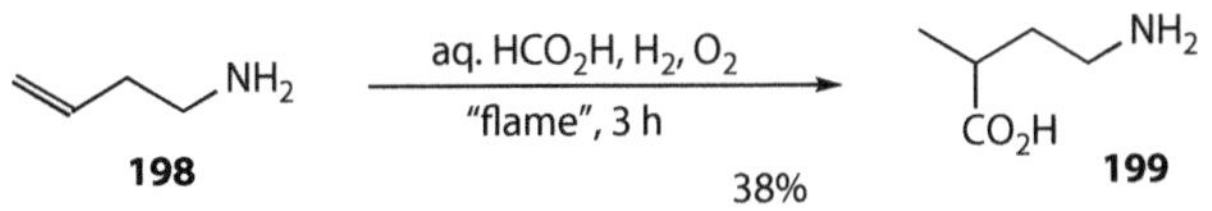

In a different approach, an oxidative addition procedure added an amine to an alkene-carboxylic acid. When ***200*** was treated with *n*-butylamine and a rhodium catalyst, 11-(*N*-butylamino)undecanoic acid (***201***) was obtained.[178] This rhodium

[176] Namoto, S.; Harada, K. *Chem. Lett.* ***1985***, 145.
[177] Namoto, S.; Harada, K. *Chem. Lett.* ***1985***, 185.
[178] Drent, E.; Breed, A.J.M. *Eur. Pat.* EP 457,386 [*Chem. Abstr.* ***1992***, *116*: P83212h].

catalysis method allowed an amine to add to an alkene in an anti-Markovnikov manner, as shown, rather than the Markovnikov addition observed in the formation of ***138***.

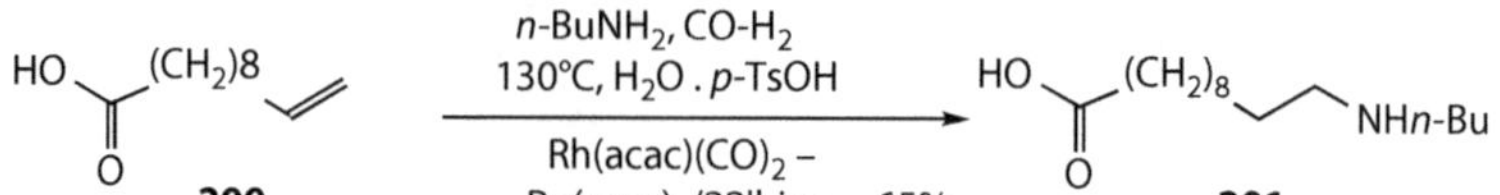

200 **201**

A formal oxidative addition with weak nucleophiles such as amides is possible if radical conditions are used.[179] The reaction of *N*-butyl acetamide and methylpent-4-enoate, in the presence of di-*tert*-butyl peroxide at 160°C, gave methyl 5-(*N*-acetyl-*N*-butylamino)pentanoate, ***202***.[179] A limitation to this approach is that only one alkene can be present in the starting material since the intermediate amido radical will add to an alkene moiety whether it is a conjugated alkene or an unconjugated alkene.

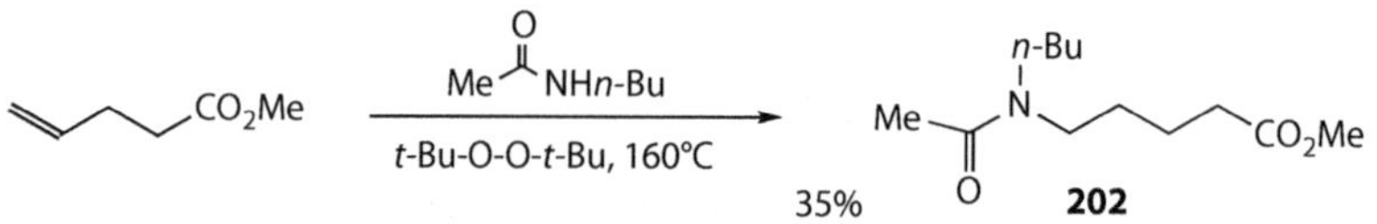

202

Alkenes can be oxidatively cleaved by a variety of methods, making them carboxyl surrogates from the standpoint of synthesis. Addition of allylmagnesium bromide to α-fluoro-acetonitrile gave imine ***203*** and in situ reduction with methanolic sodium borohydride gave ***204***.[180] Protection of the amino group as the Cbz derivative allowed oxidative cleavage of the alkene moiety with $KMnO_4$ to give the carboxylic acid, 3-amino-4-fluorobutanoic acid (***205***).

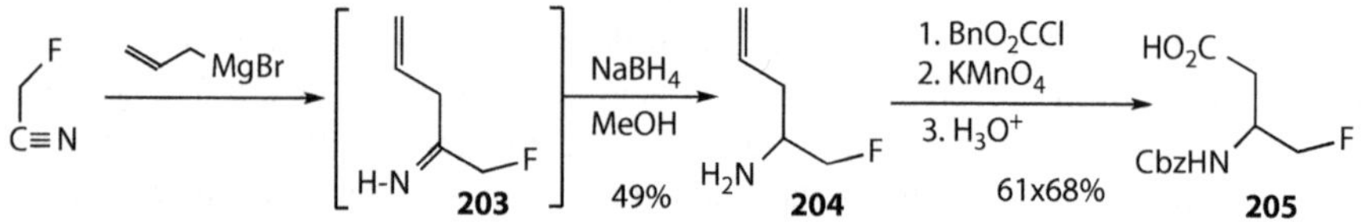

203 **204** **205**

Another strategy used an oxidative coupling and involved an initial addition of hydrocyanic acid (HCN) to alkyne ***206*** in the presence of the nickel(0) catalyst, giving a 4:1 mixture of carboxylic acids ***207:208***.[181] Hydrolysis of the phthalimidoyl group in ***207*** gave 2-aminomethylbut-2-enoic acid (***209***).

---

[179] Nikishin, G.I.; Mustafaev, R.I. *Dokl. Akad. Nauk. SSSR* ***1963***, *152*, 879 (Engl. 784).
[180] Bey, P.; Jung, M.; Gerhart, F. *Eur. Pat. Appl.* 24,965 [*Chem. Abstr.* ***1981***, *95*: P62710b].
[181] Jackson, W.R.; Perlmuter, P.; Smallridge, A.J. *Tetrahedron Lett.* ***1988***, *29*, 1983.

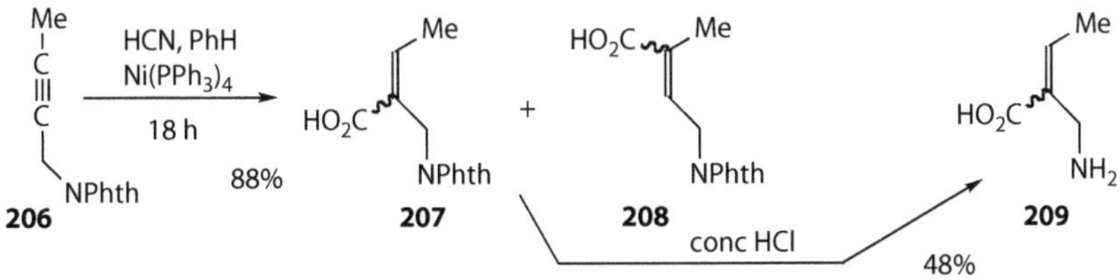

The hydrolysis of lactams to give an amino acid[182] is well known (see Chapter 2, Section 2.1). The hydrolysis of γ-butyrolactam, δ-valerolactam, ε-caprolactam, and enantholactam in strong aqueous solutions of potassium hydroxide, for example, gave the corresponding non-α-amino acid.[183] This hydrolysis reaction is important when used in conjunction with the oxidation of cyclic amines to lactams, typically using a mixture of $RuCl_3$ and $NaIO_4$. Similarly, RuO_4, generated in situ from $RuO_2/NaIO_4$,[184] oxidized 4-substituted-*N*-Boc proline to the corresponding lactam.[185] Indeed, the ruthenium oxidation of pyrrolidine to 2-pyrrolidinone is well known.[186] More highly functionalized or substituted amines may be used as precursors. When 3-hydroxyproline derivative ***210*** was treated with a mixture of ruthenium oxide and sodium periodate, lactam ***211*** was formed in >95% yield.[187] This reagent was also used for the conversion of cyclic amino acids to amino dicarboxylic acids.[188] In all cases, subsequent hydrolysis of the lactam would give the corresponding amino acid, completing the sequence that uses cyclic amines as precursors to amino acids.

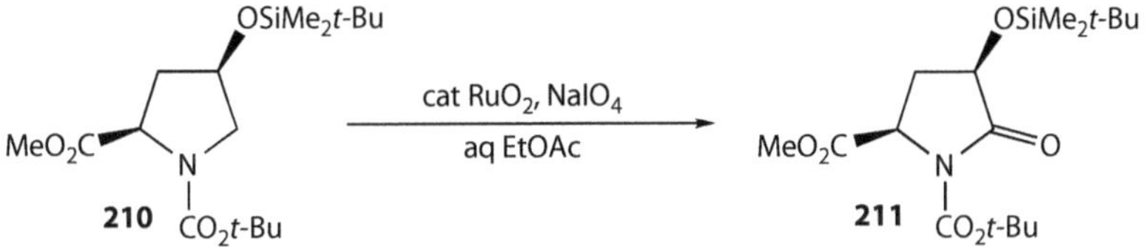

It has been reported that the ruthenium tetroxide oxidation of cyclic *N*-acylamines by a 10% $NaIO_4$ aqueous solution containing *tert*-butanol led to endocyclic C–N bond cleavage and direct formation of the Boc-protected ω-amino acid in good yields.[189] Oxidation of *N*-Boc pyrrolidine, for example, gave a 79% yield of *N*-Boc-4-aminobutanoic acid.

---

182 Testa, B.; Mayer, J.M. *Hydrolysis in Drug and Prodrug Metabolism: Chemistry, Biochemistry, and Enzymology*. Wiley-VCH, Hoboken, NJ, ***2003***, chap. 5.

183 See Vinnik, M.I.; Moieseyev, Y.V. *Tetrahedron* ***1963***, *19*, 1441.

184 Qiu, X.L.; Qing, F.L. *J. Org. Chem.* ***2003***, *68*, 3614; (b) Zhang, X.; Schmitt, A.C.; Jiang, W. *Tetrahedron Lett.* ***2001***, *42*, 5335.

185 (a) Sharma, N.K.; Ganesh, K.N. *Tetrahedron Lett.* ***2004***, *45*, 1403. Also see (b) Yoshifuji, S.; Kamane, M. *Chem. Pharm. Bull.* ***1995***, *43*, 1617; (c) Kamane, M.; Yoshifuji, S. *Tetrahedron Lett.* ***1992***, *33*, 8103;

186 See Miller, R.A.; Li, W.; Humphrey, J.L. *Tetrahedron Lett.* ***1996***, *37*, 3429; Zhang, X.; Schmitt, A.C.; Jiang, W. *Tetrahedron Lett.* ***2001***, *42*, 5335; Sharma, N.K.; Ganesh, K.N. *Tetrahedron Lett.* ***2004***, *45*, 1403.

187 Zhang, X.; Schmitt, A.C.; Jiang, W. *Tetrahedron Lett.* ***2001***, *42*, 5335.

188 (a) Yoshifuji, S.; Kamane, M. *Chem. Pharm. Bull.* ***1995***, *43*, 1617; (b) Kamane, M.; Yoshifuji, S. *Tetrahedron Lett.* ***1992***, *33*, 8103; (c) Tanaka, K.; Sawanishi, H. *Tetrahedron Asym.* ***1998***, *9*, 71.

189 Kaname, M.; Yoshifuji, S.; Sashida, H. *Tetrahedron Lett.* ***2008***, *49*, 2786.

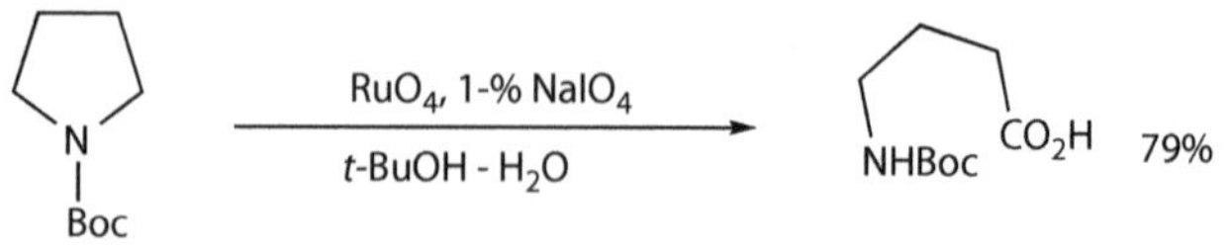

Lactams are also formed from cyclic amines by reaction with $KMnO_4$ in the presence of benzyltriethylammonium chloride.[190] Hypervalent iodine reagents may be used as the oxidant. A simple example is the oxidation of pyrrolidine with iodosylbenzene to give pyrrolidin-2-one,[191] which is readily hydrolyzed to give GABA.

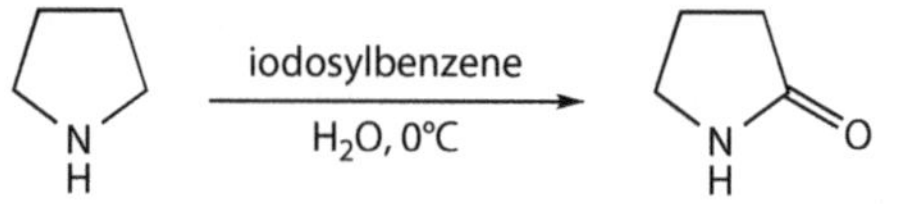

1.5.2 Rearrangement Reactions

There are several synthetic approaches that rely on a molecular rearrangement to incorporate functionality appropriate to the formation of an amino acid. One classical reaction is the Curtius rearrangement,[192,193] in which an azido-ketone rearranges to an isocyanate. Subsequent hydrolysis converts the isocyanate to an amine. This approach can be applied to the synthesis of amino acids. When 2,2,3,3-tetramethylbutanedioic acid (***212***) was converted to azido-ketone ***213***, heating led to formation of isocyanate ***214***[194] via a Curtius rearrangement. Subsequent reaction with hydrochloric acid gave 2,2,3-trimethyl-3-aminopropanoic acid, ***215***.

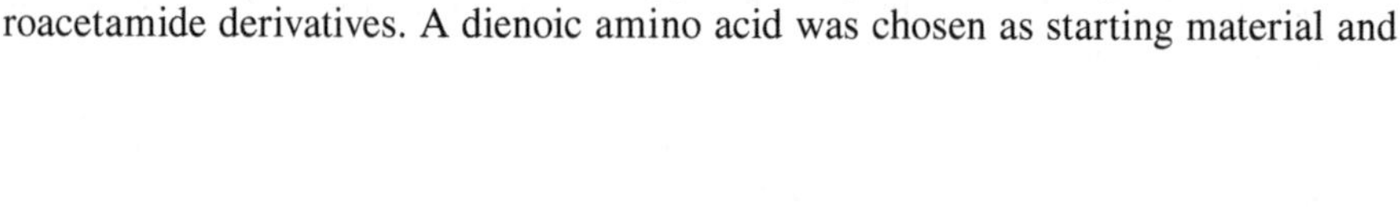

212 **213** **214** **215**

A useful modification of this basic approach targets the rearrangement of trichloroacetamide derivatives. A dienoic amino acid was chosen as starting material and

[190] Markgraf, J.H.; Stickney, C.A. *J. Heterocyclic Chem.* ***2000***, *37*, 109.

[191] (a) Moriarty, R.M.; Vaid, R.K.; Duncan, M.P.; Ochiai, M.; Inenaga, M.; Nagao, Y. *Tetrahedron Lett.* ***1988***, *29*, 6913; (b) Tada, N.; Miyamoto, K.; Ochiai, M. *Chem. Pharm. Bull.* ***2004***, *2*, 1143.

[192] (a) Curtius, T. *J. Prakt. Chem.* ***1894***, *50*, 275; (b) Smith, P.A.S. *Org. React.* ***1946***, *3*, 337; (c) Saunders, J.H.; Slocombe, R.J. *Chem. Rev.* ***1948***, *43*, 203.

[193] See (a) Smith, M.B. *Organic Synthesis*, 3rd ed. Wavefunction, Inc./Elsevier, Irvine, CA/London, England, ***2010***, pp. 192, 1051; (b) Smith, M.B. *March's Advanced Organic Chemistry*, 7th ed. John Wiley & Sons, Hoboken, NJ, ***2013***, pp. 1361–1362.

[194] Shadbolt, R.S.; Stephens, F.F. *J. Chem. Soc. C* ***1971***, 1665. Also see Schirlin, D.; Baltzer, S.; Heydt, J.G.; Jung, M.J. *J. Enzyme Inhib.* ***1987***, *1*, 243 [*Chem. Abstr.* ***1987***, *107*: 213975m].

dienoate ***216*** was converted to ***217***. Thermal rearrangement led to ***218***[195] and deprotection followed by oxidation gave ***219***.

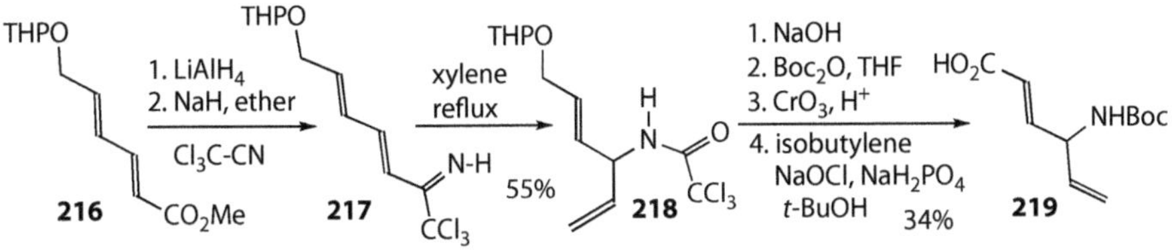

[195] Bey, P.; Gerhart, F.; Jung, M. *J. Org. Chem.* ***1986***, *51*, 2835.

2 Cyclic Precursors

The preparation of acyclic molecules from cyclic precursors is a well-known strategy.[1] The ability to control relative stereochemistry and regiochemistry in the reactions of cyclic molecules is important since cleavage of that ring transfers that stereochemistry or regiochemistry to the acyclic product.[2] Several methods will be described in this chapter that allow the synthesis of amino acids, particularly substituted amino acids, from various cyclic precursors.

2.1 FROM LACTAMS OR IMIDES

2.1.1 From β-Lactams

As first noted in Chapter 1, Section 1.5.1, the hydrolysis of lactams is a convenient source of amino acids. This transformation is, of course, illustrative of the conversion of a cyclic precursor to an acyclic amino acid. β-Lactams (2-azetidinone derivatives) are an interesting class of compounds in their own right, and hydrolysis leads to 3-aminopropanoic acid derivatives (β-alanine derivatives). The basic hydrolysis of a β-lactam such as 4-ethyl-2-azetidinone (*1*) gave 3-aminopentanoic acid (**2**),[3] for example. It is therefore appropriate to briefly discuss the synthesis of β-lactams.

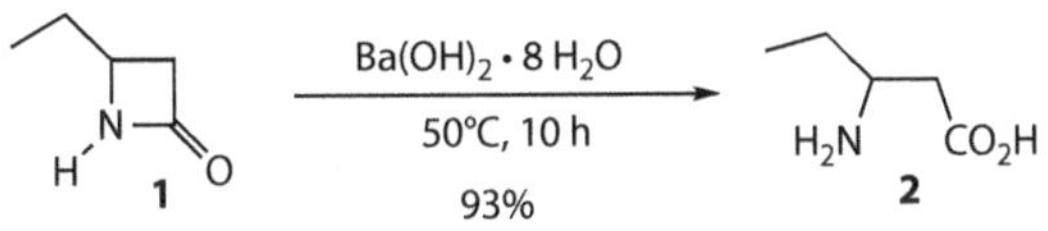

A common method for preparing β-lactams is a thermal [2+2]-cycloaddition[4] of an isocyanate with an alkene. Chlorosulfonyl isocyanate (CSI, $O{=}C{=}N{-}SO_2Cl$) is a common synthetic precursor, and in one example it reacted with 2-methyl-2-butene to give *3*, which was hydrolyzed with HCl to give 2,3-dimethyl-3-aminobutanoic acid, *4*.[5] A similar reaction with 2,3-dimethyl-2-butene gave 2,2,3-trimethyl-3-aminobutanoic acid; 2-methyl-2-pentene gave 2,2-dimethyl-3-aminopentanoic acid; and 2,4,4-trimethyl-1-pentene gave 3,5,5-trimethyl-3-aminohexanoic acid.[5] A variation

---

[1] For example, see Smith, M.B. *Organic Synthesis*, 3rd ed. Wavefunction, Inc./Elsevier, Irvine, CA/London, England, ***2010***, pp. 563–564.

[2] For example, see Smith, M.B. *Organic Synthesis*, 3rd ed. Wavefunction, Inc./Elsevier, Irvine, CA/London, England, ***2010***, pp. 551–564.

[3] Haug, Th.; Lohse, F.; Metzger, K.; Batzer, H. *Helv. Chim. Acta* ***1968***, *51*, 2069.

[4] See (a) Smith, M.B. *Organic Synthesis*, 3rd ed. Wavefunction, Inc./Elsevier, Irvine, CA/London, England, ***2010***, pp. 1076–1097; (b) Smith, M.B. *March's Advanced Organic Chemistry*, 7th ed. John Wiley & Sons, Hoboken, NJ, ***2013***, pp. 1040–1051.

[5] Graf, R. *J. L. Ann. Chem.* ***1963***, *661*, 111.

of this reaction removed the *N*-chlorosulfonyl group from the β-lactam by treatment with thiophenol and pyridine.[6]

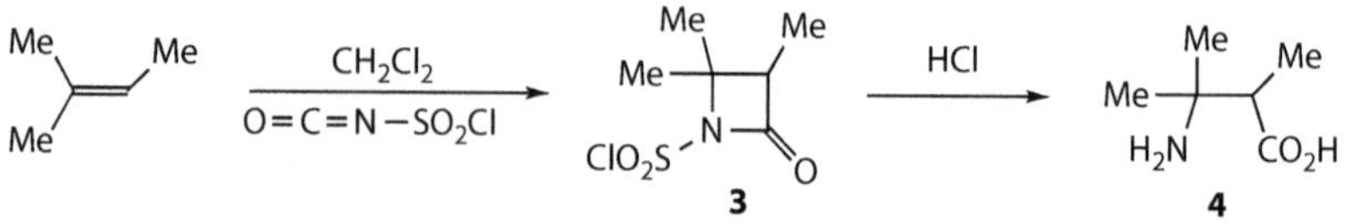

There are many examples of this type of cycloaddition-hydrolysis route to amino acids. Reaction of CSI with allenes leads to alkylidene β-lactams.[7] Hydrolysis followed by hydrogenation leads to the corresponding amino acid. Allene ***5***, for example, reacted with CSI to give ***6***. Treatment with HCl gave 2-(1-amino-2-methylethyl)-4-methylpent-2-enoic acid (***7***), and catalytic hydrogenation of the alkenyl unit gave a quantitative yield of 3-amino-3-methyl-2-(2-methylpropyl)-butanoic acid, ***8***.[7] It is clear that this methodology provides a route to both unsaturated and saturated β-amino acids. Several 2-alkyl-3-methyl-3-aminobutanoic acid derivatives were prepared by this method, and other allenes can be used in reactions with CSI.[7,8]

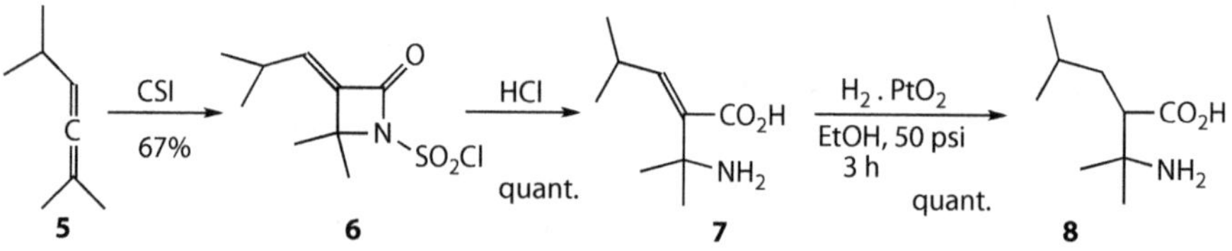

Just as allenes can be condensed with isocyanates, other dienes also react, eventually leading to alkenyl amino acids. The condensation of chlorosulfonyl isocyanate (CSI) and 1,3-pentadiene is an example that gave ***9***.[9] Removal of the sulfonyl group and hydrolysis gave methyl 3-aminohex-4-enoate (***10***), a synthetic intermediate for the preparation of daunosamine.[9] Similar reaction with 1,3-butadiene and *N*-benzyl isocyanate led to 3-aminopent-4-enoic acid (23% overall yield).[10]

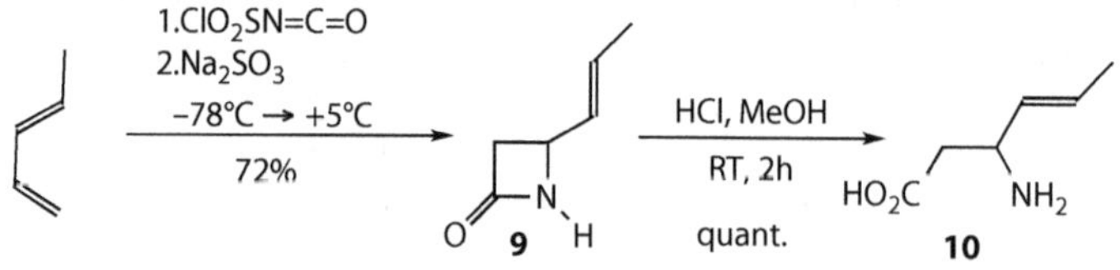

[6] Moriconi, E.J.; Kelly, J.F. *Tetrahedron Lett.* ***1968***, 1435.
[7] Moriconi, E.J.; Kelly, J.F. *J. Org. Chem.* ***1968***, *33*, 3036.
[8] See Graf, R. *J. L. Ann. Chem.* ***1963***, *661*, 111.
[9] (a) Hauser, F.M.; Rhee, R.P. *J. Org. Chem.* ***1981***, *46*, 227; (b) Moriconi, E.J.; Meyer, W.C. *J. Org. Chem.* ***1971***, *36*, 2841; (c) Hauser, F.M.; Rhee, R.P.; Ellenberger, S.R. *J. Org.Chem.* ***1984***, *49*, 2236.
[10] Arbuzov, B.A.; Zobova, N.N. *Dokl. Akad. Nauk. SSSR* ***1966***, *170*, 1317 (Engl. 993).

Alkynes are also reactive partners in β-lactam forming reactions. A [2+2]-cycloaddition reaction[11] of phenylacetylene with benzyl isocyanate gave 3-amino-3-phenylprop-2-enoic acid (***11***), after hydrolysis of the initially formed unsaturated β-lactam.[12]

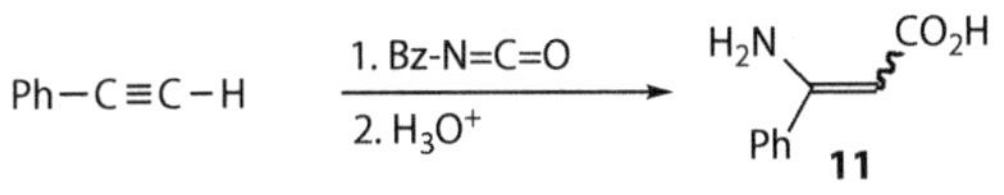

An alternative route to β-lactams[13] reacts ketenes with imines in a [2+2]-cycloaddition[14] using this strategy. Acid chlorides are excellent precursors to β-lactams via conversion to a ketene and subsequent reaction with an appropriate imine. Acid chlorides that possess a proton on the α-carbon will form a ketene under certain reaction conditions, usually upon treatment with a tertiary amine,[15] and the ketene is the actual reaction partner with the imine. An example is the reaction of acid chloride ***12*** with imine ***13*** in the presence of triethylamine, generating the ketene *in situ,* and subsequent cycloaddition gave lactam ***14***.[16]

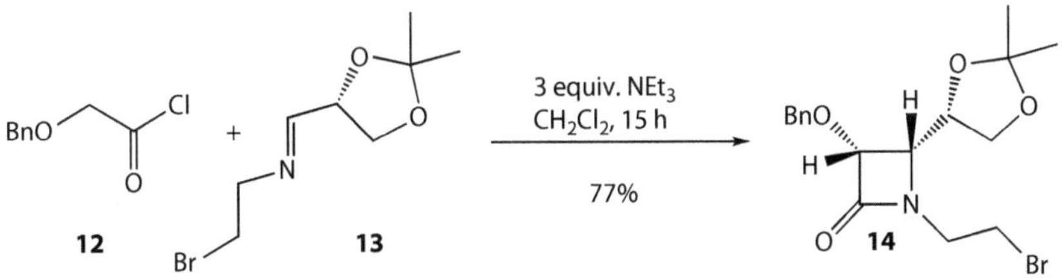

Conjugated imines are useful partners in this reaction, introducing an alkene moiety into the final amino acid. The reaction of 2-methoxyacetyl chloride and azadiene ***15*** gave β-lactam ***16***.[17] Removal of the aryl-protecting group followed by basic hydrolysis led to 3-amino-2-methoxyhex-4-enoic acid, ***17***.

[11] See (a) Smith, M.B. *Organic Synthesis*, 3rd ed. Wavefunction, Inc./Elsevier, Irvine, CA/London, England, ***2010***, pp. 1076–1097; (b) Smith, M.B. *March's Advanced Organic Chemistry*, 7th ed. John Wiley & Sons, Hoboken, NJ, ***2013***, pp. 1040–1051.

[12] Arbuzov, B.A.; Zobova, N.N. *Dokl. Akad. Nauk. SSR* ***1967***, *172*, 845 (Engl. 1037).

[13] See (a) Brown, M.J. *Heterocycles* ***1989***, *29*, 2225; (b) Isaacs, N.S. *Chem. Soc. Rev.* ***1976***, *5*, 181; (c) Mukerjee, A.K.; Srivastava, R.C. *Synthesis* ***1973***, 327; (d) Sandhu, J.S.; Sain, B. *Heterocycles* ***1987***, *26*, 777.

[14] See (a) Smith, M.B. *Organic Synthesis*, 3rd ed. Wavefunction, Inc./Elsevier, Irvine, CA/London, England, ***2010***, pp. 1076–1097; (b) Smith, M.B. *March's Advanced Organic Chemistry*, 7th ed. John Wiley & Sons, Hoboken, NJ, ***2013***, pp. 1040–1051.

[15] Brady, W.T. *Tetrahedron* ***1981***, *37*, 2949.

[16] Van Brabandt, W.; Vanwalleghem, M.; D'hooghe, M.; De Kimpe, N. *J. Org. Chem.* ***2006***, *71*, 7083.

[17] Manhas, M.S.; Hegde,V.R.; Wagle, D.R.; Bose, A.K. *J. Chem. Soc. Perkin Trans. I* ***1985***, 2045.

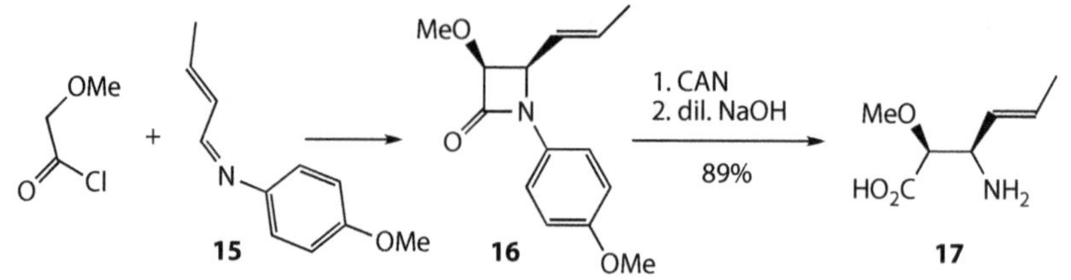

15 16 17

A ketene intermediate is not required in order to generate the β-lactam. In the case of ***19***, ketene formation is not possible. The reaction of ***18*** and ***19***, in the presence of pyridine, generated an intermediate where the nitrogen atom displaced the bromine moiety to give lactam ***20***.[18] Subsequent reaction with HCl gave 2,2-diethyl-3-aminopropanoic acid, ***21***. Similarly, 2,2-dipropyl- and 2,2-dibutyl-3-aminopropanoic acids were prepared from the appropriate acid chloride and ***19***.[18]

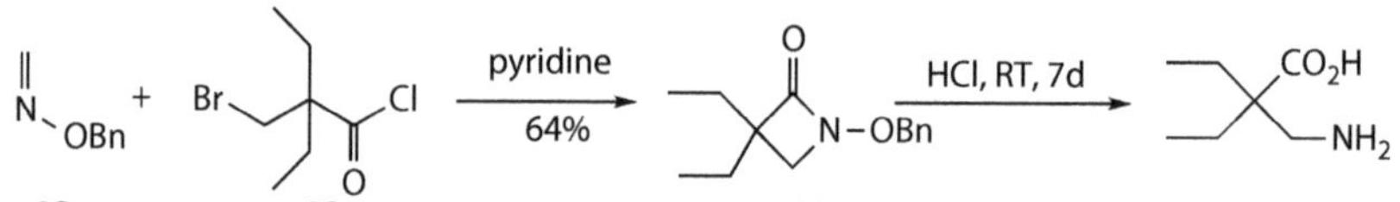

18 19 20 21

Another variation in the cycloaddition reaction employed the reaction of an imine with a ketene silyl acetal. The $TiCl_4$-catalyzed reaction of ***22*** (derived from propanal and phenethylamine) and ***23*** led to ***24***.[19] This cycloaddition is probably mediated by a nitrogen-$TiCl_3$ intermediate.[19] Hydrolysis of the β-lactam and reduction of the *N*-benzylic group with hydrogen and a palladium catalyst gave 2,2-dimethyl-3-aminopentanoic acid, ***25***. Using a chiral, nonracemic phenethylamine-protecting group for nitrogen, 2,2-dimethyl-3-aminohexanoic acid, 2,2,4-trimethyl-3-aminopentanoic acid, 2,2-dimethyl-3-aminoheptanoic acid, and 2,2,5-trimethyl-3-aminohexanoic acid were prepared with asymmetric induction ranging from 44 to 78%ee.[19]

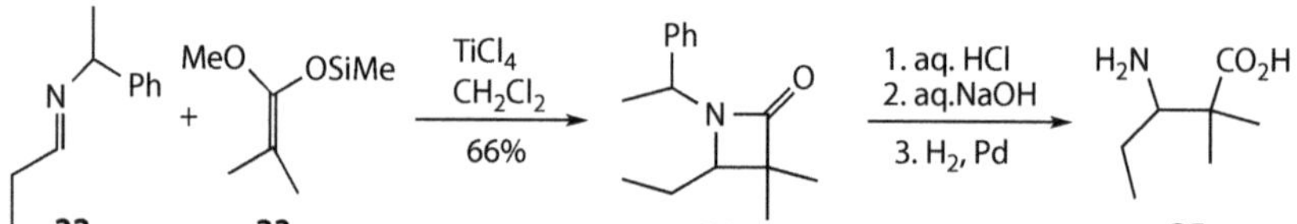

22 23 24 25

2.1.2 From Other Lactams

ω-Amino carboxylic acids are generated by the hydrolysis of lactams other than β-lactams, of course. An example is the conversion of caprolactam (***26***) to 6-aminohexanoic acid,[20] which is an *Organic Syntheses* preparation.[21] The hydrolysis of

---

[18] Nicolaus, B.J.R.; Bellasio, E.; Pagani, G.; Testa, E. *Gazz. Chim. Ital.* ***1963***, *93*, 618.
[19] Ojima, I.; Inaba, S. *Tetrahedron Lett.* ***1980***, 2077.
[20] (a) Salzbecher, M. *Mfg. Chemist* ***1962***, *33*, 463; (b) Tandara, M. *Kem. Ind.* ***1969***, *18*, 713.
[21] Eck, J.C. *Org. Synth. Coll.* ***1943***, *2*, 28.

2-pyrrolidinone (*27*) to 4-aminobutanoic acid (GABA) is another simple example.[22] Many variations of this reaction have appeared in the patent literature.[23] Other methods for hydrolysis have been developed, hydrolysis procedures under basic conditions.[24]

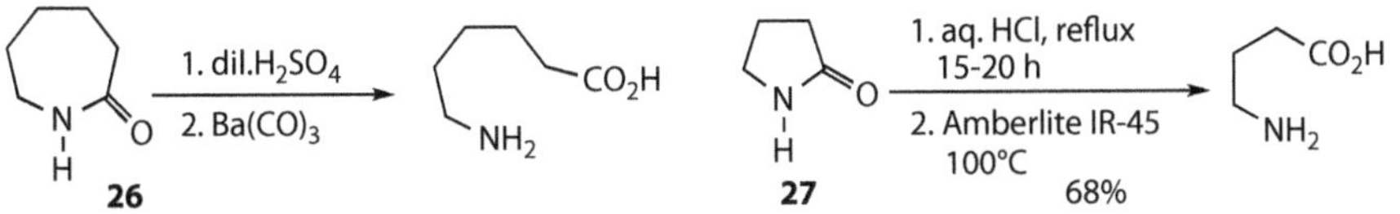

26 **27**

Substituted amino acids are generated from the corresponding lactam, as shown by the hydrolysis of 3-ethyl-2-pyrrolidinone (*28*) to give 2-ethyl-4-aminobutanoic acid (*29*).[25] Several substituted derivatives were prepared in this work, including 2-methyl-4-aminobutanoic acid, 3-ethyl-4-aminobutanoic acid, 3-propyl-4-aminobutanoic acid, 3-(1-methylethyl)-4-amino-butanoic acid, 2,3-diethyl-4-aminobutanoic acid, 2-ethyl-5-aminopentanoic acid, 3-methyl-5-aminopentanoic acid, 3-ethyl-5-amino-pentanoic acid, 3-propyl-5-aminopentanoic acid, and 3-(1-methylethyl)-5-aminopentanoic acid.[22] The next example illustrates an alternative method for the preparation of lactams in which ***30*** reacted with the sodium malonate anion (generated *in situ*) to give ***31***.[26] Aqueous hydrolysis of the ester was accompanied by decarboxylation, and subsequent catalytic hydrogenation (over Raney nickel) of the nitro group gave an amino-ester, which spontaneously cyclized to give lactam ***32***. Acid hydrolysis gave 3-methyl-3-phenyl-4-aminobutanoic acid (***33***).[26a]

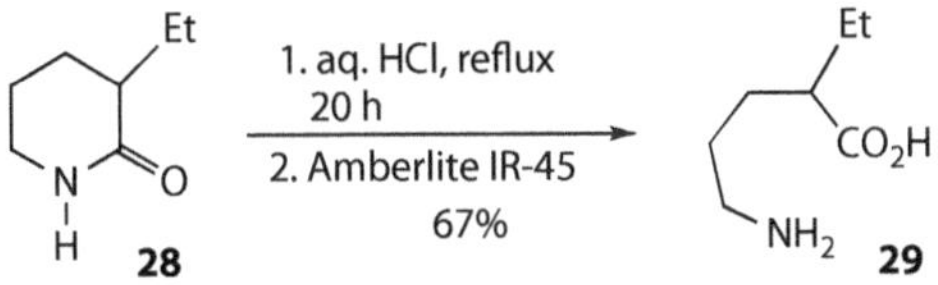

28 **29**

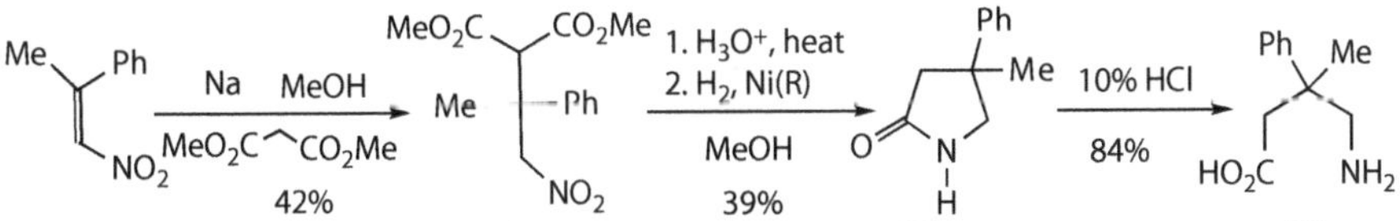

30 **31** **32** **33**

Chiral lactam precursors can be prepared from diazoamide precursors, and they serve as precursors to chiral amino acids. CH insertion of the diazo group in ***34***,

[22] Colonge, J.; Pouchol, J.-M. *Bull. Soc. Chim. Fr.* ***1962***, 598.

[23] See, for example, (a) Tandara, M. *Ger. Offen.* 1,905,530 [*Chem. Abstr.* ***1967***, *67*: 2960v]; (b) Bosnalijek-Tvornica, *Fr. Demande* 2,027,102 [*Chem. Abstr.* ***1971***, *75*: 21427h]; (c) Grigorescu, E.; Selmiciu, I.; Sauciuc, T.; Rusu, G.; Dumitrache, M.; Haulica, I. *Rom.* 51,420 [*Chem. Abstr.* ***1970***, *70*: 105980t].

[24] Flynn, D.L.; Zelle, R.E.; Grieco, P.A. *J. Org. Chem.* ***1983***, *48*, 2424.

[25] Cologne, J.; Pouchol, J.-M. *Bull. Soc. Chim. Fr.* ***1962***, 598.

[26] (a) Vasil'eva, O.S.; Zobacheva, M.M.; Smirnova, A.A.; Perekalin, V.V. *Zh. Org. Khim.* ***1978***, *14*, 1420 (Engl. 1326); (b) Sminova, A.A.; Perekalin, V.V.; Scherbakov, V.A. *Zh. Org. Khim.* ***1968***, *4*, 2245 (Engl. 2166); (c) Belostotskaya, I.S.; Dyumaev, K.M. *Sintez Prirodn. Soedin. Ikh. Analogov Fragmentov Akad.Nauk. SSSR Obshch. Tekhn. Khim.* ***1965***, 212 [*Chem. Abstr.* ***1966***, *65*: 5431f].

catalyzed by a chiral rhodium complex,[27] gave the *N*-protected lactam ***35*** [BTMSM = bis(trimethylsilylmethyl); L = MEOX[28]]. Deprotection generated lactam ***36***, which was hydrolyzed to give 4-amino-3-phenylbutanoic acid with good enantioselectivity.[29] See Chapter 5 for other asymmetric amino acid syntheses.

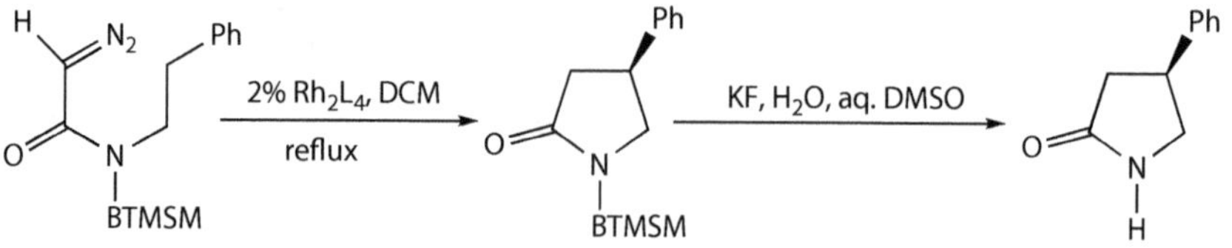

34 **35** **36**

Hydroxy-substituted amino acids are prepared from an appropriately substituted lactam. Sodium borohydride reduction of the ketone moiety in lactam ***37*** gave ***38***. Subsequent basic hydrolysis of the lactam gave 4-hydroxy-5-aminopentanoic acid, ***39***.[30,31] A different approach to hydroxy amino acids used the allylic bromination of 3-methyl-2-butenoic acid (***40***) to give a mixture of ***41*** and ***42***.[32] Reaction of ***42*** with ammonia led to formation of lactam ***43***, which was hydrolyzed to ***44***.

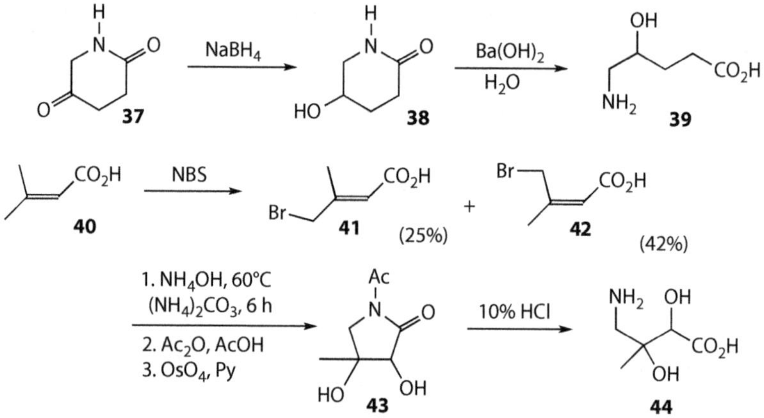

Amino acids can be prepared by the mild hydrolysis of *N*-alkenyl lactams. Lactam ***45*** was first converted to iminium salt ***46*** (a lactim ether salt) by reaction with triethyloxonium tetrafluoroborate (Meerwein's reagent).[33] Dissolution in neutral water gave

[27] (a) See Smith, M.B. *Organic Synthesis* 3rd ed. Wavefunction, Inc./Elsevier, Irvine, CA/London, England, ***2010***, pp. 1326–1331; (b) Smith, M.B. *March's Advanced Organic Chemistry*, 7th ed. John Wiley & Sons, Hoboken, NJ, ***2013***, pp. 692–693. See also, pp. 1051–1059.

[28] See (a) Doyle, M.P.; Dyatkin, A.B.; Roos, G.H.P.; Canas, F.; Pierson, D.A.; van Basten, A.; Mueller, P.; Polleux, P.J. *Am. Chem. Soc.* ***1994***, *116*, 4507; (b) Zalatan, D.N.; Du Bois, J. *J. Am. Chem. Soc.* ***2008***, *130*, 9220.

[29] Wee, A.G.H.; Duncan, S.C.; Fan, G.-J. *Tetrahedron: Asymmetry* ***2006***, *17*, 297.

[30] Herdeis, C. *Synthesis* ***1986***, 232.

[31] Herdeis, C. *Arch. Pharm.* (Weinheim) ***1983***, *316*, 719.

[32] Onda, M.; Konda, Y.; Ōmura, S.; Hata, T. *Chem. Pharm. Bull.* ***1971***, *19*, 2013.

[33] (a) Meerwein, H. *Org. Synth. Coll.* ***1973***, *5*, 1080; (b) Borch, R.F. *Tetrahedron Lett.* ***1968***, 61.

ethyl 4-aminobutanoate (***47***) in 80% yield,[34] constituting a mild hydrolysis method for converting lactams to amino acids. Indeed, iminium salt intermediates such as ***46*** are more labile than the lactam, and ring opening occurs under very mild conditions. The alkenyl group effectively functions as a trigger for the hydrolysis reaction.

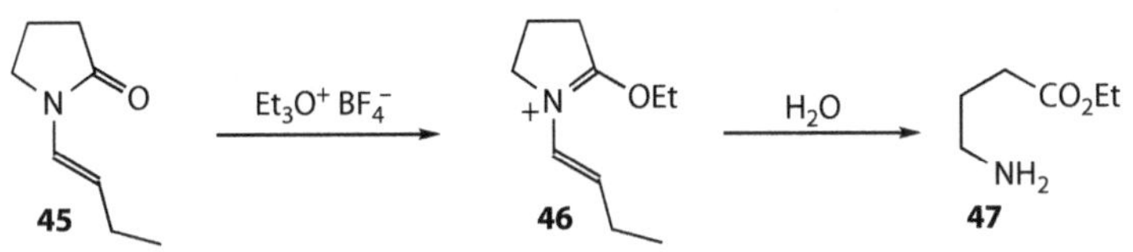

The alkenyl group in ***46*** is not essential for hydrolysis, but the lactim ether moiety is. Indeed, lactim ethers are lactam derivatives that may be hydrolyzed to give the corresponding amino acid.[35] The reaction of 2-benzyl-2-pyrrolidinone (***48***) with diethyl sulfate, for example, gave lactim ether ***49*** (3-benzyl-2-ethoxy-1-pyrroline). Subsequent dissolution in water for less than 1 hour gave amino-ester ***51*** (ethyl 2-benzyl-4-aminobutanoate).[35] Although ***49*** can be isolated, that step is unnecessary and the crude product from the first reaction can simply be dissolved in water after treatment with HCl_{gas} (which generates ***50*** in situ). Larger-ring lactams

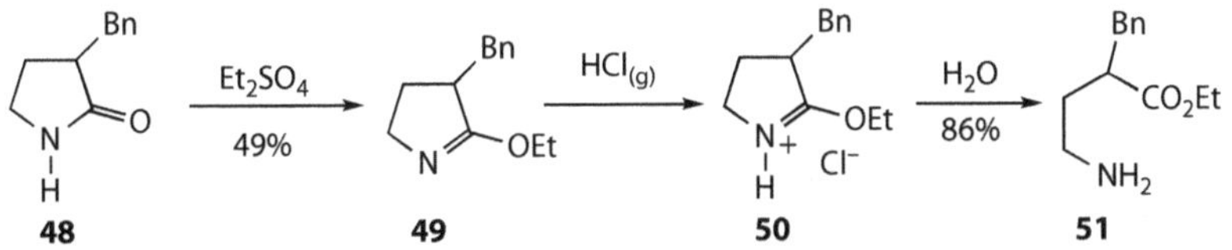

can also be converted to lactim ethers using triethyloxonium tetrafluoroborate (vide supra, giving the iminium salt directly). This method was used to prepared ethyl aminobutanoic, aminopentanoic, aminohexanoic, aminooctanoic, aminodecanoic, and aminododecanoic esters, including ethyl 4-methyl-4-aminobutanoate, ethyl 2-methyl-4-aminobutanoate, as well as ethyl 5-aminopentanoate, ethyl 6-aminohexanoate, ethyl 7-aminoheptanoate, and ethyl 12-aminododecanoate.[35] Hydrolysis of lactim ethers was known in the literature prior to the work just discussed. The work of Smith and coworkers found that the lactim iminium salt hydrolyzed under milder conditions than lactim ethers. The work of Bartlett and coworkers converted 2-pyrrolidinone to the lactim ether 2-methoxypyrroline (***52***) by treatment with dimethyl sulfate.[36] This methoxy-imine was brominated[37] with *N*-bromosuccinimide to give ***53***.[38] Reaction with potassium acetate gave ***54***, and acid hydrolysis led to 2-hydroxy-4-aminobutanoic acid (***55***).[36]

---

[34] Zezza, C.A.; Kwon, T.W.; Sheu, J.L.; Smith, M.B. *Heterocycles* ***1992***, *34*, 1325.
[35] Menezes, R.; Smith, M.B. *Synth. Commun.* ***1988***, *18*, 1625.
[36] Wick, A.E.; Bartlett, P.A.; Dolphin, D. *Helv. Chim. Acta* ***1971***, *54*, 513.
[37] Yamada, Y.; Emori, T.; Kinoshita, S.; Okada, H. *Agric. Biol. Chem.* ***1973***, *37*, 649.
[38] Yamada, Y.; Okada, H. *Agric. Biol. Chem.* ***1976***, *40*, 1437.

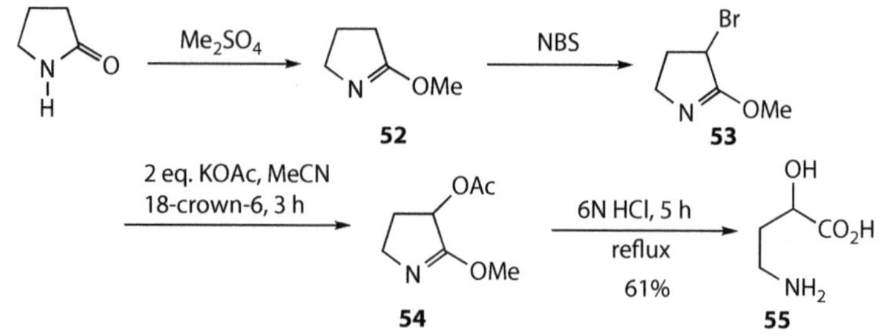

2.1.3 From Imides

An imide can be hydrolyzed to an amino acid similarly to hydrolysis of a lactam. A simple example of this reaction is the hydrolysis of glutarimide to give 4-oxo-pentanoic acid (5-amino levulinic acid, ***56***).[39] This latter compound showed relatively high herbicidal activity.[39]

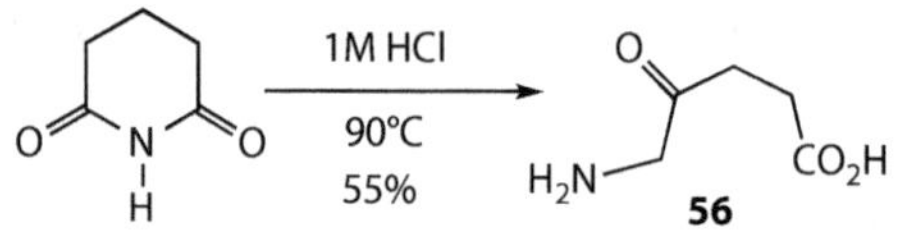

One carbonyl group of an imide is subject to reaction with certain nucleophiles that will generate a lactam. In one example, the reaction of allylmagnesium bromide and *N*-methylsuccinimide (***57***) led to lactam ***58***. Subsequent treatment with base gave lactam diene ***59***.[40] Flitsch and coworkers hydrogenated the diene to give *N*-methyl-5-propyl-2-pyrrolidinone (***60***), and acid hydrolysis gave 4-(*N*-methylamino)heptanoic acid, ***61***.[40]

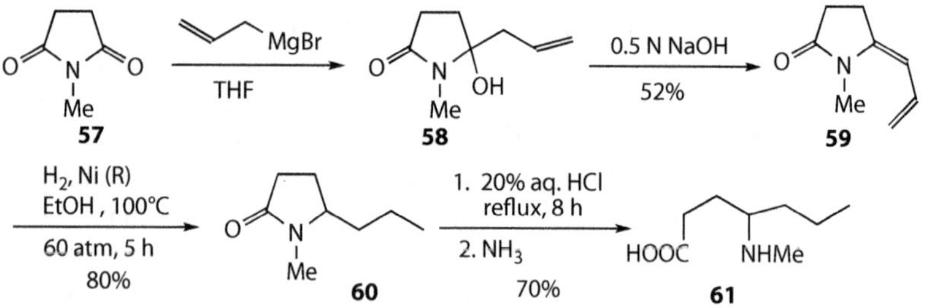

2.2 FROM KETONES

Cyclic ketones can be refunctionalized to produce amino acids. Refunctionalization is typically followed by cleavage of the ring, or in some cases, cleavage is followed by refunctionalization.

---

[39] Suzuki, K.; Takeya, H. *Jpn. Kokai Tokkyo Koho* JP 03 72,450 [*Chem. Abstr.* ***1991***, *115*: P91665c].
[40] Flitsch, W. *J. L. Ann. Chem.* ***1965***, *684*, 141.

2.2.1 Direct Cleavage of Cyclic Ketones

A useful method for the preparation of amino acids begins with direct cleavage of a cyclic ketone. A cyclic ketone such as cyclopentanone can be oxidatively cleaved with sulfuric acid. When this cleavage was followed by treatment with nitrosyl sulfate, an oximino acid (***62***) was formed.[41] Catalytic hydrogenation of the oximino group gave 4-aminobutanoic acid (GABA). Similar treatment of cycloheptanone gave 6-amino-hexanoic acid, and cyclononanone led to 8-aminooctanoic acid.

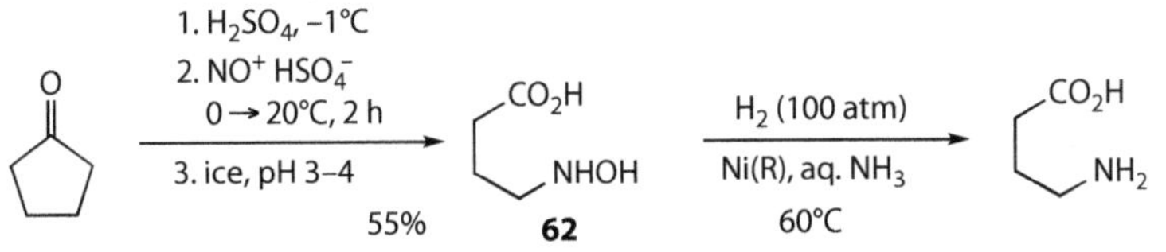

Cyclic α-nitroketones are important sources of amino acids since the nitro group can be reduced to an amino group (see Chapter 1, Section 1.1.3). Cyclohexene was treated with N_2O_4, and the product was oxidized with chromium trioxide to give 2-nitrocyclohexanone (***63***).[42] When ***63*** was treated with aqueous bicarbonate, oxidative cleavage of the ring gave 6-nitro-hexanoic acid (***64***). Catalytic hydrogenation of the nitro group generated 6-aminohexanoic acid.[42,43] Ammonium formate, in the presence of methanolic Pd on carbon, is also effective for the reduction of a nitro group to an amine.[44] Both methyl 4-aminobutanoate and methyl 3-aminopropanoate were produced by reduction of the appropriate nitro-ester with this reagent.[44]

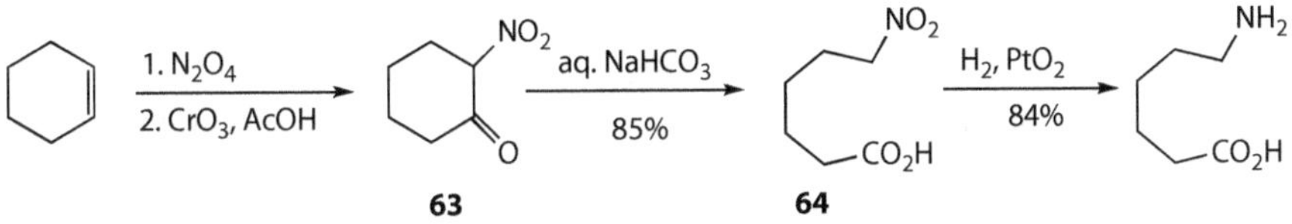

An alternative preparation of 2-nitro-ketones was reported by Lang, who reacted cyclodecene with nitrous oxide and air to give oxime ***65***.[45] Hydrolysis with aqueous HCl converted the oxime moiety to a ketone moiety in ***66***. Oxidative cleavage occurred upon treatment with aqueous NaOH and gave 12-nitrododecanoic acid, ***67***. Catalytic hydrogenation of the nitro group gave 12-aminododecanoic acid.[45]

---

[41] Lafont, P.; Thiers, M. *Fr.* 1,349,281 [*Chem. Abstr.* ***1964***, *60*: P13145b].

[42] Matlack, A.S.; Breslow, D.S. *J. Org. Chem.* ***1967***, *32*, 1995.

[43] (a) Tanaka, I.; Uehara, H.; Oosaki, F.; Yamagata, M.; Tanaka, S.; Takeshita, T. *Jpn.* 71 09,810 [*Chem. Abstr.* ***1971***, *75*: 98933c]; (b) Tanaka, I.; Yamagata, M.; Uehara, H. *Jpn.* 70 29,965 [*Chem. Abstr.* ***1971***, *74*: 63924c].

[44] Ram, S.; Ehrenkaufer, R.E. *Tetrahedron Lett.* ***1984***, *25*, 3415.

[45] Lang, A. *Ger. Offen.* 2,062,928 [*Chem. Abstr.* ***1971***, *75*: 110690f]. Also see Ballini, R.; Papa, F.; Abate, C. *Eur. J. Org. Chem.* ***1999***, 87.

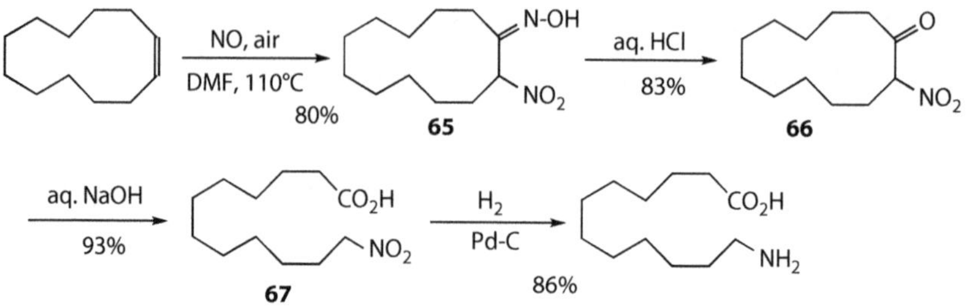

2.2.2 Via Rearrangement Reactions

As illustrated in the previous sections, cyclic ketones are excellent starting materials for the synthesis of amino acids. An important strategy converts cyclic ketones directly to lactams or imines, which can then be hydrolyzed to the corresponding amino acid. There are two classical reactions of cyclic ketones that lead to a lactam via a rearrangement: the Beckmann rearrangement and the Schmidt rearrangement.

The Beckmann rearrangement[46] is a classical method for converting a cyclic ketone to a lactam that treats the oxime of a cyclic ketone with a strong acid such as sulfuric acid or polyphosphoric acid (PPA). Reagents such as phosphorus pentachloride (PCl_5) can also be used. In a simple example, the oxime of cyclohexanone was treated with sulfuric acid to give caprolactam. Subsequent hydrolysis gave 6-aminohexanoic acid. A modification was used in the Beckmann rearrangement of cyclohexanone oxime. The resulting product was trapped with potassium benzene sulfinate to give an *O*-benzenesulfonate lactim ether (***68***). Subsequent reaction of ***68*** with the morpholino enamine of cyclohexanone (***69***) was followed by ring opening to give 1-amino-7-oxododecanoic acid (***70***).[47] Oximes of several different cyclic ketones were similarly treated to produce a variety of oxo-amino acids, including 7-oxo-13-aminotridecanoic acid, 7-oxo-14-aminotetradecanoic acid, 7-oxo-15-aminopentadecanoic acid, and 7-oxo-16-aminohexadecanoic acid.[47]

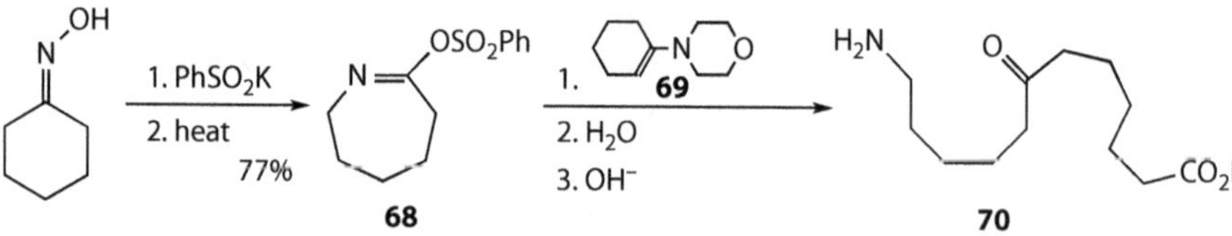

Cyclic ketones can be converted directly to a lactam by treatment with a mixture of sodium azide and an acid, which generates hydrazoic acid (HN_3) in situ, in what

[46] (a) Beckmann, E. *Ber.* ***1886***, *19*, 988; (b) Donaruma, L.G.; Heldt, W.Z. *Org. React.* ***1960***, *11*, 1; (c) see Smith, M.B. *Organic Synthesis*, 3rd ed. Wavefunction, Inc./Elsevier, Irvine, CA/London, England, ***2010***, pp. 188–190; (d) Smith, M.B. *March's Advanced Organic Chemistry*, 7th ed. John Wiley & Sons, Hoboken, NJ, ***2013***, pp. 1365–1367.

[47] Hünig, S.; Gräßmann, W.; Meuer, V.; Lücke, E.; Brenninger, W. *Chem. Ber.* ***1967***, *100*, 3039.

is known as the Schmidt reaction.[48] The general procedure uses sodium azide in an acidic medium to generate hydrazoic acid. As with the Beckmann rearrangement, the lactam products can be hydrolyzed to the corresponding amino acid. When 2-isobutylcyclopentanone (*71*) was reacted with polyphosphoric acid and sodium azide, 5-isobutyl-2-piperidone (*72*) was obtained.[49] Acid hydrolysis gave a near-quantitative yield of 7-methyl-5-aminooctanoic acid (*73*). In general, the rearrangement leads to attachment of nitrogen to the more highly substituted carbon of the cyclic ketone.

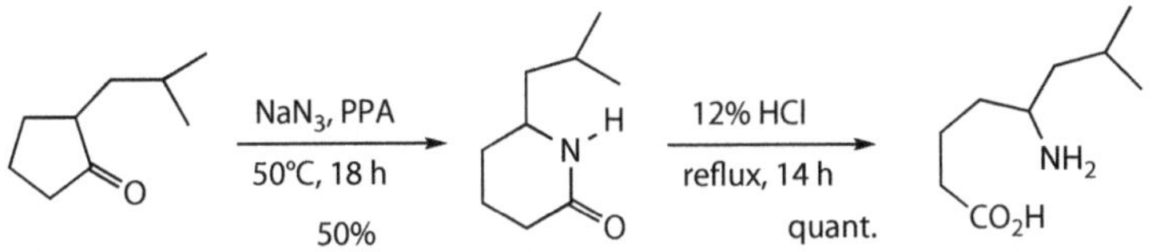

71 **72** **73**

Although there is usually a preference for one isomer, regioisomeric mixtures of lactams are commonly produced by Schmidt rearrangement[48] (and also Beckmann rearrangement[46]) when two carbon atoms with differing substituents are adjacent to the carbonyl group in the starting material. The reaction of 2-methyl-cyclohexanone and sodium azide, in the presence of polyphosphoric acid, provides an example that generates two regioisomers.[50] The two lactam products were 7-methylhexahydroazepin-2-one (7-methyl caprolactam, *74*) in 54% yield and 3-methylhexahydroazepin-2-one (3-methyl caprolactam, *75*) in 10% yield.[51] Acid hydrolysis followed by chromatographic separation gave the mixture of 6-aminoheptanoic acid and 2-methyl-7-aminoheptanoic acid.

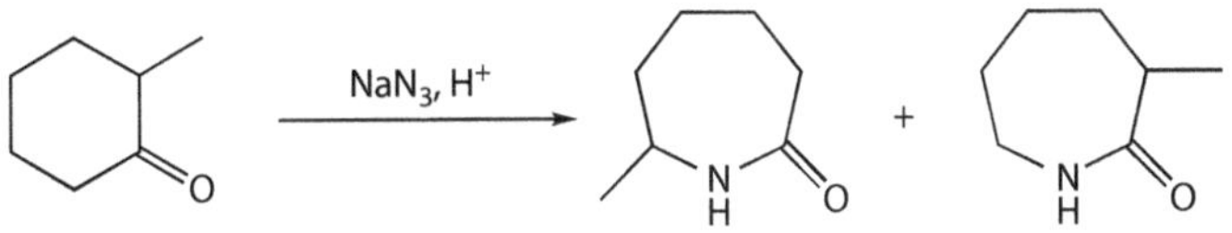

74 **75**

2.3 FROM LACTONES AND ANHYDRIDES

Lactones and anhydrides each react with ammonia or amines to give nitrogen derivatives. In some cases, lactones are converted to lactams and anhydrides to imides. In these cases, hydrolysis will open the ring and generate an amino acid (see Chapter 1,

[48] (a) Schmidt, R.F. *Ber.* ***1924***, *57*, 704; (b) Wolff, H. *Org. React.* ***1946***, *3*, 307; (c) Koldobskii, G.I.; Terreshchenko, G.F.; Gerasimova, E.S.; Bagal, L.I. *Russ. Chem. Rev.* ***1971***, 835; (d) Beckwith, A.L.J., in *Chemistry of Amides*, Zabicky, J., ed. Interscience, London, ***1970***, pp. 137–145; (e) see Smith, M.B. *Organic Synthesis*, 3rd ed. Wavefunction, Inc./Elsevier, Irvine, CA/London, England, ***2010***, pp. 190–192; (f) Smith, M.B. *March's Advanced Organic Chemistry*, 7th ed. John Wiley & Sons, Hoboken, NJ, ***2013***, pp. 1363–1365.

[49] Chimiak, A. *Bull. Acad. Pol. Sci. Ser. Sci. Chim.* ***1969***, *17*, 197.

[50] Conley, R.T. *J. Org. Chem.* ***1958***, *23*, 1330.

[51] Overberger, C.G.; Parker, G.M. *J. Polym. Sci. A-1* ***1968***, *6*, 513.

Section 1.5.1 and Section 2.1). In other cases, the lactone or anhydride may be converted to an amino acid directly or in situ. Reaction of five- and six-membered ring lactones with amines tends to give the lactam, whereas the same reaction with larger-ring lactones leads to the acyclic amino acid. Anhydrides show similar behavior.

2.3.1 From Lactones

Just as amines react with esters to form amides, lactones react with amines to give lactams or amino acid derivatives directly. When β-propiolactone (***76***) reacted with ammonia, for example, 3-aminopropanamide (***77***) was obtained.[52] A similar reaction with methylamine gave no ***77b*** but rather a 97% yield of GABA. Dimethylamine, however, converted ***76*** to ***77c*** in 79% yield, along with 13% of GABA.[52]

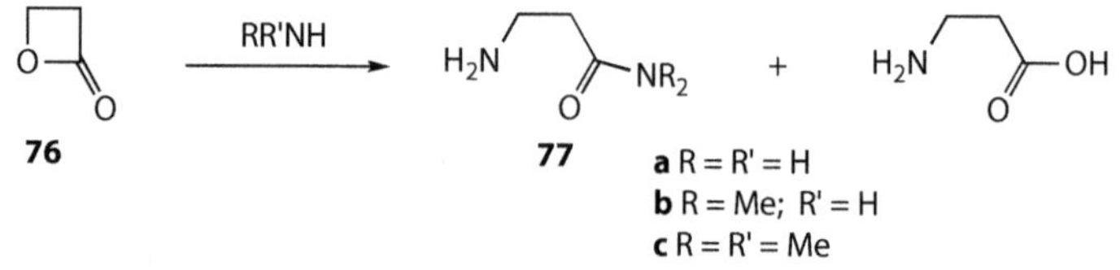

76 **77**

With five-membered ring lactones in particular, reaction with ammonia or an amine tends to give a lactam rather than the amino acid. When 3-benzyl-γ-butyrolactone (***78***) was heated with ammonia, 4-benzyl-2-pyrrolidinone (***79***) was isolated.[53] Subsequent hydrolysis led to 3-benzyl-4-aminobutanoic acid (***80***), an anti-spasmodic sedative.[53]

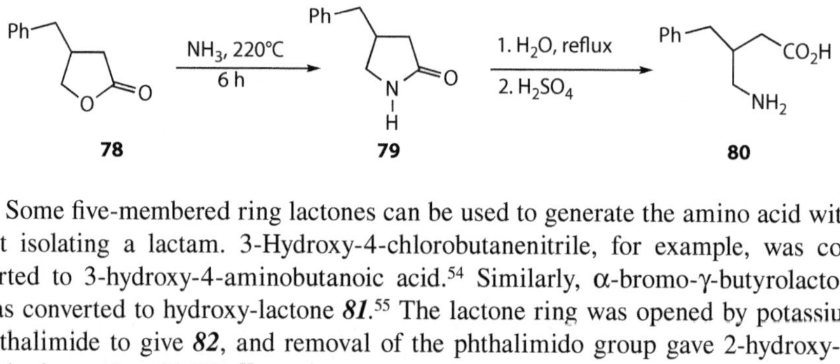

78 **79** **80**

Some five-membered ring lactones can be used to generate the amino acid without isolating a lactam. 3-Hydroxy-4-chlorobutanenitrile, for example, was converted to 3-hydroxy-4-aminobutanoic acid.[54] Similarly, α-bromo-γ-butyrolactone was converted to hydroxy-lactone ***81***.[55] The lactone ring was opened by potassium phthalimide to give ***82***, and removal of the phthalimido group gave 2-hydroxy-4-aminobutanoic acid (***55***).[55]

---

[52] Gresham, T.L.; Jansen, J.E.; Shaver, F.W.; Bankert, R.A.; Fiedorek, F.T. *J. Am. Chem. Soc.* ***1951***, *73*, 3168.
[53] Nabeta, S.; Kojima, S.; Sago, K. *Jpn.* 71 03,767 [*Chem. Abstr.* ***1971***, *74*: P140928m].
[54] Kurono, M.; Shigeoka, S.; Sakai, Y.; Itoh, H. *Ger. Offen.* 2,323,043 [*Chem. Abstr.* ***1974***, *80*: P37450z].
[55] (a) Goel, O.P.; Krolls, U.; Lewis, E.P. *Org. Prep. Proceed. Int.* ***1985***, *17*, 91; (b) Saito, Y.; Hashimoto, M.; Seki, H.; Kamiya, T. *Tetrahedron Lett.* ***1970***, 4863; (c) Leuenberger, H.G.; Matzinger, P.K.; Wirz, B. *Eur. Pat. Appl.* EP 272,605 [*Chem. Abstr.* ***1988***, *109*: P188838a].

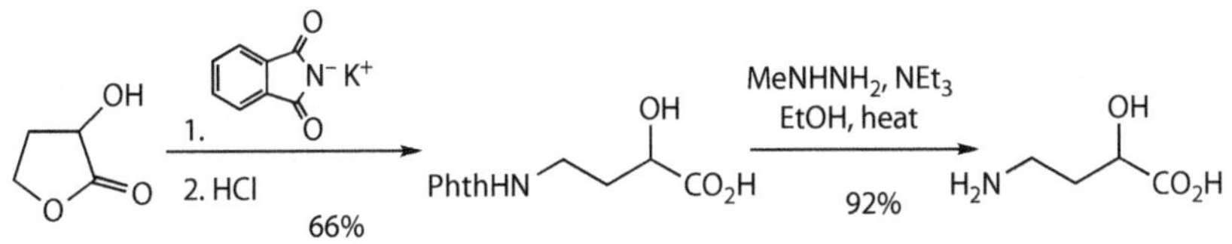

81 **82** **55**

Lactone ***83*** (paraconic acid) was converted to 4-hydroxy-3-amino-butanoic acid (***85***) via initial generation of the amide (***84***). Treatment of ***84*** with a basic solution of bromine converted the amide to the amine via a Hofmann rearrangement,[56] and acid hydrolysis gave 3-amino-4-hydroxybutanoic acid, ***85***.[57]

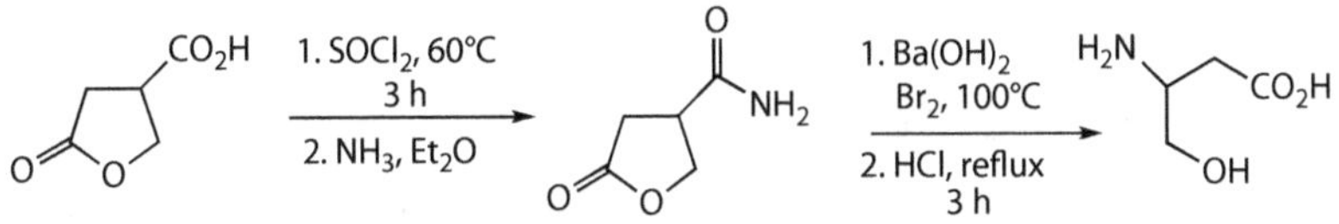

83 **84** **85**

2.3.2 From Anhydrides

Anhydrides can be converted to ω-amino acids via reaction with amines and reduction of the resultant imide or acyclic amide product. When succinic anhydride was treated with two molar equivalents of diethylamine, amido acid ***86*** was produced.[58] Reaction with triethyloxonium tetrafluoroborate[59] generated an iminium ester that was reduced with sodium borohydride to give ethyl 4-(*N*,*N*-diethylamino)butanoate (*87*) in 83% overall yield.[58] In a similar manner, maleic anhydride reacts with ammonia to give 3*Z*-aminoprop-2-enoic acid (also known as maleamic acid).[60]

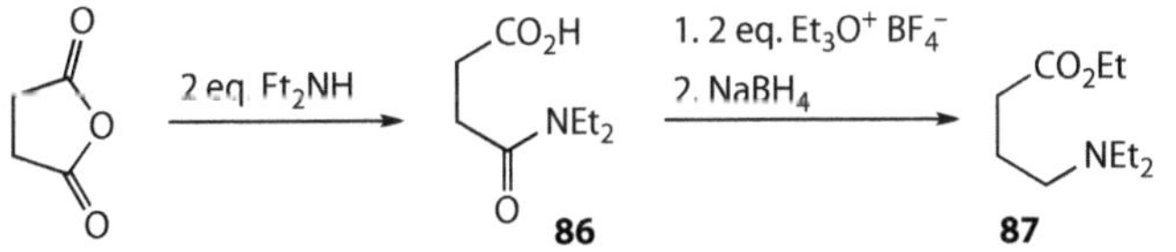

86 **87**

Both anhydrides and imides (see Section 2.1.3) have been used as precursors for substituted amino acids. β-Amino acids in particular can be prepared from these

[56] (a) Hofmann, A.W. *Ber.* ***1881***, *14*, 2725; (b) Wallis, E.S.; Lane, J.F. *Org. React.* ***1949***, *3*, 267; (c) see Smith, M.B. *Organic Synthesis*, 3rd ed. Wavefunction, Inc./Elsevier, Irvine, CA/London, England, ***2010***, pp. 190–192; (d) Smith, M.B. *March's Advanced Organic Chemistry*, 7th ed. John Wiley & Sons, Hoboken, NJ, ***2013***, pp. 1360–1361.

[57] Piskov, V.B. *Zh. Obshch. Khim.* ***1962***, *32*, 3407 (Engl. 3343).

[58] Bell, K.H.; Fullick, G. *Synth. Commun.* ***1987***, *17*, 1965.

[59] (a) Meerwein, H. *Org. Synth. Coll.* ***1973***, *5*, 1080; (b) Borch, R.F. *Tetrahedron Lett.* ***1968***, 61.

[60] (a) Abramyan, R.K.; Filippychev, G.F. *Uch. Zap. Yaroslav. Tekhnol. Inst.* ***1971***, 161 [*Chem. Abstr.* ***1973***, *78*: 42826q]; (b) Takaya, T.; Ono, T.; Okuda, Y. *Jpn. Kokai* 74 35,325 [*Chem. Abstr.* ***1974***, *81*: P119980g].

reagents. Weinreb showed that anhydride ***88*** was converted to methyl 2*R*-methyl-3-aminobutanoic acid (***89***) via formation of an intermediate azide.[61] This sequence utilized a Curtius rearrangement (see Chapter 1, Section 1.5.2)[62,63] of a diazoketone. An example is presented here so it can be compared with the anhydride reaction that was used to prepare GABA.

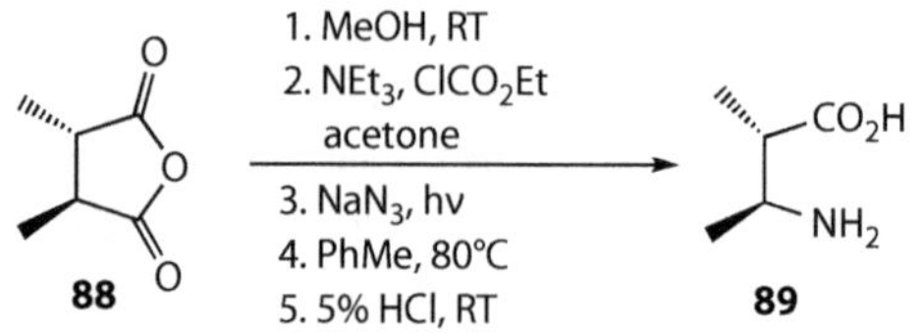

2.4 FROM HETEROCYCLES

Many heterocycles contain nitrogen, and with suitable functionality, reduction would produce an amino compound. Indeed, a variety of functionalized heterocycles have been converted to amino acids using this fundamental idea.

2.4.1 Isoxazolidines, Isoxazoles, Oxazolidines, and Oxazine Derivatives

When a nitrone reacts with a vinyl ether, an isoxazolidine is formed via a [3+2]-cycloaddition reaction.[64,65] A similar reaction with an alkyne gives an isoxazole.[64] With proper functionalization, hydrogenation can lead to ring opening and formation of an amino acid. Hydrogenation of isoxazole ***90***, for example, gave a quantitative yield of ***91***.[66] Isoxazole ***90*** was prepared from phenyl acetylene and methyl 4-nitrobutanoate.[67]

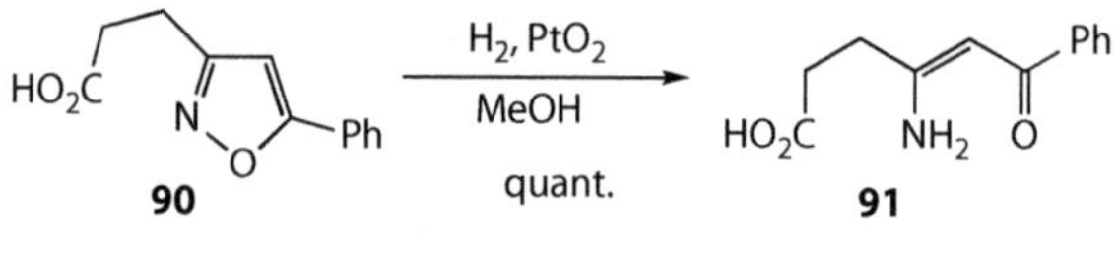

[61] Garigipati, R.S.; Morton, J.A.; Weinreb, S.M. *Tetrahedron Lett.* ***1983***, *24*, 987.

[62] (a) Curtius, T. *J. Prakt. Chem.* ***1894***, *50*, 275; (b) Smith, P.A.S. *Org. React.* ***1946***, *3*, 337; (c) Saunders, J.H.; Slocombe, R.J. *Chem. Rev.* ***1948***, *43*, 203.

[63] See (a) Smith, M.B. *Organic Synthesis*, 3rd ed. Wavefunction, Inc./Elsevier, Irvine, CA/London, England, ***2010***, pp. 192, 1051; (b) Smith, M.B. *March's Advanced Organic Chemistry*, 7th ed. John Wiley & Sons, Hoboken, NJ, ***2013***, pp. 1361–1362.

[64] Huisgen, R. *Angew. Chem. Int. Ed. Engl.* ***1963***, *2*, 565, 633.

[65] See (a) Smith, M.B. *Organic Synthesis*, 3rd ed. Wavefunction, Inc./Elsevier, Irvine, CA/London, England, ***2010***, pp. 1101–1115; (b) Smith, M.B. *March's Advanced Organic Chemistry*, 7th ed. John Wiley & Sons, Hoboken, NJ, ***2013***, pp. 1014–1020.

[66] Barco, A.; Benetti, S.; Pollini, G.P. *J. Org. Chem.* ***1979***, *44*, 1734.

[67] Stevens, R.V.; Christensen, C.G.; Edmonson, W.L.; Kaplan, M.; Reid, E.B.; Wentland, M.P. *J. Am. Chem. Soc.* ***1971***, *93*, 6629.

Isoxazolidines are produced by [3+2]-cycloaddition reactions,[68] and can also be converted to amino acids. Diethylamine was oxidized with hydrogen peroxide to give a nitrone, which reacted with ethyl vinyl ether via [3+2]-cycloaddition to give ***92***.[69] Subsequent reaction with benzyl bromide gave an ammonium salt, and the ring was opened with DABCO to give ethyl 3-(*N*-benzyl-3-ethylamino)butanoate, ***93***. Other 3-aryl aminopropanoic acid derivatives have been prepared using oxazolidine derivatives derived from vinyl acetates and hydroxylamines.[70]

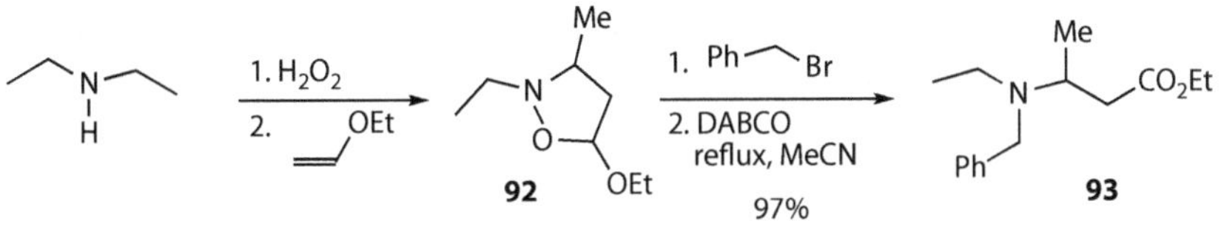

An alternative strategy prepared isoxazolidinyl esters by displacement of the halide moiety in an α-halo-acid derivative by an amine, an amine surrogate. γ-Butyrolactone reacted with triphenylphosphine and bromine to give a dibromo acid that was esterified to give ethyl 2,4-dibromobutanoate.[71] Subsequent reaction with *N*-Cbz-hydroxylamine gave ***94*** via sequential displacement of the two bromine atoms. Reaction with hydroxylamine and base, followed by catalytic hydrogenation, gave 2*S*-hydroxy-4-aminobutanoic acid (***55***).[71] Isoxazolidine-5-carboxylic acid derivatives are, in general, excellent precursors to hydroxy amino acids. Larger-ring heterocycles can also be produced by this fundamental approach. α-Bromoester ***95***, for example, was converted to perhydrooxazine-6-carboxylic acid (***96***).[72] Catalytic hydrogenation led to 2-hydroxy-5-amino-pentanoic acid (***97***, 2-hydroxy-DAVA; also see Chapter 6, Section 6.5).

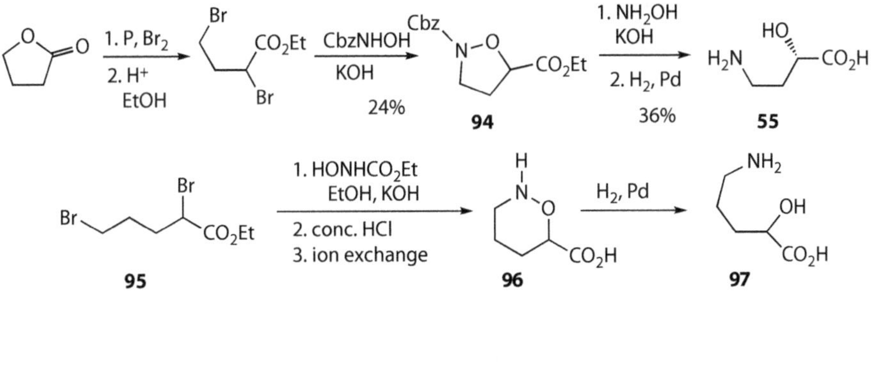

[68] See (a) Smith, M.B. *Organic Synthesis*, 3rd ed. Wavefunction, Inc./Elsevier, Irvine, CA/London, England, ***2010***, pp. 1101–1115; (b) Smith, M.B. *March's Advanced Organic Chemistry*, 7th ed. John Wiley & Sons, Hoboken, NJ, ***2013***, pp. 1014–1020.

[69] Murahashi, S.; Kodera, Y.; Hosomi, T. *Tetrahedron Lett.* ***1988***, 29, 5949. Also see Keirs, D.; Moffat, D.; Overton, K.; Tomanek, R. *J. Chem. Soc. Perkin Trans. I* ***1991***, 1041.

[70] Keirs, D.; Moffat, D.; Overton, K.; Tomanek, R. *J. Chem. Soc. Perkin Trans. I* ***1991***, 1041.

[71] Hjeds, H.; Honore, T. *Acta Chem. Scand. Ser. B* ***1978***, *B32*, 187.

[72] Falch, E.; Hedegaard, A.; Nielsen, L.; Jensen, B.R.; Hjeds, H.; Krogsgaard- Larsen, P. *J. Neurochem.* ***1986***, *47*, 898.

Other hydroxy amino acid derivatives can be prepared by a [3+2]-cycloaddition[73] route, and there are other methods that generate nitrones. Nitro compounds react with alkenes, in the presence of isocyanates, to produce dihydroisoxazole derivatives via a nitrile *N*-oxide.[74] Kametani and coworkers generated nitrile *N*-oxide ***99*** in situ from the reaction of nitro compound ***98*** and phenylisocyanate. When ***99*** was condensed with *tert*-butylcrotonate, ***100*** was formed in 58% yield, along with 14% of ***101***.[75] Isolation of ***100*** was followed by catalytic hydrogenation to give a 1:1 mixture of ***102*** and ***103***. Isocyanates also react with some heterocycles, including isoxazolones, to give ring-enlarged heterocycles (isoxazol-5-ones to 1,3-oxazin-6-ones, for example).[76] Isoxazole acids have been shown to be useful precursors to amino acids.[77]

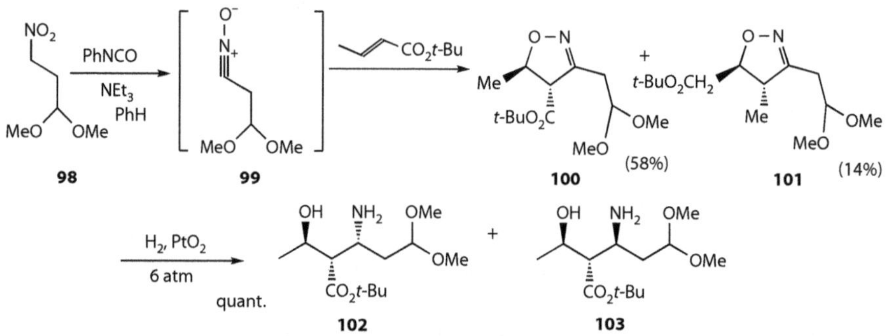

Isoxazolidin-5-ones have been used as protected amino acids, in some cases allowing an asymmetric synthesis of more substituted derivatives. Conjugated addition of the chiral lithium amide reagent ***104*** to methyl 3-phenylpropenoate was followed by cyclization to ***105***.[78] Subsequent alkylation gave ***106*** in 75% yield and >99:1 dr, and hydrogenation gave 1-alkylated 3-amino acid (***107***) as the final product. 2,2-Disubstituted amino acids were prepared by this method.[78]

[73] See (a) Smith, M.B. *Organic Synthesis*, 3rd ed. Wavefunction, Inc./Elsevier, Irvine, CA/London, England, ***2010***, pp. 1101–1115; (b) Smith, M.B. *March's Advanced Organic Chemistry*, 7th ed. John Wiley & Sons, Hoboken, NJ, ***2013***, pp. 1014–1020.

[74] Tufariello, J.J. *Accts. Chem. Res.* ***1979***, *12*, 396.

[75] (a) Kametani, T.; Nagahara, T.; Suzuki, Y.; Yokohama, S.; Huang, S.-P.; Ihara, M. *Tetrahedron*, ***1981***, *37*, 715; (b) Kametani, T.; Huang, S.-P.; Ihara, *Heterocycles* ***1979***, *12*, 1183; (b) Kanetani, T.; Nagahara, T.; Suzuki, Y.; Yokohama, S. *Heterocycles* ***1980***, *14*, 403; (d) Kametani, T.; Huang, S.-P.; Yokohama, S.; Suzuki, Y.; Ihara, M. *J. Am. Chem. Soc.* ***1980***, *102*, 2060.

[76] Risitano, F.; Grassi, G.; Foti, F.; Caruso, F.; LoVecchio, G. *J. Chem. Soc. Perkin Trans. I*, ***1979***, 1522.

[77] Auricchio, S.; Ricca, A.; DePava, O.V. *Tetrahedron Lett.* ***1980***, 857.

[78] Bentley, S.A.; Davies, S.G.; Lee, J.A.; Roberts, P.M.; Russell, A.J.; Thomson, J.E.; Toms, S.M. *Tetrahedron* **2010**, *66*, 4604.

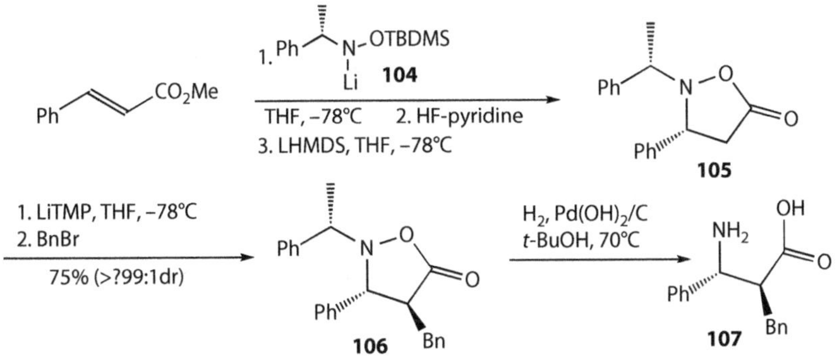

Using a different method, 1,3-amino-1,2-propanediol (***108***) was treated with phosgene to give oxazolidinone ***109***.[79] Subsequent oxidation of the primary hydroxy group with basic permanganate gave a carboxylic acid (***110***), and acid hydrolysis opened the ring to give isoserine (***111***) in good yield. Oxazinones such as ***112*** can be used to prepare amino acids. In this particular example, a palladium-catalyzed ring opening was accompanied by rearrangement to give ethyl 6-(*N*-tosylamino)-5,5-dimethylhex-3-enoate, ***113***.[80]

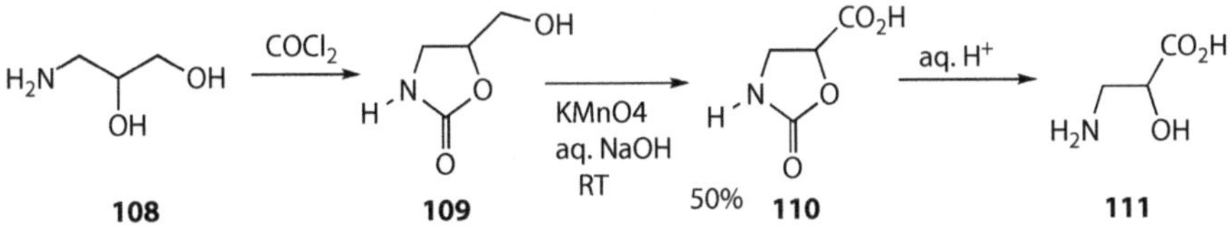

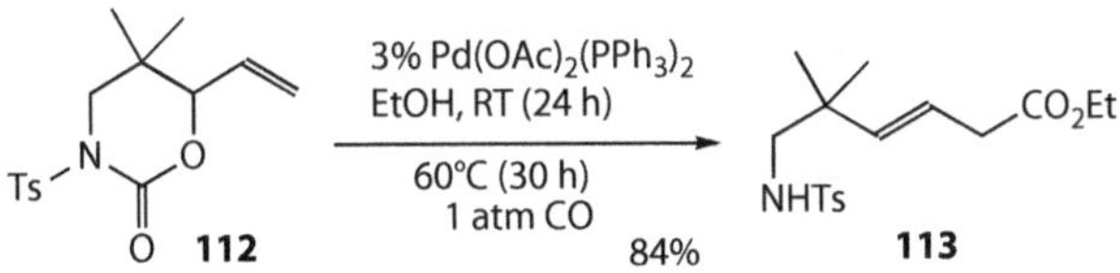

A different approach coupled ester acid chloride ***114*** with ***115***[81] to give ***116***.[82] Treatment with γ-picoline led to the rearranged product ***117*** in 84% overall yield from ***114***. Acid hydrolysis gave 8-oxo-9-aminononanoic acid, where the oxazolone moiety served as both a protected amine and a protected ketone.

---

[79] Tsuchihashi, G.; Iriuchijima, S.; Mori, Y. *Jpn. Kokai* 76 16,660 [*Chem. Abstr.* ***1976***, *85*: P123901g].
[80] Bando, T.; Tanaka, S.; Fugami, K.; Yoshida, Z.; Tamaru, Y. *Bull. Chem. Soc. Jpn.* ***1992***, *65*, 97.
[81] Stewart, J.M.; Woolley, D.W. *J. Am. Chem. Soc.* ***1956***, *78*, 5336.
[82] Zav'yalov, S.I.; Kravchenko, N.E. *Izv. Akad. Nauk. SSSR Ser. Khim.* ***1984***, 422 (Engl. 384).

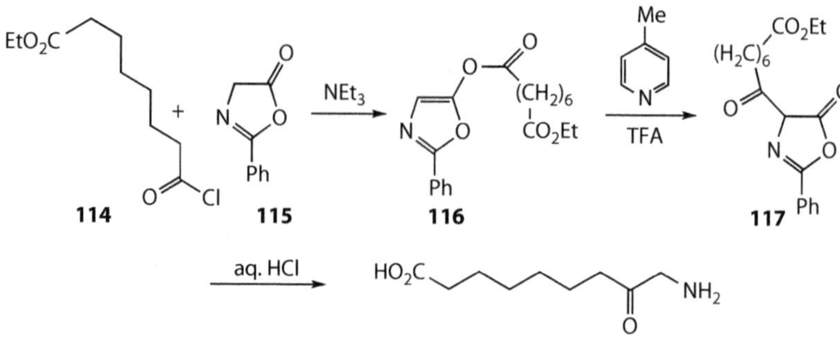

More exotic heterocycles have been used for amino acid synthesis. For example, chiral pyrimidinone ***118*** was prepared from an amino-oxazolidine, allowing conjugate addition by an organocuprate to give derivatives such as ***119***.[83] Subsequent alkylation gave ***120***, and catalytic hydrogenation followed by vigorous hydrolysis gave chiral β-amino acid ***121***. Other alkylated β-amino acids were prepared by this method.

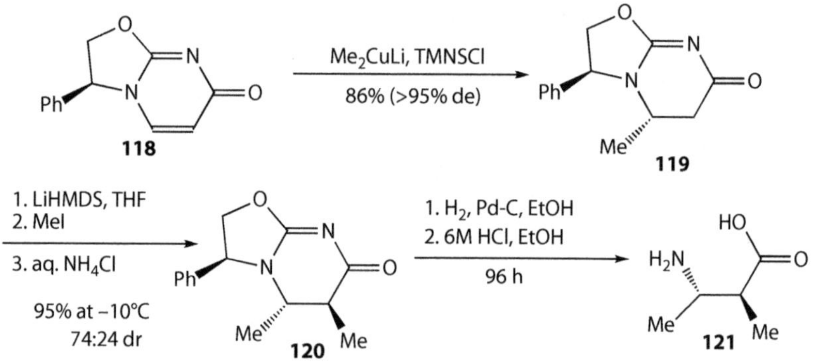

2.4.2 Pyrazole Derivatives

Pyrazole derivatives such as ***122*** (a pyrazoline) are formed by a [3+2]-cycloaddition[84] reaction of an alkene with a diazoalkane.[85] Basic hydrolysis of ***122*** with aqueous sodium hydroxide gave 2-methyl-3-aminopropanoic acid (***123***).[86] Similarly, 2-methyl-3-aminobutanoic acid was prepared in 52% yield, 2-ethyl-3-aminopropanoic acid was prepared in 81% yield, and, 2-methyl-3-aminopentanoic acid was prepared in 52% yield, all from the appropriate pyrazoline.[86]

---

[83] Agami, C.; Cheramy, S.; Dechoux, L.; Melaim, M. *Tetrahedron* ***2001***, *57*, 195.

[84] See (a) Smith, M.B. *Organic Synthesis*, 3rd ed. Wavefunction, Inc./Elsevier, Irvine, CA/London, England, ***2010***, pp. 1101–1115; (b) Smith, M.B. *March's Advanced Organic Chemistry*, 7th ed. John Wiley & Sons, Hoboken, NJ, ***2013***, pp. 1014–1020.

[85] (a) Schneider, M.P.; Goldbach, M. *J. Am. Chem. Soc.* ***1980***, *102*, 6114; (b) see Smith, M.B. *Organic Synthesis*, 3rd ed. Wavefunction, Inc./Elsevier, Irvine, CA/London, England, ***2010***, p. 1110

[86] Ioffe, B.V.; Isidorov, V.A.; Stolyarov, B.V. *Dokl. Akad. Nauk SSSR*, ***1971***, *197*, 91 (Engl. 177).

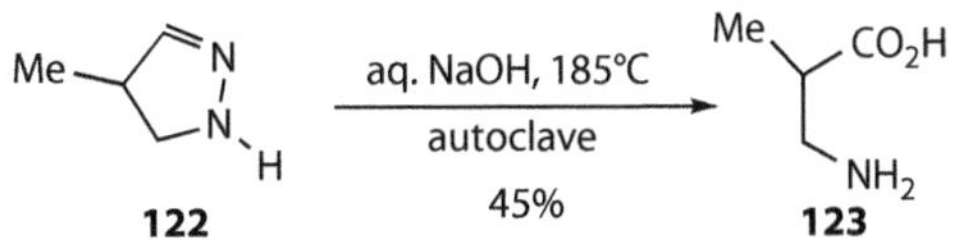

2.4.3 Pyridine, Hydropyridine, and Pyridone Derivatives

Several saturated and unsaturated derivatives of pyridine and pyridone can serve as synthetic precursors to amino acids. The ring in *N*-benzoylpiperidine (***124***) was oxidatively cleaved with potassium permanganate to give ***125***.[87] Hydrolysis of the amide moiety gave 5-aminopentanoic acid.

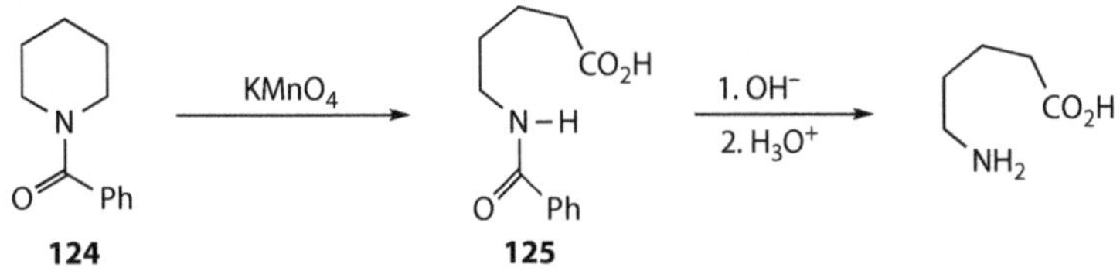

Ring cleavage can occur by many routes, depending on the attached substituents. The reaction of 1,4,5,6-tetrahydronicotinamide (***126***) with hydroxylamine, for example, gave a transient isoxazol-5-one (***127***). This heterocycle spontaneously opened under the reaction conditions to give 5-amino-2-cyanopentanoic acid (***128***) as the final product.[88]

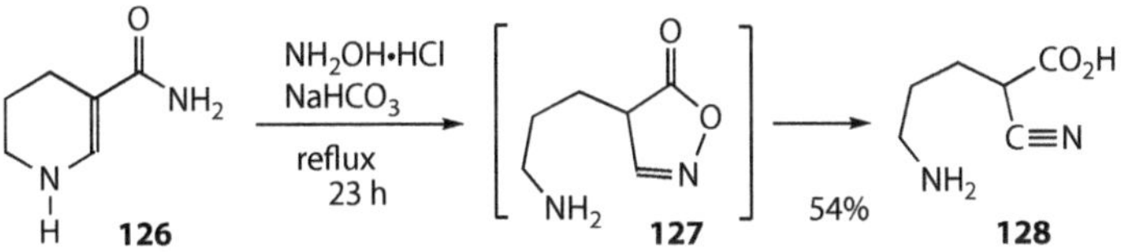

A pyridone route to 5-aminopentanoic acids began with the conversion of 2-pyridone (***129***) to 5-hydroxy-2-piperidone (***130***)[89] by an Elbs oxidation.[90] Catalytic hydrogenation converted the pyridone ring to a lactam (***131***), and acid hydrolysis gave 5-amino-4-oxopentanoic acid, ***132***.[89,91]

---

[87] (a) Schotten, C. *Chem. Gesel. Ber. Deutch.* ***1888***, *21*, 2240; (b) Schotten, C. *Chem. Gesel. Ber. Deutch.* ***1884***, *17*, 2546.

[88] Quin, L.D.; Pinion, D.O. *J. Org. Chem.* ***1970***, *35*, 31310.

[89] Franck, B.; Stratmann, H. *Heterocycles*, ***1981***, *15*, 919; (b) Franck, B. *Angew Chem.* ***1982***, *94*, 327.

[90] (a) Elbs, K. *J. Prakt. Chem.* ***1893***, *48*, 179; (b) Haines, A.H. *Methods for the Oxidation of Organic Compounds*. Academic Press, London, ***1985***, p. 174; (c) Sethna, S.M. *Chem. Rev.* ***1951***, *49*, 91; (d) see Smith, M.B. *Organic Synthesis*, 3rd ed. Wavefunction, Inc./Elsevier, Irvine, CA/London, England, ***2010***, p. 267.

[91] Herdeis, C.; Dimmerling, A. *Arch. Pharm.* (Weinheim) ***1984***, *317*, 304.

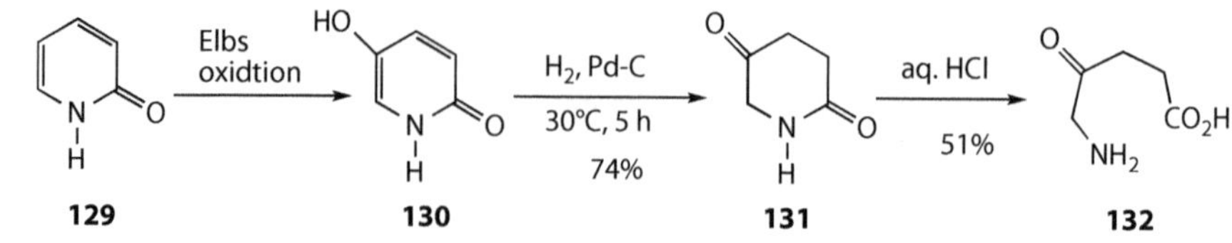

Cyclic imines other than reduced pyridine derivatives can be utilized in a reaction sequence that prepares amino acids. Further, functionalization of a cyclic amine, followed by conversion to the corresponding lactam and hydrolysis, leads to an amino acid if desired. Preparation of aryldiazonium salt ***133*** in four steps allowed a palladium-catalyzed Heck coupling[92] of ***133*** with *N*-Boc-3-pyrroline (***134***) to give ***135***.[93] Hydrolysis with aqueous acid gave ***136***, which was oxidized to 2-pyrrolidinone ***137*** with tetrapropylammoniumperruthenate (TPAP). Hydrolysis with 6N HCl led to lactam ***138*** (known as rolipam). Rolipram was originally developed as an antidepressant[94] but has been shown to be a selective inhibitor of phosphodiesterase IV (PDE4), which is related to pathological processes of the central nervous system.[95] In this case, the target was the lactam, but ring opening the 4-aminobutanoic acid derivative is also possible by heating ***137*** with 6N HCl at reflux for several hours.[93]

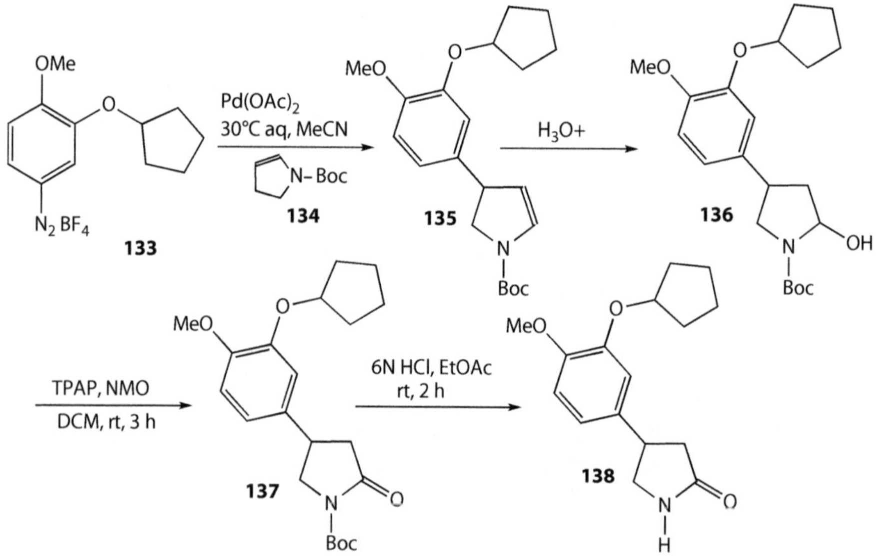

[92] See Smith, M.B. *Organic Synthesis*, 3rd ed. Wavefunction, Inc./Elsevier, Irvine, CA/London, England, ***2010***, pp. 1236–1239.

[93] Garcia, A.L.L.; Carpes, M.J.S.; de Oca, A.C.B.M.; dos Santos, M.A.G.; Santana, C.C.; Correia, C.R.D. *J. Org. Chem.* ***2005***, *70*, 1050.

[94] (a) Seildelman, D.; Schmiechen, R.; Paschelke, G.; Muller, B. (Schering A.G., Berlin). *Ger. Pat.* 2413935/3, ***1974*** [*Chem. Abstr.* ***1976***, *84*: 30878u]. (b) Schmiechen, R.; Horowski, R.; Palenschat, D.; Paschelke, G.; Wachtel, H.; Kehr, W. (Schering A.G., Berlin). *U.S. Pat.* 4193926, ***1976***.

[95] (a) Beavo, J.A.; Conti, M.; Heaslip, R. *J. Mol. Pharmacol.* ***1994***, *46*, 399. (b) Brackeen, M.F.; Cowan, D.J.; Stafford, J.A.; Schoenen, F.J.; Veal, J.M.; Domanico, P.L.; Rose, D.; Strickland, A.B.; Verghese, M.; Feldman, P.L. *J. Med. Chem.* ***1995***, *38*, 4848. (c) Seika, M. *Drugs Future* ***1998***, *23*, 108.

2.4.4 From Uracil Derivatives and Cyclic Carbamates

Uracil derivatives contain both an amine and a carboxyl group in the structure, and they are effective β-amino acid surrogates. An initial cyclization reaction of the alkene moiety of a conjugated acid such as 4-methyl-2-pentenoic acid with urea gave dihydrouracil derivative ***139***.[96] Acid hydrolysis led to cleavage of the ring and formation of 4-methyl-3-aminopentanoic acid, ***140***. Substituted dihydrouracils have been converted to 2-alkyl, 2,3-dialkyl, 3-alkyl, 3,3-dialkyl, and 2,3,3-trialkyl-3-aminopropanoic acid derivatives.[97]

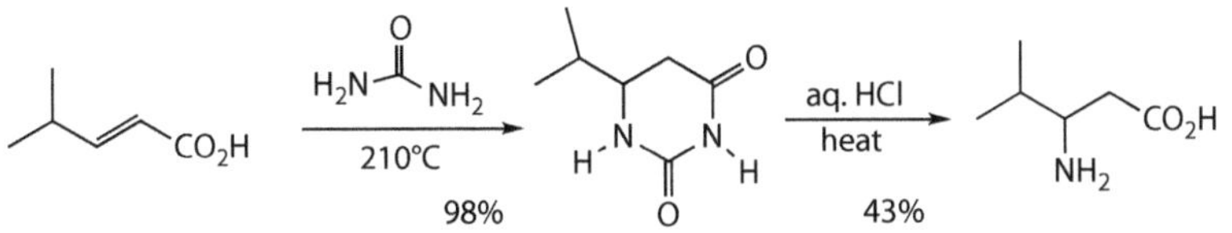

139 **140**

Dihydrouracils are actually perhydropyrimidinones (see ***141***), and they can be prepared from β-alanine, as pointed out above. Once in hand, formation of the enolate anion of ***141*** by treatment with lithium diisopropylamide (LDA) was followed by alkylation and hydrolysis to give substituted β-alanine derivative ***142*** with high enantioselectivity.[98]

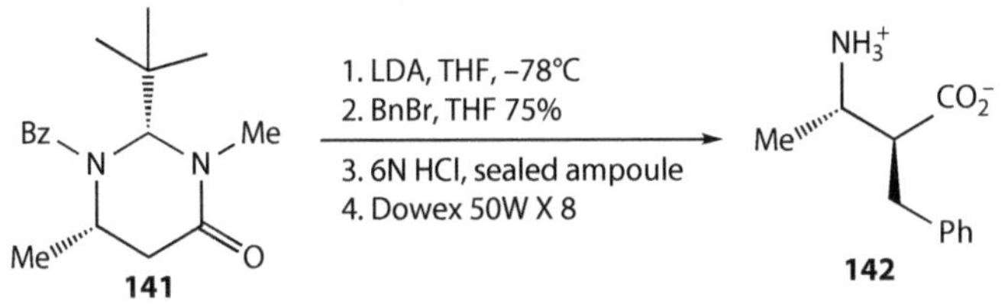

141 **142**

Cyclic five-membered ring urethane derivatives (imidazolidin-2-ones) are also available, and they serve as precursors to amino acids when appropriately functionalized. In one specialized synthesis, dethiobiotin (***143***) was treated with hot HCl to give 7,8-diaminononanoic acid (7,8-diaminopelargonic acid, ***144***).[99]

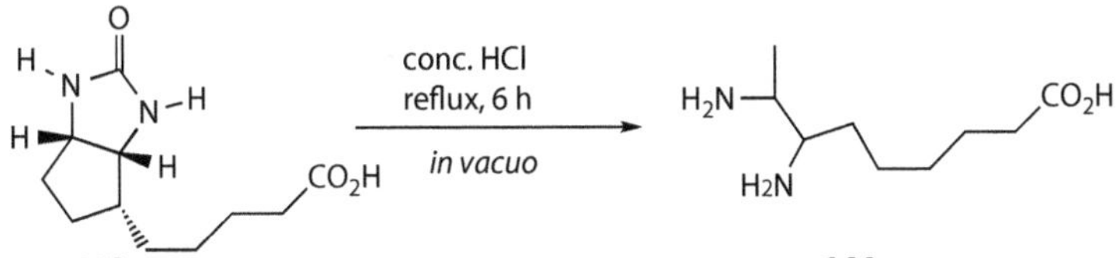

143 **144**

---

[96] Birkofer, L.; Storch, I. *Chem. Ber.* ***1953***, *86*, 529.
[97] Rachina, V.; Blagoeva, I. *Synthesis* ***1982***, 967.
[98] Juaristi, E.; Escalante, J.; Lamatsch, B.; Seebach, D. *J. Org. Chem.* ***1992***, *57*, 2396. Also see Juaristi, E.; Escalante, J. *J. Org. Chem.* ***1993***, *58*, 2282.
[99] Suyama, T.; Tsugawa, R.; Okumura, S.; Kanao, S. *Yakugaku Zasshi* ***1968***, *88*, 223 [*Chem. Abstr.* ***1968***, *69*: 51501w].

2.4.5 From Thiophene, Thiazolidine, and Oxathiolone Derivatives

The thiophene ring has been used as a template to "hold" a carbonyl moiety and an amino moiety (or its surrogate) at predetermined relative positions. Reductive cleavage of the thiophene ring "liberates" the amino acid.[100] When ***145*** was treated with Raney nickel in aqueous ammonium hydroxide, for example, reduction of the nitro moiety as well as the thiophene ring gave 4-aminoheptanoic acid (***146***, also known as 4-aminoenanthic acid). In a similar manner, 4-aminohexanoic acid was produced in 46% yield.[101,102] Oxime-substituted thiophene derivatives can be used, as in the conversion of ***147*** to 9-aminoundecanoic acid (***148***).[101]

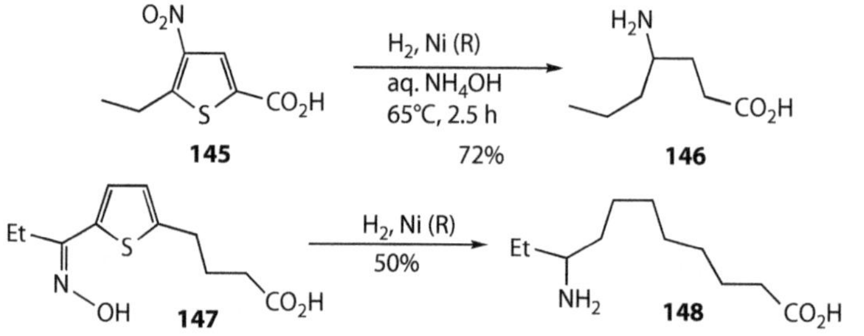

Other sulfur-containing heterocycles serve as useful precursors. When hydrothiazole ***149*** was condensed with malonic acid, thiazolidine ***150*** was formed.[103] Heating to reflux in water was sufficient to open the thiazolidine ring and give 3-amino-4-methyl-4-mercaptopentanoic acid, ***151***.

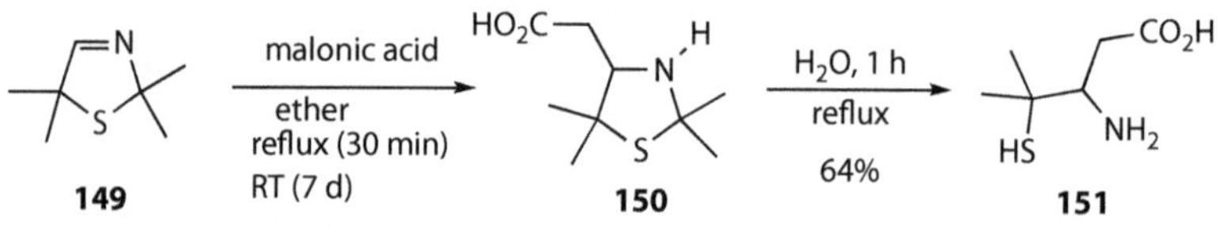

2.5 DIELS-ALDER STRATEGIES

A very different approach from those cited in previous sections uses cyclic precursors for the preparation of acyclic amino acids. In work by Krief and coworkers, a Diels-Alder

[100] Fabrichnyi, B.P.; Zurabyan, S.E.; Shalavina, I.F.; Gol'dfarb, Ya.L. *Izv. Akad. Nauk. SSSR Ser Khim.* ***1967***, 2102

[101] (a) Gol'dfarb, Ya.L.; Fabrichnyı, B.P.; Shalavina, I.F. *Tetrahedron*, ***1962***, *18*, 21; (b) Fabrichnyı, B.P.; Krasnyanskaya, E.A.; Shalavina, I.F.; Gol'dfarb, Ya.L. *Zh. Obshch. Khim.* ***1963***, *33*, 2697 (Engl. 2627).

[102] Fabrichnyi, B.P.; Shalavina, I.F.; Gol'dfarb, Ya.L. *Zh. Obshch. Khim.* ***1964***, *34*, 3878 (Engl. 3938).

[103] Martens, J.; Kintscher, J.; Arnold, W. *Synthesis* ***1991***, 497.

reaction[104] between methyl acrylate and ***152*** gave pyran ***153***.[105] Catalytic hydrogenation opened the ring and gave methyl 5-(*N,N*-diethylamino)-4-methylpentanoate, ***154***.

Oxazines are another class of heterocycles that have been converted to amino acids. A dihydro-1,2-oxazine is available by a Diels-Alder reaction of dienyl esters with nitroso derivatives, which react as a dienophile. The reaction of methyl sorbate (***155***) and 1-chloro-1-nitrosocyclohexane (***156***) gave ***157***, for example.[106] Catalytic hydrogenation opened the ring, and hydrolysis of the ester moiety gave 5-amino-2-hydroxyhexanoic acid, ***158***.

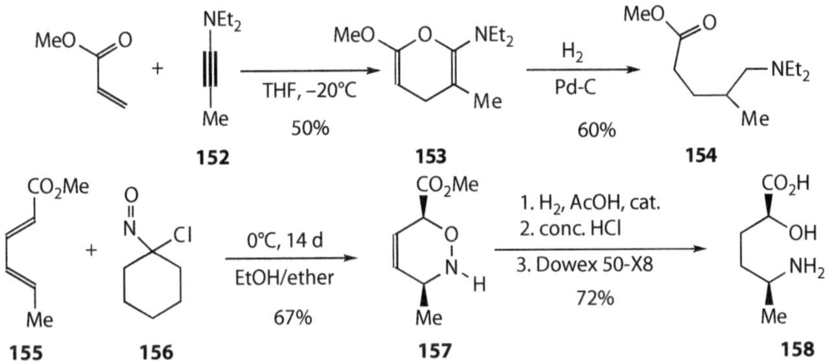

---

104 (a) Desimoni, G.; Tacconi, G.; Barco, A.; Pollini, G.P. *Natural Product Synthesis through Pericyclic Reactions*. American Chemical Society, Washington, DC, ***1983***; (b) see Smith, M.B. *Organic Synthesis*, 3rd ed. Wavefunction, Inc./Elsevier, Irvine, CA/London, England, ***2010***, pp. 1013–1075; (c) Smith, M.B. *March's Advanced Organic Chemistry*, 7th ed. John Wiley & Sons, Hoboken, NJ, ***2013***, pp. 1020–1039.

105 Ficini, J.; Krief, A. *Tetrahedron Lett.* ***1970***, 885.

106 Belleau, B.; Au-Young, Y.-K. *J. Am. Chem. Soc.* **1963**, *85*, 64.

3 Conjugate Addition Reactions

Amines and many amine surrogates are nucleophiles in conjugate addition reactions (the Michael reaction).[1] It is therefore not surprising that they add to conjugated acid derivatives to give amino acids. This strategic approach to their synthesis is significant, and for that reason it has been segregated into a separate chapter.

3.1 AMMONIA AND AMINE NUCLEOPHILES

Ammonia is sufficiently nucleophilic that addition to the double bond of a conjugated ester is possible, without attack at the acyl carbon, although the conjugate addition tends to be reversible. Amines undergo similar conjugate addition. In both cases, the product is a 3-aminopropanoic acid derivative (a β-amino acid derivative).

The most common method for producing substituted β-amino acids is via conjugate addition to conjugated acid derivatives, usually esters. This reaction is sometimes called an aza-Michael reaction.[2] A chemoenzymatic variation led to an enantioselective synthesis of β-amino acids.[3] Just as ammonia reacts as a nucleophile in a Michael addition with α,β-unsaturated carbonyl derivatives, amines add to conjugated acids such as acrylic acid to give β-amino acids (3-aminopropanoic acids).[4] Methylamine, aniline, dimethylamine, pyrrolidine, and morpholine add to acrylic acid to give the corresponding 3-*N*-substituted or *N*,*N*-disubstituted aminopropanoic acid derivative.[4]

Since ammonia or an amine react as bases, an acid-base reaction with conjugate carboxylic acids is inevitable, and this fact can limit the yield and utility of the reaction. Yields are often better when conjugated esters are used as a substrate rather than the acid itself, as shown by the addition of diethylamine to ethyl acrylate, which gave ethyl 3-(*N*,*N*-diethyl-amino)propionate (***1***) in 87% yield.[5] Similarly, *tert*-butylamine added to give an 87% yield of ***2***.[6] The reaction can also be done using an

---

[1] (a) Michael, A. *J. Prakt. Chem.* ***1887***, *35*, 379; (b) Bergmann, E.D.; Gingberg, D.; Pappo, R. *Org. React.* ***1959***, *10*, 179; (c) Perlmutter, P. *Conjugative Addition Reactions in Organic Synthesis*. Pergamon Press, Oxford, ***1992***; (d) see Smith, M.B. *Organic Synthesis*, 3rd ed. Wavefunction, Inc./Elsevier, Irvine, CA/London, England, ***2010***, pp. 877–888; (e) Smith, M.B. *March's Advanced Organic Chemistry*, 7th ed. John Wiley & Sons, Hoboken, NJ, ***2013***, pp. 943–949.

[2] See (a) Yamagiwa, N.; Qin, H.; Matsunaga, S.; Shibasaki, M. *J. Am. Chem. Soc.* ***2005***, *127*, 13419; (b) Munro-Leighton, C.; Blue, E.D.; Gunnoe, T.B. *J. Am. Chem. Soc.* ***2006***, *128*, 1446; (c) Smith, M.B. *March's Advanced Organic Chemistry*, 7th ed. John Wiley & Sons, Hoboken, NJ, ***2013***, p. 946.

[3] Strompen, S.; Weiß, M.; Ingram, T.; Smirnova, I.; Gröger, H.; Hilterhaus, L.; Liese, A. *Biotechn. Bioeng.* ***2012***, *109*, 1479.

[4] Jolidon, S.; Meul, T. *Eur. Pat. Appl.* EP 144,980 [*Chem. Abstr.* ***1986***, *105*:43325p].

[5] Weisel, C.A.; Taylor, R.B.; Mosher, H.S.; Whitmore, F.C. *J. Am. Chem. Soc.* ***1945***, *67*, 1071.

[6] Robinson, J.B.; Thomas, J. *J. Chem. Soc.* ***1965***, 2270.

alcohol solvent such as ethanol rather than using the amine, neat.[7] Aromatic amines such as aniline add to acrylate esters,[8] as do a wide variety of other amines.[9]

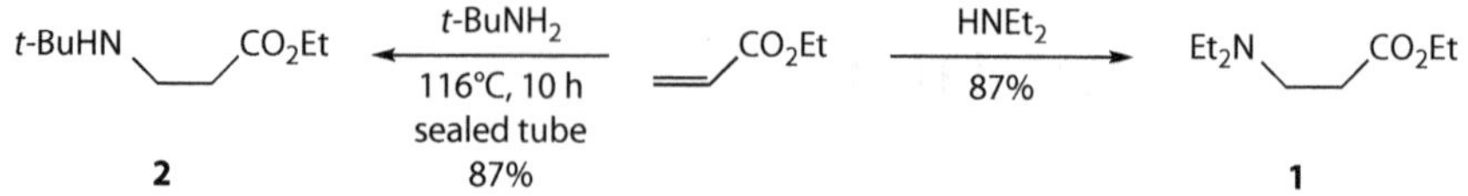

Michael addition of *N*-substituted amines[10] is a rather general method to produce amino acids with alkyl substituents. In one example, addition of benzylamine to ethyl hexenoate produced *N*-benzyl-3-aminohexanoate (*3*) in 75% yield.[11] In other work, the reaction of ethyl cinnamate with piperidine gave ethyl 3-phenyl-3-piperidino propanoate,[12] and another example reacted methylamine with ethyl 4-phenylbut-2-enoate (*4*) to give ethyl 4-phenyl-3-methylaminobutanoate, ***5***.[13]

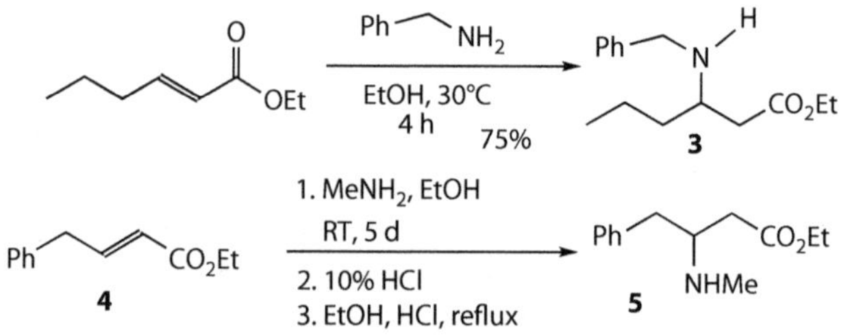

Functionalized amino acids can also be prepared by this methodology. Hydroxy amino acids were prepared by initial protection of the alcohol moiety in ***6***, allowing conjugate addition of benzylamine to give *7*.[14] Removal of the silyl-protecting group with tetrabutylammonium fluoride gave methyl 3-(*N*-benzylamino)-2-(1-hydroxyethyl) propanoate, ***8***. When the reaction was done using methanol as a solvent, a 3.5:1 *syn:anti* mixture 15 was obtained in 99% yield. When the reaction was done in tetrahydrofuran (THF), however, a 1:3.9 ratio of *syn:anti* ***8*** was obtained, in 77% yield.[14]

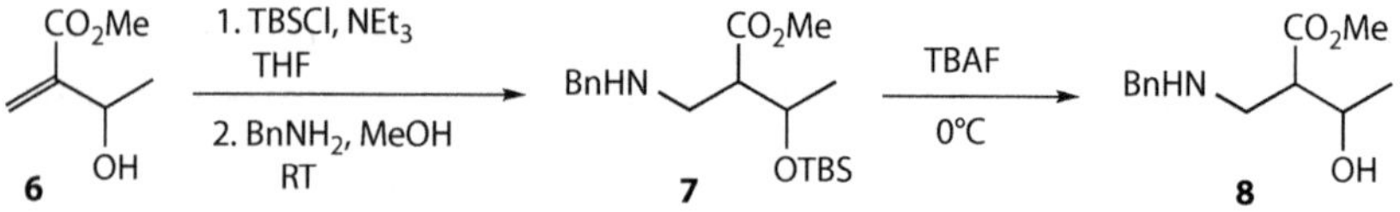

[7] Frankhauser, R.; Grob, C.A.; Krasnobajew, V. *Helv. Chim. Acta* ***1966***, *49*, 690.
[8] Simonova, N.I. *Tr. Leningr. Inst. Kinoinzhenerov* ***1961***, 19 [*Chem. Abstr.* ***1964***, *60*: 4127b].
[9] Röhnert, H. *Archiv. Pharm.* ***1963***, *296*, 257.
[10] (a) Michael, A. *J. Prakt. Chem.* ***1887***, *35*, 379; (b) Bergmann, E.D.; Gingberg, D.; Pappo, R. *Org. React.* ***1959***, *10*, 179; (c) Perlmutter, P. *Conjugative Addition Reactions in Organic Synthesis*. Pergamon Press, Oxford, ***1992***; (d) see Smith, M.B. *Organic Synthesis*, 3rd ed. Wavefunction, Inc./Elsevier, Irvine, CA/London, England, ***2010***, pp. 877–888; (e) Smith, M.B. *March's Advanced Organic Chemistry*, 7th ed. John Wiley & Sons, Hoboken, NJ, ***2013***, pp. 943–949.
[11] Habermehl, G.; Andres, H. *Liebigs Ann. Chem.* ***1977***, 800.
[12] Pacheco, H.; Dreux, M.; Beauvillain, A. *Bull. Soc. Chim. Fr.* ***1962***, 1379.
[13] Teotino, U.M. *J. Org. Chem.* ***1962***, *27*, 1906.
[14] Perlmutter, P.; Tabone, M. *Tetrahedron Lett.* ***1988***, *29*, 949.

Enantioselective variations of the Michael reaction are known.[15] The conjugate addition of a chiral amine to methyloct-2-enoate gave ***9***, and deprotection led to methyl 3-aminooctanoate, ***10***. The copper-catalyzed arylation of ***10*** led to ***11***, which allowed the total synthesis of ***12***, (+)-angustureine.[16]

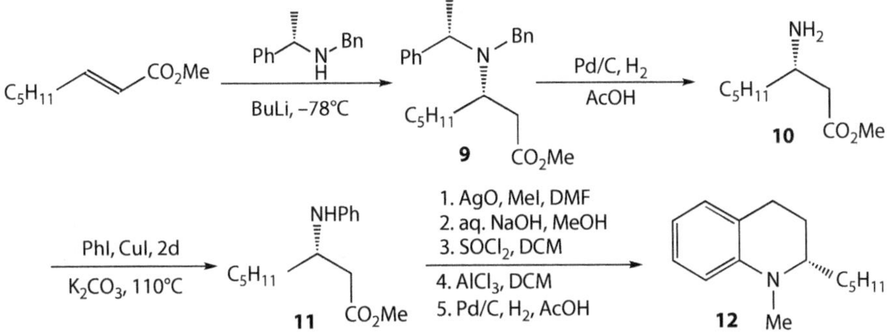

A similar approach used chiral lithium amide ***13*** as the nucleophilic species, and conjugate addition led to ***14***.[17] Subsequent catalytic hydrogenation gave ***15*** in 83% yield and >97%de. It is noted that the use of lithium amides as homochiral ammonia equivalents in

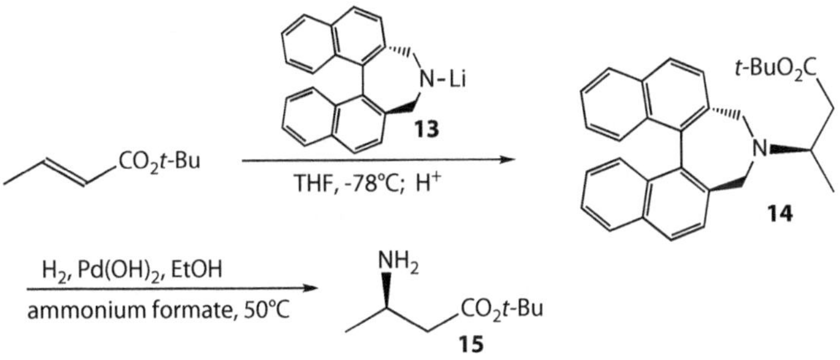

conjugate addition reactions of conjugated esters is an *Organic Syntheses* preparation.[18] 3-Amino acids with substituents can be prepared by the conjugate addition of lithium amides to conjugate esters. In the case of ***16***, this approach led to ***17*** in good yield and modest asymmetric induction.[19]

---

15 See Smith, M.B. *Organic Synthesis*, 3rd ed. Wavefunction, Inc./Elsevier, Irvine, CA/London, England, ***2010***, pp. 878.

16 Lin, X.-F.; Li, Y.; Ma, D.-W. *Chin. J. Chem.* ***2004***, *22*, 932. For the isolation of angustureine, see Jacquemond-Coliet, I.; Hannedouche, S.; Favre, N.; Fourasté, I.; Moulis, C. *Phytochemistry* ***1999***, *51*, 1167.

17 (a) Davies, S.G.; Smith, A.D.; Price, P.D. *Tetrahedron Asym.* ***2005***, *16*, 2833. Also see (b) Davies, S.G.; Garrido, N.M.; Kruchinin, D.; Ichihara, O.; Kotchie, L.J.; Price, P.D.; Price Mortimer, A.J.; Russell, A.J.; Smith, A.D. *Tetrahedron Asym.* ***2006***, *17*, 1793; (c) Langenhan, J.M.; Gellman, S.H. *J. Org. Chem.* ***2003***, *68*, 6440.

18 Davies, S.G.; Fletcher, Ai.M.; Roberts, P.M. *Org. Synth.* ***2010***, *87*, 143.

19 Langenhan, J.M.; Gellman, S.H. *J. Org. Chem.* ***2003***, *68*, 6440.

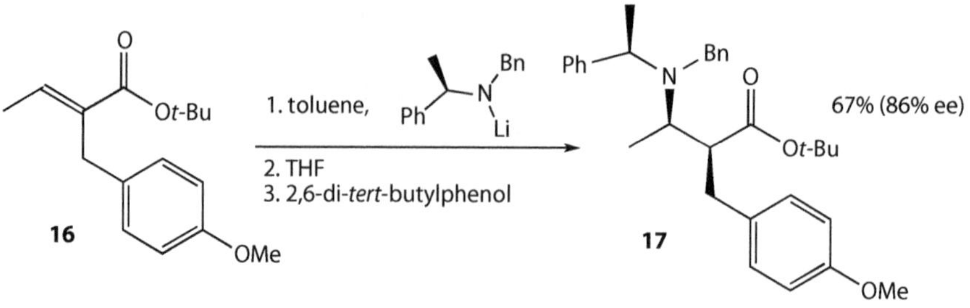

16 17

Conjugate addition of amines to alkynyl esters or alkynyl acids will give 3-amino-propenoic acid derivatives. When ***18*** reacted with the amine unit of ethyl glycine in methanolic HCl, ***19*** was formed. Catalytic hydrogenation of the enamino-ester gave ***20***.[20] The use of R–C≡C–CO_2Et (R = *n*-C_4F_9, C_6F_{13}, C_7F_{15}, or C_8F_{17}) led to the preparation of not only the 3-aminooctanoic acid derivatives, but also the heptanoic, decanoic, dodecanoic, and undecanoic acid derivatives (all as their polyfluoro analogs).[20a]

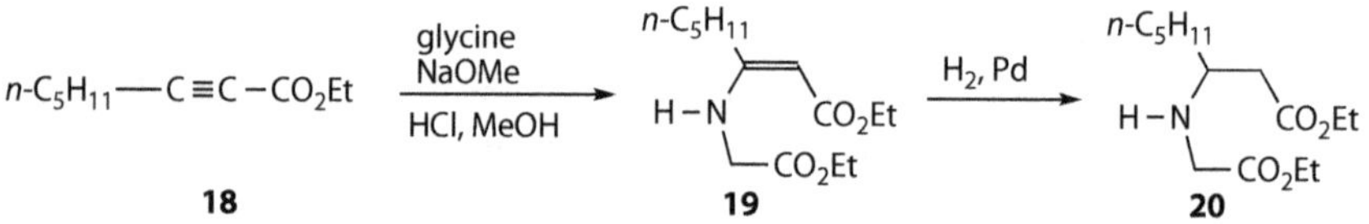

18 19 20

Catalytic hydrogenation of enamino-esters such as ***19*** or an enamino acid is a convenient method to produce the corresponding saturated β-amino acid, and thereby extend the synthetic utility of the methodology. The use of chiral hydrogenation catalysts has the potential to produce amino acids with high enantiomeric purity. Many catalysts and additives are available.[21] Conjugate imido-ester ***21*** was prepared from 4-chloroacetophenone in three steps. Asymmetric catalytic hydrogenation led to the imido-ester, and acid hydrolysis gave the important amino acid baclofen (***22***; see Chapter 6, Section 6.3) with good enantioselectivity for the overall process.[22]

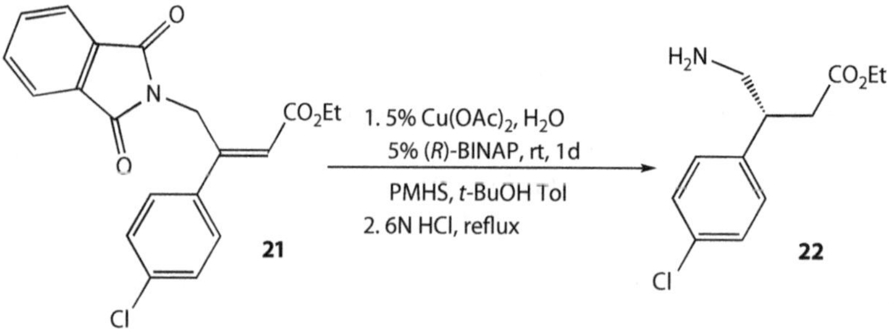

21 22

---

[20] (a) Chauvin, A.; Fabron, J.; Aityahia, M.O.; Pastor, R.; Cambon, R. *Tetrahedron*, ***1990***, *46*, 6705; (b) Haddach, M.; Pastor, R.; Riess, J.G. *Tetrahedron Lett.* ***1990***, *31*, 1989.

[21] Zhu, W.; Ma, D. *Org. Lett.* ***2003***, *5*, 5063.

[22] Deng, J.; Hu, X.-P.; Huang, J.-D.; Yu, S.-B.; Wang, D.-Y.; Duan, Z.-C.; Zheng, Z. *J. Org. Chem.* ***2008***, *73*, 6022. Also see Deng, J.; Duan, Z.-C.; Huang, J.-D.; Hu, X.-P.; Wang, D.-Y.; Yu, S.-B.; Xu, X.-F.; Zheng, Z. *Org. Lett.* ***2007***, *9*, 4825; Deng, J.; Duan, Z.-C.; Huang, J.-D.; Hu, X.-P.; Wang, D.-Y.; Yu, S.-B.; Xu, X.-F.; Zheng, Z. *Org. Lett.* ***2008***, *10*, 3379.

A photochemically induced radical addition reaction of amines has been reported. Photolysis of 2-aminopropane in the presence of ethyl crotonate led to an amino-ester (***23***), which spontaneously cyclized to 4,5,5-trimethyl-2-pyrrolidinone, ***24***.[23] Aqueous acid hydrolysis liberated 3,4-dimethyl-4-aminopentanoic acid, ***25***. Amine radicals (the unpaired electron is on N) add in a conjugate manner to α,β-unsaturated esters, giving an amino acid derivative.

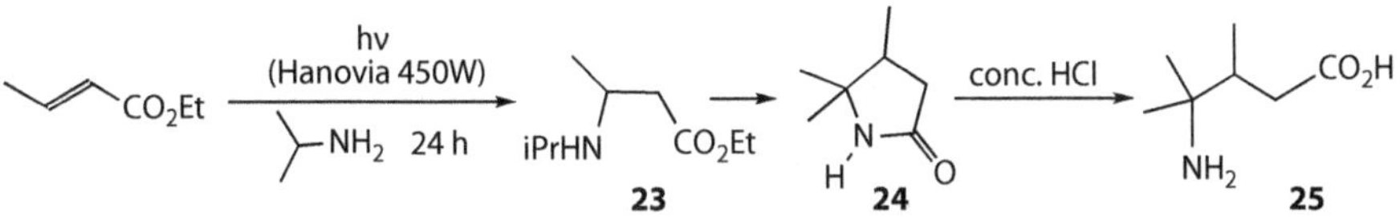

23 **24** **25**

Esters are not the only acid derivatives that participate in conjugate addition reactions. Nitriles may be hydrolyzed to the corresponding acid, so conjugated nitriles can be utilized to prepare amine acids. Amines add to conjugated nitriles, and the resulting cyanoamine can be hydrolyzed to amino acids. In such a case, the nitrile moiety is a carboxylic acid surrogate (see Chapter 1, Section 1.3.1). A simple example is the reaction of benzylamine with but-2-enenitrile (***26***) to give ***27***. Acid hydrolysis of the nitrile moiety gave *N*-benzyl-3-aminobutanoic acid (***28***), and removal of the benzyl group by hydrogenation with a palladium catalyst gave 3-aminobutanoic acid, ***29***.[24]

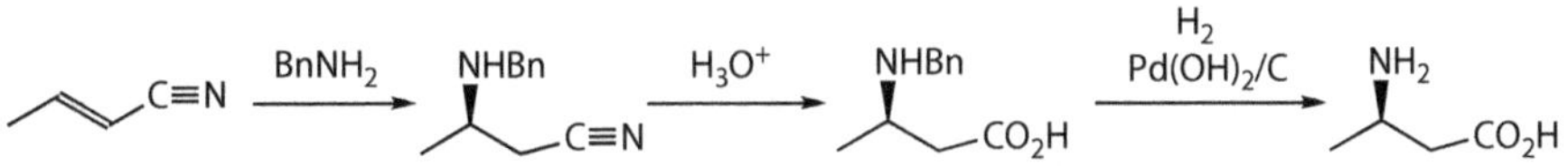

26 **27** **28** **29**

Another example heated methacrylonitrile (***30***) with ammonia in aqueous NaOH (in an autoclave) to give 2-methyl-3-aminopropanoic acid (***31***),[25] after acid hydrolysis. These two examples illustrate that placement of an alkyl substituent can be determined in part by the substituent pattern on the conjugated starting material.

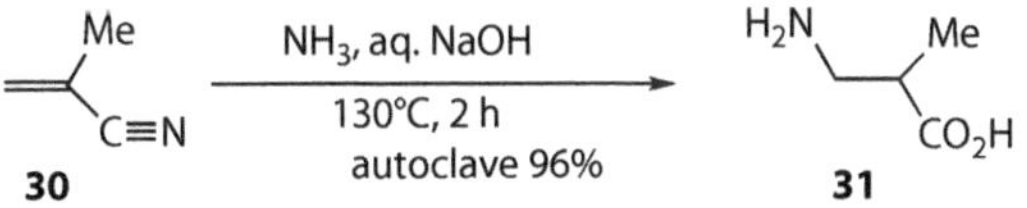

30 **31**

[23] Pfau, M.; Dulou, R. *Bull. Soc. Chim. Fr.* ***1967***, 3336.

[24] Furukawa, M.; Okawara, T.; Terawaki, Y. *Chem. Pharm. Bull.* ***1977***, *25*, 1319.

[25] Mekhtiev, S.I.; Safarov, Yu.D.; Makhmedov, R.; Tagiev, R.B. *Azeb. Khim. Zh.* ***1981***, 21 [*Chem. Abstr.* ***1981***, *95*: 62618c]. Also see Szlompek-Nesteruk, D. *Prezemysl. Chem.* ***1965***, *44*, 85 [*Chem. Abstr.* ***1965***, *62*:16366f].

3.2 CARBON NUCLEOPHILES

3.2.1 Nitro Enolates and Nitrile Enolates

Certain amine surrogates can add to conjugated carbonyl systems if that surrogate can be converted to a carbanionic species, typically an enolate anion. Addition of these species to conjugated acid derivatives will give an amino acid, but a second reaction is required to convert the surrogate to the amino group. This section will describe reactions of specialized enolate anions that participate in Michael additions,[26] but it is noted that Chapter 4 is devoted to enolate anion reactions of all types that will produce an amino acid.

Nitroalkanes undergo conjugate addition reactions with electron-deficient alkenes under the right conditions.[27] There are several variations. Nitroalkanes react with a suitable base to form the corresponding nitro enolate anion, which readily adds to conjugated systems. The reaction of ethyl crotonate and nitroethane in the presence of the base, 1,3-diazabicyclo[5.4.0]-undecane (DBU), led to nitro-ester ***30*** in 74% yield.[28] The conjugate addition was proceeded by formation of the nitroethane enolate anion in situ. Catalytic hydrogenation of the nitro group in ***32*** and acid hydrolysis of the ester produced 3-methyl-4-aminobutanoic acid, ***33***. The use of different conjugated esters led to formation of 3-ethyl- (59% overall yield), 3-propyl- (51% overall yield), 3-isopropyl- (62% overall yield), 3-butyl- (56% overall yield), 3-*sec*-butyl- (60% overall yield), 3-isobutyl- (58% overall yield), and 3-*tert*-butyl-4-aminobutanoic acid (31% overall yield).[28] Organocatalysts such as optically active pyrrolidines have been used to facilitate the conjugate addition of nitroalkanes with good enantioselectivity.[29] Mesitylcopper/(*R*)-DTBM-Segphos precatalysts have been used for the conjugate addition of nitroalkanes to conjugated thioamides to give γ-nitrothioamides with good enantioselectivity.[30] Cinchona alkaloid-derived chiral bifunctional thiourea organocatalysts have been used for the conjugate addition of nitromethane to chalcones with high enantioselectivity.[31]

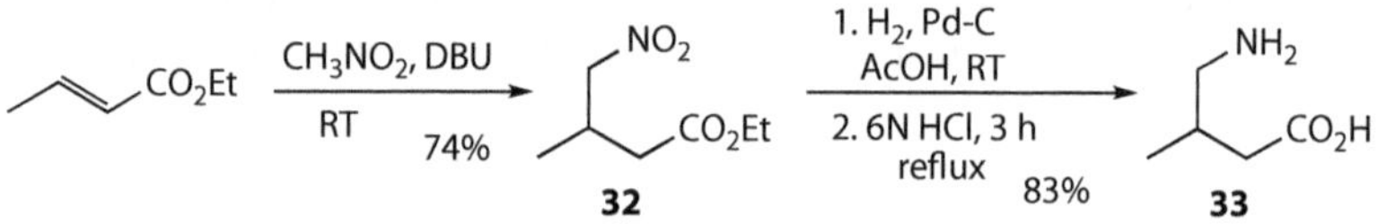

32 **33**

In conjugate addition reactions of nitroalkanes, the nitro group of the nitroalkane can react further with the conjugate addition product to generate an isoxazolidine

[26] (a) Michael, A. *J. Prakt. Chem.* ***1887***, *35*, 379; (b) Bergmann, E.D.; Gingberg, D.; Pappo, R. *Org. React.* ***1959***, *10*, 179; (c) Perlmutter, P. *Conjugative Addition Reactions in Organic Synthesis.* Pergamon Press, Oxford, ***1992***; (d) see Smith, M.B. *Organic Synthesis*, 3rd ed. Wavefunction, Inc./Elsevier, Irvine, CA/London, England, ***2010***, pp. 877–888; (e) Smith, M.B. *March's Advanced Organic Chemistry*, 7th ed. John Wiley & Sons, Hoboken, NJ, ***2013***, pp. 943–949.

[27] Ballini, R.; Bosica, G.; Fiorini, D.; Palmieri, A.; Petrini, M. *Chem. Rev.* **2005**, *105*, 933.

[28] Andruszkiewicz, R.; Silverman, R.B. *Synthesis*, ***1989***, 953.

[29] Jensen, K.L.; Poulsen, P.H.; Donslund, B.S.; Morana, F.; Jørgensen, K.A. *Org. Lett.* **2012**, *14*, 1516; Gotoh, H.; Ishikawa, H.; Hayashi, Y. *Org. Lett.* **2007**, *9*, 5307.

[30] Ogawa, T.; Mouri, S.; Yazaki, R.; Kumagai, N.; Shibasaki, M. *Org. Lett.* **2012**, *14*, 110.

[31] Vakulya, B.; Varga, S.; Csámpai, A.; Soós, T. *Org. Lett.* **2005**, *7*, 1967.

such as ***34*** (also see Chapter 2, Section 2.4.1). When nitromethane was treated with triethyloxonium tetrafluoroborate (Meerwein's reagent)[32] and then with methyl crotonate, ***34*** was formed.[33] Heating to 100°C led to quantitative formation of ***35***. Catalytic hydrogenation of the oxime moiety gave 4-amino-2-hydroxy-3-methylbutanoic acid, ***36***.

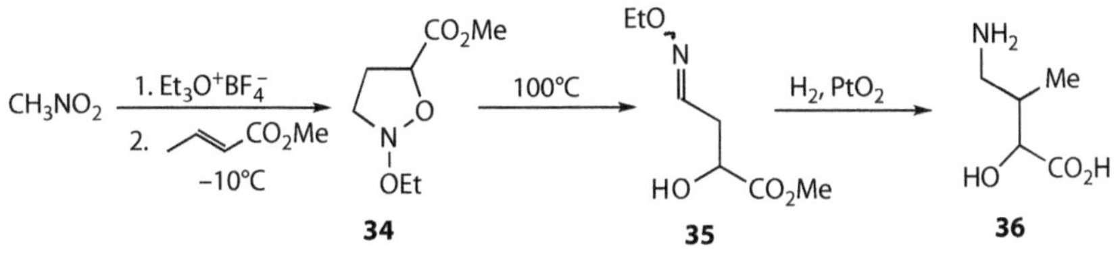

Nitrile enolate anions constitute a class of amine surrogate (see Chapter 1, Section 1.3.1) that may be used in conjugate addition reactions. Enolate anions of nitriles are generated by treatment of an alkyl nitrile with a suitable base, and undergo Michael addition[34] with an appropriate conjugated ester to give cyano-esters. Subsequent reduction of the cyano group generates the corresponding amino derivatives. Lactams are sometimes formed during this process when the initial amino-ester can form a five- or six-membered ring. This process is illustrated by the reaction of the anion of ***37*** with acrylonitrile to give ***38***.[35] Catalytic hydrogenation initially generated an amino-ester, but it spontaneously cyclized to the lactam (***39***). A hydrolysis step was therefore required to give 4-(3,5-dimethylphenyl)-5-aminopentanoic acid, ***40***.[35]

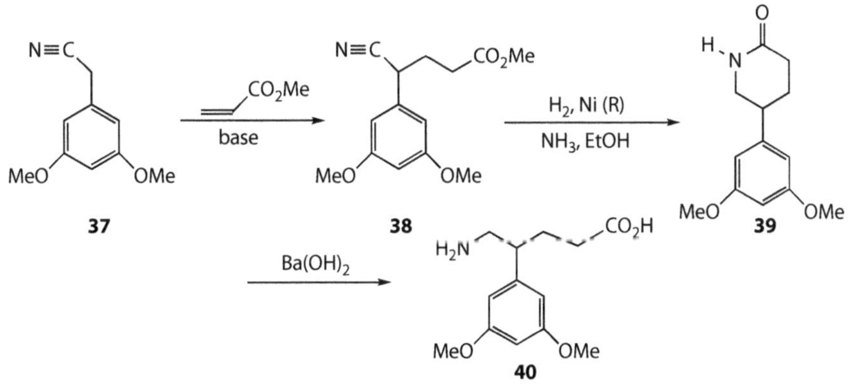

---

[32] Meerwein, H. *Org. Synth. Coll.* ***1973***, *5*, 1080.

[33] Sato, H.; Kusumi, T.; Imaye, K.; Kakisawa, H. *Chem. Lett.* ***1975***, 965.

[34] (a) Michael, A. *J. Prakt. Chem.* ***1887***, *35*, 379; (b) Bergmann, E.D.; Gingberg, D.; Pappo, R. *Org. React.* ***1959***, *10*, 179; (c) Perlmutter, P. *Conjugative Addition Reactions in Organic Synthesis*. Pergamon Press, Oxford, ***1992***; (d) see Smith, M.B. *Organic Synthesis*, 3rd ed. Wavefunction, Inc./Elsevier, Irvine, CA/London, England, ***2010***, pp. 877–888; (e) Smith, M.B. *March's Advanced Organic Chemistry*, 7th ed. John Wiley & Sons, Hoboken, NJ, ***2013***, pp. 943–949.

[35] Oine, T.; Kugita, H.; Takeda, M. *Chem. Pharm. Bull.* ***1963***, *11*, 541.

3.2.2 Addition to Conjugated Nitro Derivatives

The previous section discussed the addition of an amine surrogate to a conjugate acid derivative, but an alternative strategy adds an acid surrogate to a conjugated system that bears the nitrogen moiety. Nitroalkenes serve as such a substrate in Michael addition reactions, but the nucleophile must contain a carbonyl moiety or a carboxyl surrogate. The malonate anion is one such entity. When nitroethene was treated with dimethyl malonate under basic conditions, formation of the malonate anion and conjugate addition led to ***41***.[36] Catalytic hydrogenation of ***41*** led to an amino-ester that cyclized to 2-pyrrolidinone derivative (***42***). Acid hydrolysis with heating was accompanied by decarboxylation to give 4-aminobutanoic acid (GABA).[37]

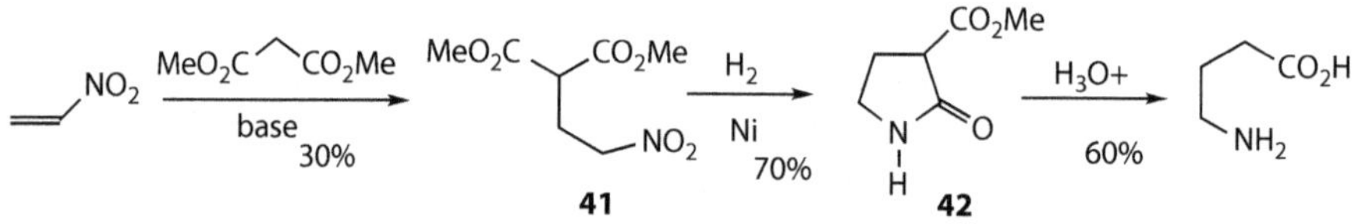

In the presence of a chiral base, good enantioselectivity is possible in this reaction. Conjugate addition of the malonate anion to conjugated nitro compound ***43***, in the presence of a chiral organocatalyst, gave ***44*** in 86% yield and 93%ee.[38] The organocatalyst contained both a thiourea and an amino moiety. Nitro diesters such as ***44*** are readily converted to chiral 4-aminobutanoic acid derivatives.

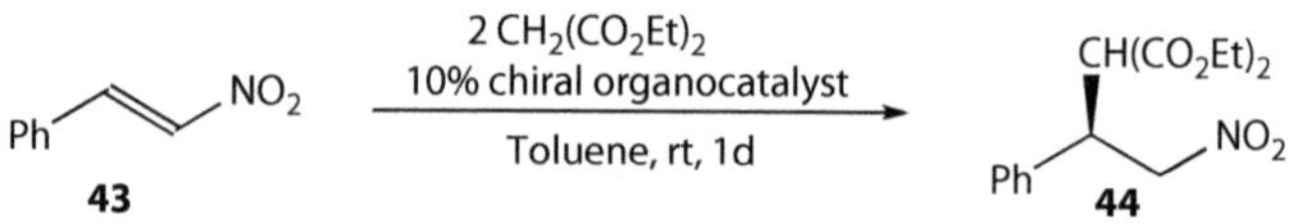

Another enantioselective synthesis of substituted aminobutanoic acid derivatives involved a conjugate addition of enolate anions to nitroalkenes in the presence of an Evans' auxiliary. The reaction of ***45*** gave ***46***, for example.[39] Hydrogenation using Raney nickel led to lactam ***48***, and protection of nitrogen followed by base hydrolysis gave ***47***.

---

[36] Smirnova, A.A.; Perekalin, V.V.; Shcherbakov, V.A. *Zh. Org. Khim.* ***1968***, *4*, 2245 (Engl. 2166).
[37] Fabrochnyi, B.P.; Shalavina, I.F.; Zurabyan, S.E.; Gol'dfarb, Ya.L.; Kostrova, S.M. *Zh. Org. Khim.* ***1968***, *4*, 680 (Engl. 663).
[38] Okino, T.; Hoashi, Y.; Furukawa, T.; Xu, X.; Takemoto, Y. *J. Am. Chem. Soc.* ***2005***, *127*, 119.
[39] Brenner, M.; Seebach, D. *Helv. Chim. Acta* ***1999***, *82*, 2365.

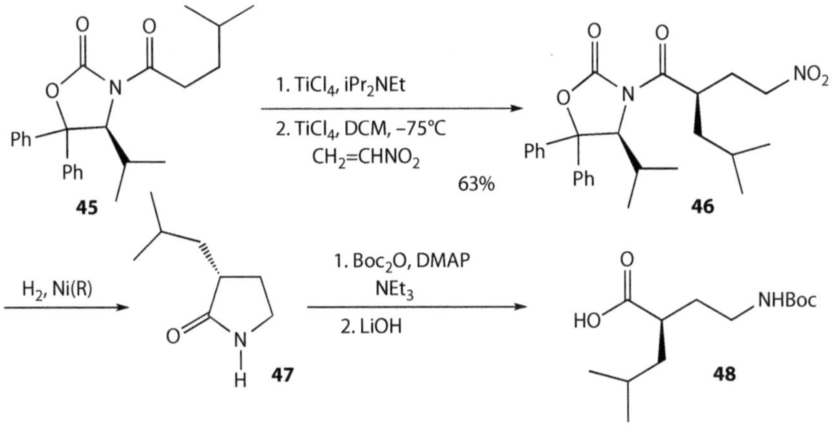

3.2.3 Addition of Enamines

Enamines are nucleophilic species that contain an amine moiety. Dienoic acid derivatives such as ***50*** were prepared by the condensation of enamino-esters (see ***49***) and ethyl propiolate.[40] Changing the 3-methyl group in the alkenyl amino acid to phenyl or hydrogen led to the synthesis of other dienoic acid derivatives. In this study, the *N*-alkyl group was modified from benzyl to hydrogen or methyl in order to produce other derivatives.[40]

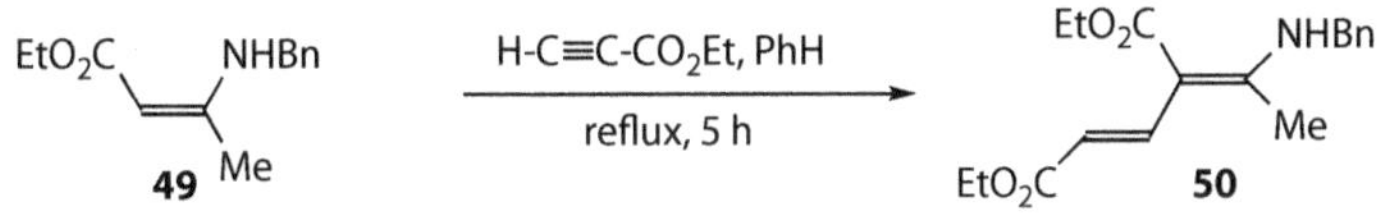

[40] Anghedide, N.; Draghici, C.; Raileanu, D. *Tetrahedron* ***1974***, *30*, 623.

4 Enolate Anion and Related Reactions

An important general strategy for the preparation of amino acids involves generating a carbanion from an acid derivative and subsequent reaction with another suitably functionalized derivative. This reaction may be the conjugate addition discussed in Chapter 3, Section 3.2.2, but alkylation or acyl addition reactions may also be used. When appropriate functionality is present, these reactions constitute a useful route to non-α-amino acids.

4.1 ACID, ESTER, AND MALONATE ENOLATE ANION REACTIONS

The reaction of an ester bearing an α-hydrogen atom with a nonnucleophilic base such as lithium diisopropylamide (LDA) generates the corresponding enolate anion.[1] Modern techniques allow generation of both mono- and dianions of carboxylic acids. Such enolate anions undergo C-alkylation and C-condensation reactions.

An example of an ester enolate alkylation reaction first treated methyl 2-methylpropanoate with lithium diisopropylamide to generate the enolate anion, and then with 4-bromobutanenitrile to give ***1***.[2] Catalytic hydrogenation of the cyano group gave methyl 6-amino-2,2-dimethylhexanoate (**2**). In this case, the nitrile was the amine surrogate and the ester was the acid precursor.

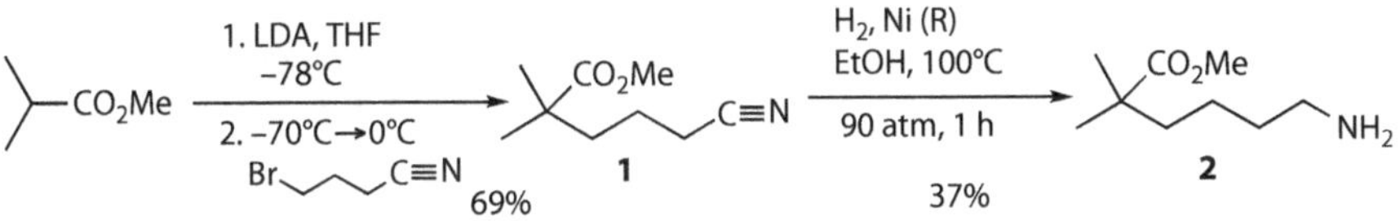

The enolate alkylation reaction that generated ***1*** used cyano as a nitrogen surrogate (see Chapter 1, Section 1.1.3). Other nitrogen surrogates may also be used in enolate alkylation reactions. An example is the reaction of the sodium enolate of diethyl 2-methyl malonate with phthalimide derivative ***3***. This displacement reaction was followed by removal of the phthalimidoyl group, hydrolysis of the ester moieties, and decarboxylation to give 2-methyl-6-aminohexanoic acid (***4***).[3] Phthalimide ***1*** was prepared by reaction of 1,4-dibromobutane with potassium phthalimide.[3] The length

---

[1] See Smith, M.B. *Organic Synthesis*, 3rd ed. Wavefunction, Inc./Elsevier, Irvine, CA/London, England, ***2010***, pp. 823–829.

[2] (a) Cefelín, P.; Lochman, L.; Stehlícek, J. *Coll. Czech. Chem. Commun.* ***1973***, *38*, 1339; (b) Cefelín, P.; Lochman, L. *Czech.* 161,459 [*Chem. Abstr.* ***1976***, *85*: P143778a].

[3] Overberger, C.G.; Parker, G.M. *J. Polym. Sci. A-1* ***1968***, *6*, 513. Also see Böhme, H.; Broese, R.; Eiden, F. *Chem. Ber.* ***1959***, *92*, 1258.

of the carbon chain and the substituent pattern in the final amino acid is determined by the halo-phthalimide precursors.

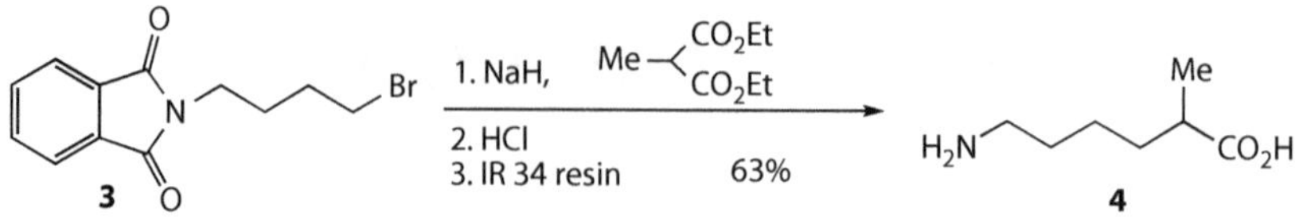

One advantage of using a malonate derivative is the ability to incorporate either one or two alkyl groups at the α-position via subsequent alkylation. Dibenzyl malonate was methylated via reaction of its sodium enolate with iodomethane, and then later condensed with α-bromo-ketone ***5*** to give ***6***.[4] Catalytic hydrogenation deprotected the esters to give the diacid, and heating led to decarboxylation and formation of 6-(*N*-Boc amino)-3,8-dimethyl-5-oxononanoic acid, *7*. Malonate anions also react with α-amidoalkyl-*p*-tolyl(phenyl)sulfones in the presence of a base to give β-amino diacid derivatives.[5] Subsequent hydrolysis, with deprotection and decarboxylation, gave the amino acid.

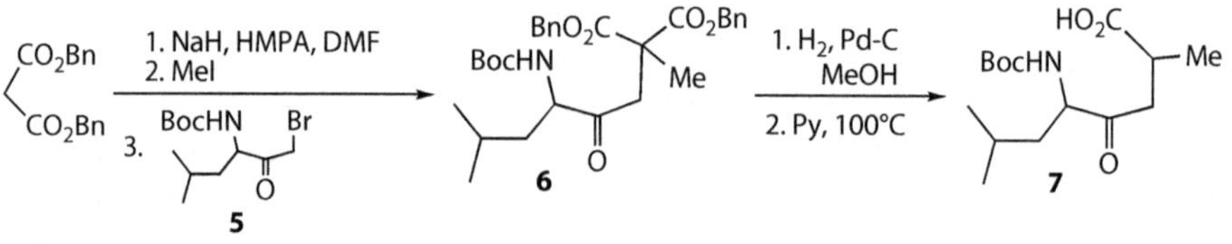

Ester enolate anions react with other electrophilic substrates, including imino chlorides. Perfluorinated β-amino acid derivatives were prepared by condensation of an ester enolate anion with a trifluoromethyl imino chloride to give enamino-esters such as ***8***.[6] Subsequent reduction gave the amino-ester, ***9***. β-Trifluoroalkyl β-amino acids were also prepared via a base-catalyzed [1,3]-proton shift reaction.[7]

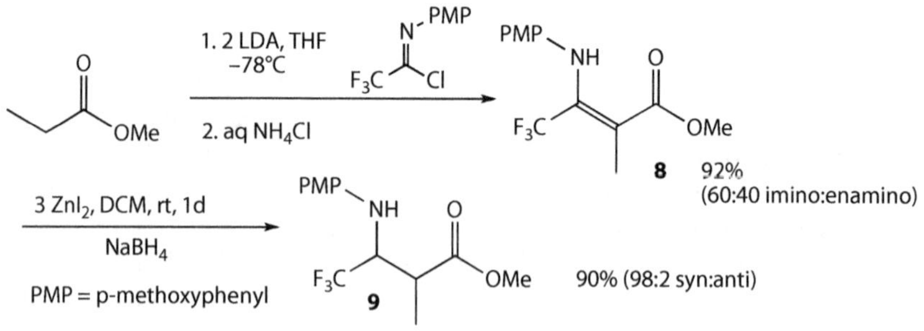

[4] Harbeson, S.L.; Rich, D.H. *J. Med. Chem*, ***1989***, *32*, 1378.

[5] Nejman, M.; Śliwińska, A.; Zwierzak, A. *Tetrahedron* ***2005***, *61*, 8536.

[6] Fustero, S.; Pina, B.; García de la Torre, M.; Navarro, A.; Ramírez de Arellano, C.; Simón, A. *Org. Lett.* ***1999***, *1*, 977.

[7] Soloshonok, V.A.; Kukhar, V.P. *Tetrahedron* ***1996***, *52*, 6953.

Both aldehydes and ketones can be condensed with malonate derivatives, but ammonia can be used as a base.[8] Reaction of butanal with ammonia gave the imine, and in the presence of the half-ester of malonic acid ethyl 3-aminohexanoate (***10***) was formed in 77% yield.[9] 2,2-Disubstituted β-amino acids were also prepared by this method.[9,10] Ketones also react, but

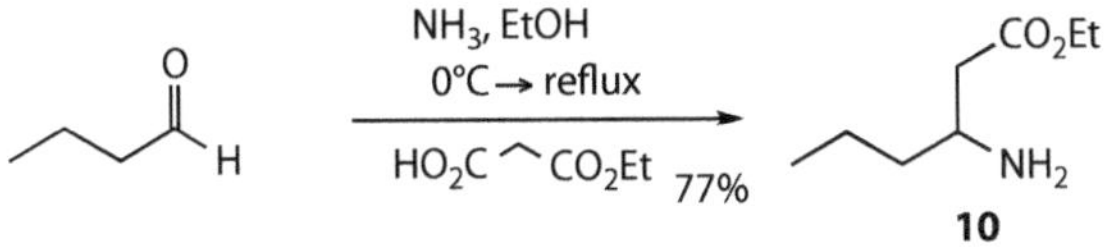

10

the yields are often lower. When 3-amino-esters such as ***10*** were treated with base, a β-lactam was formed. An example is the reaction of ***11*** with methylmagnesium bromide to give β-lactam ***12***.[8] Note that the precursor in the reaction was a ketone. Hydrolysis led to the β-alanine derivative (a 3-amino propanoic acid).

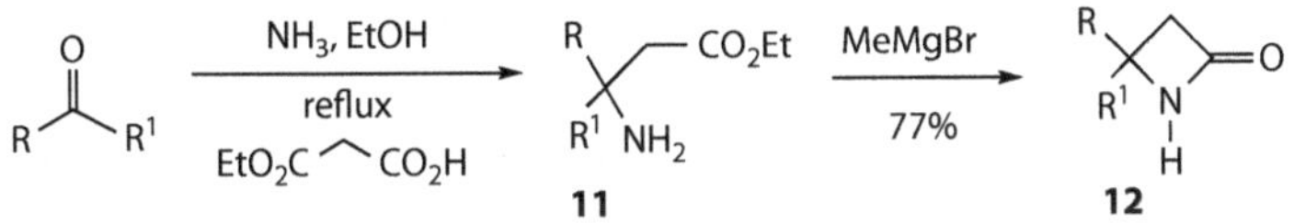

11 **12**

Aziridine derivatives function as amine surrogates because they can be opened by malonate enolates, as in the condensation of ***13*** with the enolate anion of diethyl 2-fluoromalonate to give ***14***. Subsequent treatment with HCl gave 4-amino-2-fluorobutanoic acid, ***15***.[11]

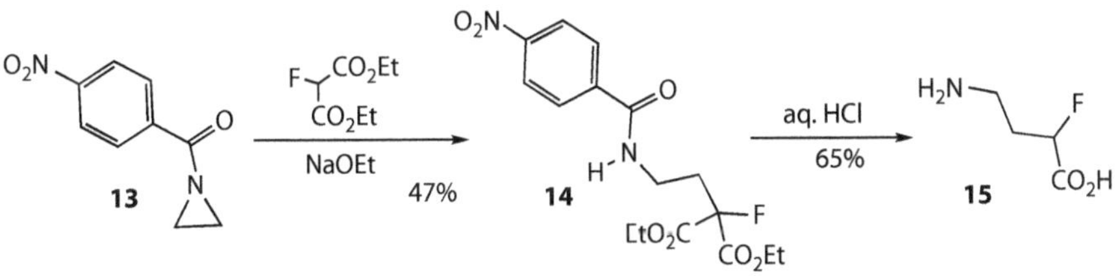

13 **14** **15**

In another aziridine ring-opening reaction, the dianion of diphenylacetic acid (2-phenylpropanoic acid) was generated by reaction of sodium naphthalenide, and subsequent reaction with *N*-benzylaziridine to give 4-(*N*-benzoylamino)-2,2-diphenylbutanoic acid (or 4-(*N*-benzoylamino)-2-methyl-2-phenylbutanoic acid) after hydrolysis (***16***, R = Ph, Me respectively).[12]

[8] Rodionow, W.M.; Federowa, A.M. *Arch. Pharmaz. Ber. Dtsch. Pharmaz. Ges.* ***1928***, *266*, 116.
[9] Testa, E.; Fontanella, L.; Aresi, V. *J. L. Ann. Chem.* ***1964***, *673*, 60.
[10] Chernova, N.G.; Rybkina, E.I.; Berlin, A.Ya. *Zh. Obshch. Khim.* ***1964***, *34*, 2129 (Engl. 2142).
[11] Buchanan, R.L.; Pattison, F.L.M. *Can. J. Chem.* ***1965***, *43*, 3466.
[12] Stamm, H.; Weiss, R. *Synthesis* ***1986***, 395.

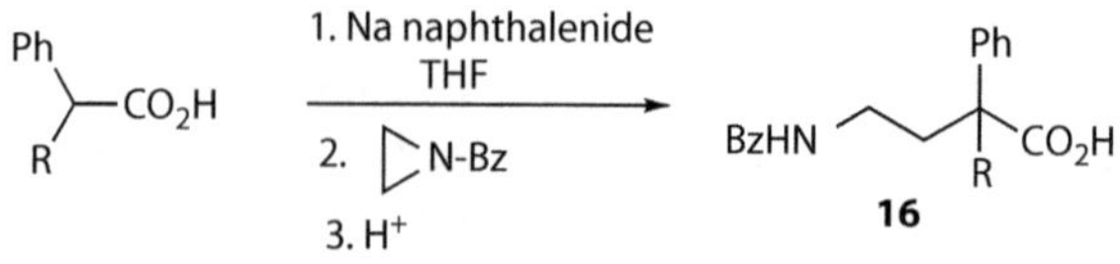

Acetoacetic acid esters (β-keto-esters) are another readily available source of carboxylate enolate anions. Further, β-keto-esters are convenient sources of oxo-alkenyl amino acid. An example is the formation of the *bis*-anion of ***17*** (formed by reaction of ***17*** with sodium hydride and then butyllithium) allowed condensation with benzonitrile (at the methyl carbon of ***17***) to give primarily ***18***, along with some of the cyclized product, ***19***.[13]

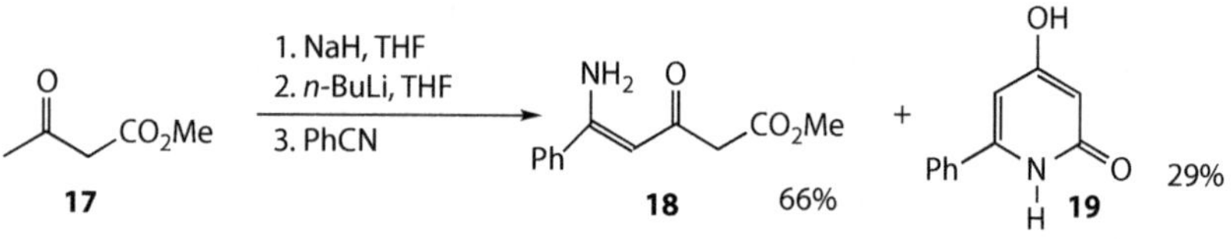

Knoevenagel type[14] condensation reactions generally include the condensation of α-cyano-esters (and some other malonate analogs) with an aldehyde or ketone. An example is the condensation of naphthalene-1-carbaldehyde (***20***) with ethyl cyanoacetate to give ***21***.[15] Subsequent catalytic hydrogenation of the nitrile moiety gave ethyl 3-amino-2-(1-naphthylmethyl)propanoate, ***22***.

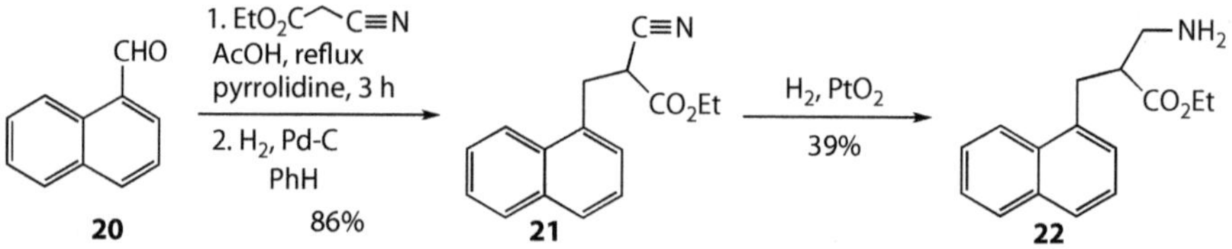

Enolate anions derived from cyano-esters react with alkyl halides, as in the conversion of ***23*** to ***24*** via reaction with sodium hydroxide and 1-bromo-butane.[16] Catalytic hydrogenation gave ethyl 2-butyl-2-methyl-3-amino-propanoate, ***25***. Similarly, 2,2-dimethyl-3-aminopropanoate, 2-methyl-2-ethyl-3-aminopropanoate, and 2-methyl-3-propyl-3-aminobutanoate were prepared in 64, 71, and 69% yields, respectively.[16]

---

[13] Huckin, S.N.; Weiler, L. *Can. J. Chem.* ***1974***, *52*,1343.

[14] (a) Japp, F.R.; Streatfeild, F.W. *J. Chem. Soc.* ***1883***, *43*, 27; (b) Knoevenagel, F. *Ber.* ***1896***, *29*, 172; (c) Knoevenagel, F. *Ber.* ***1898***, *31*, 730; (d) Jones, G. *Org. React.* ***1967***, *15*, 204; (e) see Smith, M.B. *Organic Synthesis*, 3rd ed. Wavefunction, Inc./Elsevier, Irvine, CA/London, England, ***2010***, p. 829; (f) Smith, M.B. *March's Advanced Organic Chemistry*, 7th ed. John Wiley & Sons, Hoboken, NJ, ***2013***, pp. 1157–1160.

[15] Harada, H.; Iizuka,K.; Kamijo, T.; Akahane, K.; Yamamoto, R.; Nakano, Y.; Tsubaki, A.; Kubota, T.; Shimaoka, I.; Umeyama, H.; Kiso, Y. *Chem. Pharm. Bull.* ***1990***, *38*, 3042.

[16] Eisenbach, C.D.; Lenz, R.W. *Macromolecules* ***1976***, *9*, 227.

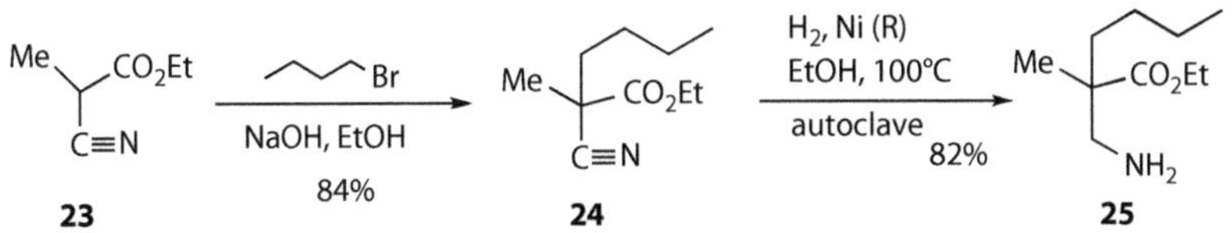

When malonic acid is used as the enolate anion partner, with pyridine as the base, the condensation reaction with aldehydes is known as the Doebner condensation.[17] The reaction of phthalimido aldehyde ***26***[18] with malonic acid and pyridine is an example.[17] The initial reaction was followed by treatment with aqueous sulfuric acid, which induced decarboxylation, and ***27*** was obtained.[19] Catalytic hydrogenation of the alkene moiety gave ***28***, and subsequent reaction with aqueous hydrazine removed the phthalimidoyl group to give 4-aminobutanoic acid (***29***).[19] It is noted that hydrolysis of ***28*** followed by treatment with hydrazine gave 4-amino-2-butenoic acid, but in only 7% yield.[19]

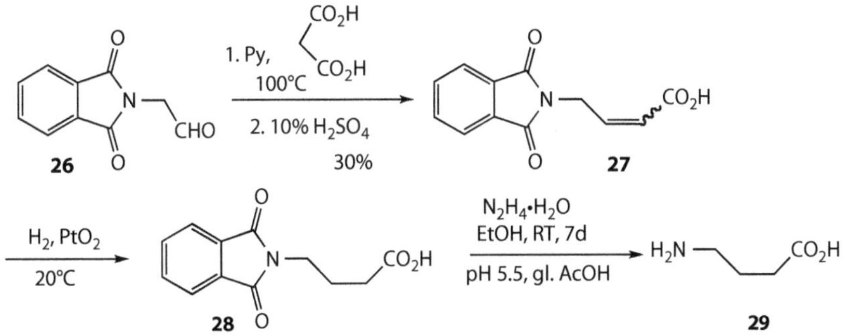

A conjugate addition strategy related to the Knoevenagel reaction[20] used the enolate anion of diethyl 2-acetamidopropanoate (a functionalized malonate) in a reaction with ketene ***30***[21] to give ***31***.[22] In this particular example, both the amino and carboxyl moieties of the final amino acid were present in the malonate starting material. Aqueous acid hydrolysis of the ester groups and the amide was accompanied by decarboxylation to give 2-amino-3-methylenebutanedioic acid, ***32***.[22] This is

---

[17] (a) Doebner, O. *Ber.* ***1900***, *33*, 2140; (b) see Smith, M.B. *Organic Synthesis*, 3rd ed. Wavefunction, Inc./Elsevier, Irvine, CA/London, England, ***2010***, p. 830; (c) Smith, M.B. *March's Advanced Organic Chemistry*, 7th ed. John Wiley & Sons, Hoboken, NJ, ***2013***, p. 1158.

[18] Radde, E. *Ber.* ***1922***, *55*, 3174.

[19] Balenovic, K.; Jambresic, I.; Urbas, B. *J. Org. Chem.* ***1954***, *19*, 1589.

[20] (a) Japp, F.R.; Streatfeild, F.W. *J. Chem. Soc.* ***1883***, *43*, 27; (b) Knoevenagel, F. *Ber.* ***1896***, *29*, 172; (c) Knoevenagel, F. *Ber.* ***1898***, *31*, 730; (d) Jones, G. *Org. React.* ***1967***, *15*, 204; (e) see Smith, M.B. *Organic Synthesis*, 3rd ed. Wavefunction, Inc./Elsevier, Irvine, CA/London, England, ***2010***, p. 829; (f) Smith, M.B. *March's Advanced Organic Chemistry*, 7th ed. John Wiley & Sons, Hoboken, NJ, ***2013***, pp. 1157–1160.

[21] (a) Lange, R.W.; Hansen, H.J. *Org. Synth.* ***1982*** *62*, 202; (b) Rosenthal, R.W.; Schwartzman, L.H. *J. Org. Chem.* ***1959***, *24*, 836.

[22] Paik, Y.H.; Dowd, P. *J. Org. Chem.* ***1986***, *51*, 2910.

obviously a conjugate addition reaction (see Chapter 3, Section 3.2), but it is included here because of the Knoevenagel type process.[23]

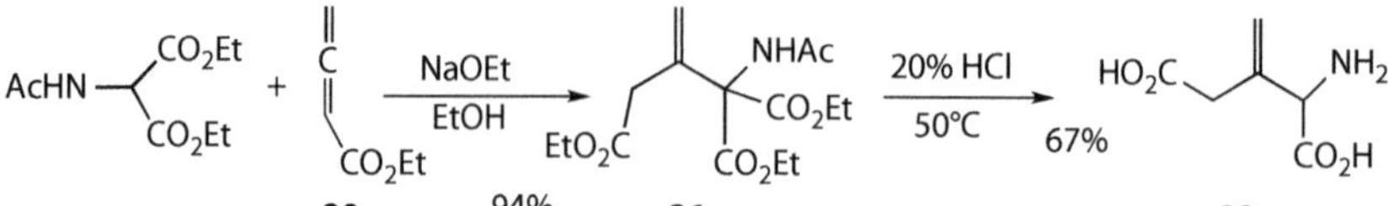

30 **31** **32**

Doebner condensation[24] with malonic acid itself sometimes leads to very poor yields of product, as seen above. One reported method that improved the reaction used nitro-ester ***33*** rather than malonic acid in a reaction with an ortho-ester and aniline, and ***34*** was produced in good yield.[25] In this variation, the amine surrogate was the nitro group (see Chapter 1, Section 1.1.3).

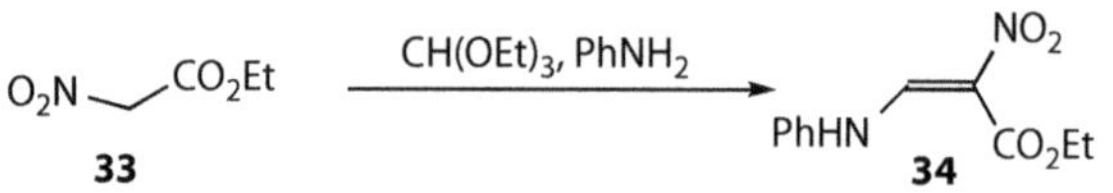

33 **34**

Imines are readily prepared by reaction of an aldehyde with an amine. These imines can then be condensed with an ester enolate to generate an amino-ester directly. An example is the condensation of ***35*** with ethyl phenylacetate in the presence of sodium ethoxide, to give ethyl 3-(*N*-phenylamino)-2,3-diphenyl-3-aminopropanoate (***36***).[26] The *N*-(4-chlorophenyl) derivative was also prepared by this method.

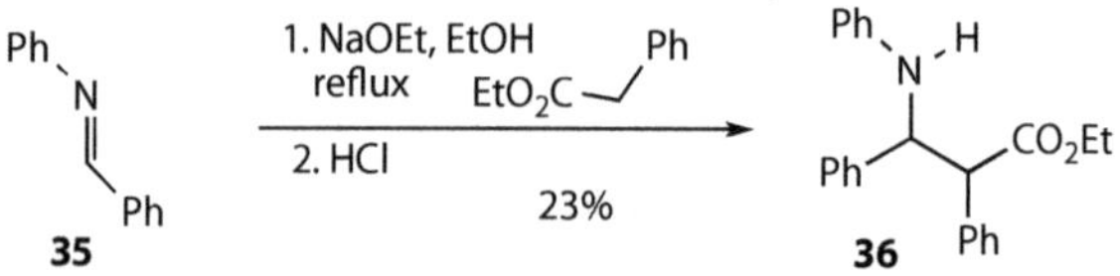

35 **36**

In an interesting approach, an imine moiety was generated in situ by reaction of a methoxy carbamate with lithium diisopropylamide. This imine was condensed with an ester enolate to give the amino-ester, protected as a carbamate. Reaction of ***37*** with lithium diisopropylamide, for example, gave imine ***38***, which reacted with the metal enolate anion shown (M = Li, Na) to give a mixture of diastereomeric

[23] (a) Japp, F.R.; Streatfeild, F.W. *J. Chem. Soc.* ***1883***, *43*, 27; (b) Knoevenagel, F. *Ber.* ***1896***, *29*, 172; (c) Knoevenagel, F. *Ber.* ***1898***, *31*, 730; (d) Jones, G. *Org. React.* ***1967***, *15*, 204; (e) see Smith, M.B. *Organic Synthesis*, 3rd ed. Wavefunction, Inc./Elsevier, Irvine, CA/London, England, ***2010***, p. 829; (f) Smith, M.B. *March's Advanced Organic Chemistry*, 7th ed. John Wiley & Sons, Hoboken, NJ, ***2013***, pp. 1157–1160.

[24] (a) Doebner, O. *Ber.* ***1900***, *33*, 2140; (b) see Smith, M.B. *Organic Synthesis*, 3rd ed. Wavefunction, Inc./Elsevier, Irvine, CA/London, England, ***2010***, p. 830; (c) Smith, M.B. *March's Advanced Organic Chemistry*, 7th ed. John Wiley & Sons, Hoboken, NJ, ***2013***, p. 1158.

[25] Knippel, E.; Knippel, M.; Michalik, M.; Kelling, H.; Kristen, H. *Z. Chem.* ***1975***, *15*, 446 [*Chem. Abstr.* ***1976***, *84*: 121068w].

[26] Simova, E.; Kurtev, B. *Monatsh. Chem.* ***1965***, *96*, 722.

ethyl 3-(*N*-methylcarbamoyl)-2-ethyl-4-methylpentanoates, ***39*** and ***40*** in a 7:3 ratio.[27] Substituted butanoic acid derivatives were also prepared in this manner, as well as 2-ethyl, 2-isopropyl, and 2-phenyl derivatives.[27,28]

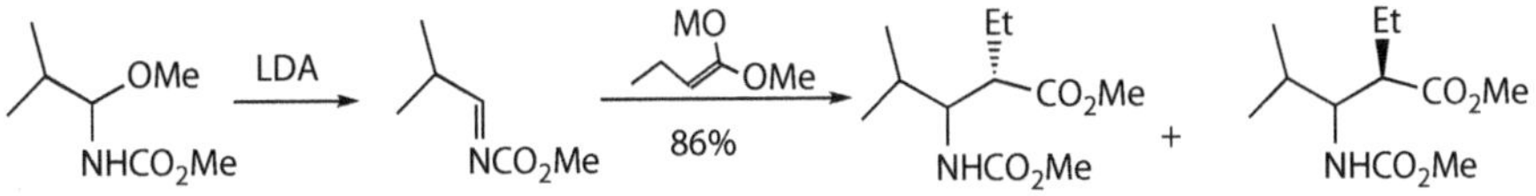

37 **38** **39** **40**

An example of a conjugate addition reaction with an amine-bearing molecule generates an α-amino radical (the unpaired electron is on carbon). The reaction of butylamine and butyl acrylate, in the presence of di-*tert*-butyl peroxide, gives butyl 4-aminoheptanoate, ***41***.[29] In a similar manner, cyclohexylamine was heated with butyl acrylate in the presence of di-*tert*-butyl peroxide to produce 1-amino-1-(butoxycarbonylethyl)cyclohexane.[29]

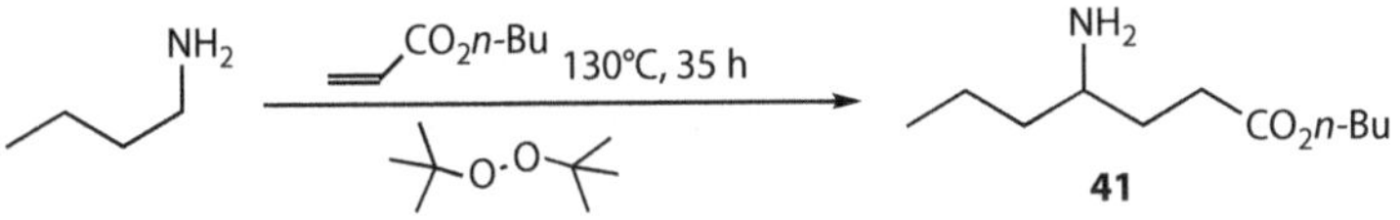

41

4.2 NITRILE ENOLATE ANIONS

The previous section presented at least one example of the use of nitrile-ester enolate anions to prepare amino acids. The enolate of alkyl mono-nitriles has been used in both alkylation reactions and condensation reactions. An example of alkylation reacted the enolate anion of 2-phenylacetonitrile with ethyl bromoacetate to give nitrile ***42***, and subsequent reduction of the cyano moiety leads to ***43***.[30]

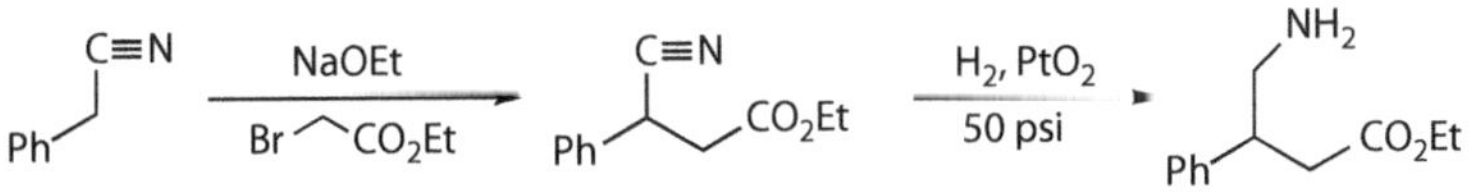

42 **43**

The sodium salt of succinimide reacted with α,ω-dibromides such as ***44*** to give ***45***.[31] In a second step, the enolate anion of α-phenyl nitrile ***46*** displaced the bromide moiety in ***44*** to give ***47***. Acid hydrolysis converted the nitrile to an acid and the imide

[27] Shono, T.; Kise, N.; Sanda, F.; Ohi, S.; Yoshioka, K. *Tetrahedron Lett.* ***1989***, *30*, 1253. Also see Shono, T.; Kise, N.; Sanda, F.; Ohi, S.; Tsubata, K. *Tetrahedron Lett.* ***1988***, *29*, 231.

[28] Shono, T.; Tsubata, K.; Okinaga, N. *J. Org. Chem.* ***1984***, *49*, 1056.

[29] Masuo, F.; Yamamoto, K. *Jpn.* 23,161 ('63) [*Chem. Abstr.* ***1964***, *60*: P2800g].

[30] Buu, N.T.; Van Gelder, N.M. *Br. J. Pharmac.* ***1974***, *52*, 401.

[31] (a) Salmon-Legagneur, F.; Neveu, C. *Bull. Soc. Chim. Fr.* ***1965***, 2270; (b) Salmon-Legagneur, F.; Neveu, C. *C. R. Acad. Sci.* ***1964***, *259*, 1878; (c) Salmon-Legagneur, F.; Neveu, C. *C. R. Acad. Sci.* ***1963***, *256*, 187.

to an amine, leading to ***48***. Reaction of ***44*** (n = 4) with ***40*** (R = H) led to a 57% yield of 2-phenyl-6-amino-hexanoic acid (***48a***).[31]

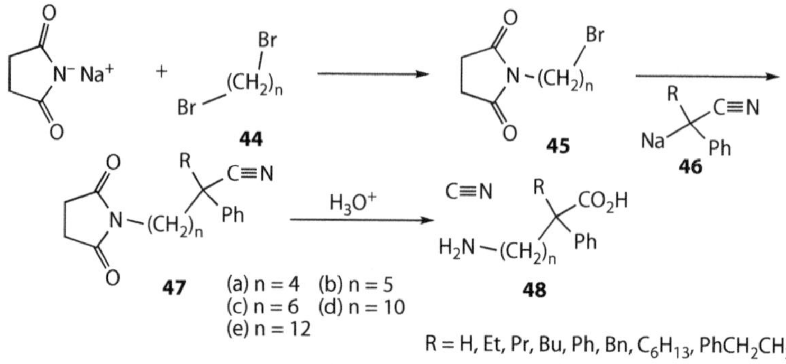

Similarly prepared were 2-phenyl-7-aminoheptanoic acid (***48b***) in 68% yield, 2-phenyl-8-amino-octanoic acid (***47c***) in 68% yield, 2-phenyl-10-aminodecanoic acid (***48d***) in 44% yield, and 2-phenyl-12-aminododecanoic acid (***48e***) in 26% yield, all with R = H.[31] Many other nitriles were used in this sequence to give a wide variety of phenyl-substituted amino acids.[31] A variety of other α-aryl and α,α-diaryl derivatives can be prepared by using substituted nitriles, including 2,2-diphenylethanenitrile, 2-phenylbutanenitrile, 2-phenylpentanenitrile, 2,3-diphenylpropanenitrile, etc.

Nitrile enolate anions react in condensation reactions with carbonyl derivatives[32] such as oxalate derivatives, which serve as a carboxyl surrogate. Reaction of 2-phenylacetonitrile and diethyl oxalate, in the presence of sodium amide, gave ***49***.[33] Catalytic hydrogenation of the cyano group and hydrolysis led to ethyl 2-phenyl-3-aminopropanoate, ***50***. Once again, the nitrile moiety was an amino methyl surrogate. Other α-aryl acetonitrile derivatives were similarly converted to the following β-alanines: 2-(3-chlorophenyl)-, 2-(4-chloro-phenyl)-, 2-(2-methylphenyl)-, 2-(4-methylphenyl)-, 2-(2-methylphenyl)-, and 2-(3-methyl-phenyl)-3-aminopropanoates.[34]

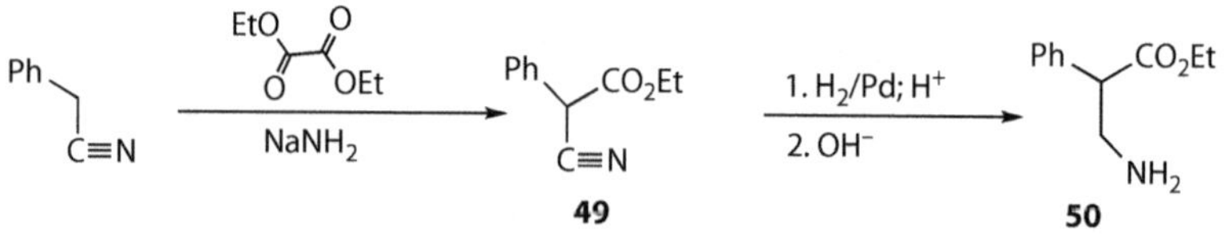

The condensation reaction of ester enolates with nitriles is another variation that can be used to prepare alkenyl-β-amino acids. In a simple example, the reaction of *tert*-butyl acetate with magnesium diisopropylamide and then propanenitrile led to

[32] See Smith, M.B. *Organic Synthesis*, 3rd ed. Wavefunction, Inc./Elsevier, Irvine, CA/London, England, ***2010***, p. 832.

[33] Cignarella, G.; Mariani, L.; Testa, E. *Gazz. Chim. Ital.* ***1965***, *95*, 831.

[34] (a) Leonard, F.; Wajngurt, A.; Klein, M.; Smith, C.M. *J. Org. Chem.* ***1961***, *26*, 4062; (b) Leonard, F. *U.S. Pat.* 3,125,583 [*Chem. Abstr.* ***1964***, *60*: P13193d].

tert-butyl 3-amino-hex-2-enoate, ***51***.[35] In further work, other derivatives were formed that include *tert*-butyl 3-aminobut-2-enoate (66%), 3-amino-4-methylpent-2-enoate (74%), 3-amino-4,4-dimethylpent-2-enoate (43%), 3-amino-4-phenylpent-2-enoate (25%), and other 4-aryl- and 2-alkyl-substituted alkenyl amino acids. Aromatic nitriles can be condensed with esters to form aryl-substituted alkenyl amino acids. The enolate anion of *tert*-butyl acetate reacted with benzonitrile, for example, to give *tert*-butyl 3-amino-3-phenylprop-2-enoate, ***52***.[36]

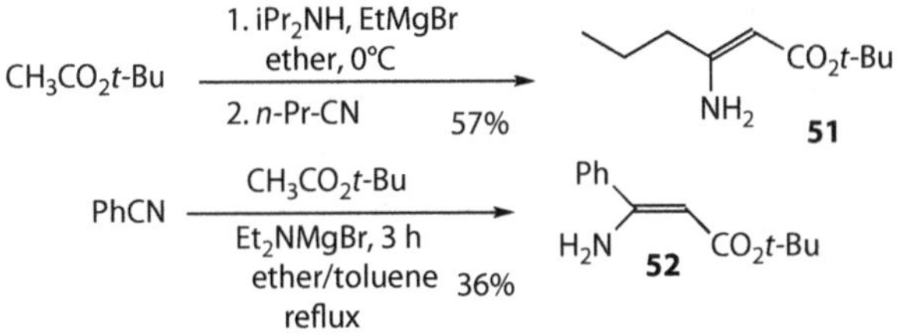

4.3 NITROALKANE ENOLATE ANIONS

The enolate anion derived from nitroalkyl derivatives[37] can function as an amine surrogate in reactions with alkyl halides that also have an ester moiety as part of the structure. In general, these anions react with a species bearing a carboxyl group or a carboxyl surrogate. Later in the reaction sequence, the nitro group can be reduced to an amine.

In a simple example, reaction of nitroethane (***53***) with LDA followed by quenching with methyl 3-chloropropanoate gave ***54***, and reduction with ammonium formate gave an 81% overall yield of ***55***.[38] Enolate anions derived from secondary nitro compounds tend to give better yields in the alkylation step than primary nitro compounds. Enolate anions of benzylic nitro compounds also tend to give better yields in alkylation reactions.

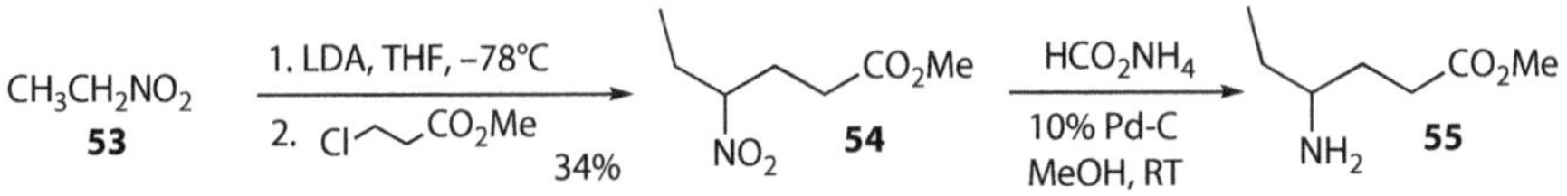

Reaction of the anion of nitroalkanes with α-halo-esters leads to β-amino acid derivatives. Nitromethane reacted with lithium diisopropylamide to form the nitro-enolate, and subsequent reaction with methyl 2-chloro-2-phenylethanoate

[35] (a) Hiyama, T.; Kobayashi, K. *Tetrahedron Lett.* ***1982***, *23*, 1597; (b) Hiyama, T.; Kobayashi, K.; Nishide, K. *Bull. Chem. Soc. Jpn.* ***1987***, *60*, 2127; (c) Hiyama, T.; Oishi, H.; Suetsugu, Y.; Nishide, K.; Saimoto, H. *Bull. Chem. Soc. Jpn.* ***1987***, *60*, 2139.

[36] Sosnovskikh, V.Ya.; Ovsyannikov, I.S.; Nikol'ski, A.L. *Zh. Org. Khim.* ***1986***, *22*, 1775 (Engl. 1595).

[37] See Smith, M.B. *Organic Synthesis*, 3rd ed. Wavefunction, Inc./Elsevier, Irvine, CA/London, England, ***2010***, p. 830.

[38] Kaji, E.; Igarashi, A.; Zen, S. *Bull. Chem. Soc. Jpn.* ***1976***, *49*, 3181.

gave ***56***.[39] Reduction of the nitro group with ammonium formate gave methyl 2-phenyl-3-aminopropanoate, ***57***.

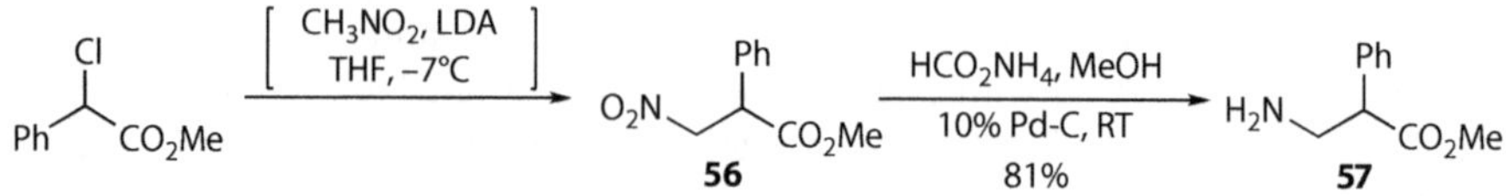

56 57

This general approach allows the synthesis of highly functionalized amino acids. Dioxolane-protected 3-nitropropanal (***58***), for example, was condensed with methyl glycolate to give ***59*** as a 1:1 mixture of *syn:anti* diastereomers.[40] Conversion to the dimethyl-*tert*-butylsilyl derivative allowed hydrogenation of the nitro group to an amino group, giving methyl 3-amino-2-hydroxybutanoate 4-carbaldehyde (protected as a dioxolane, ***60***).

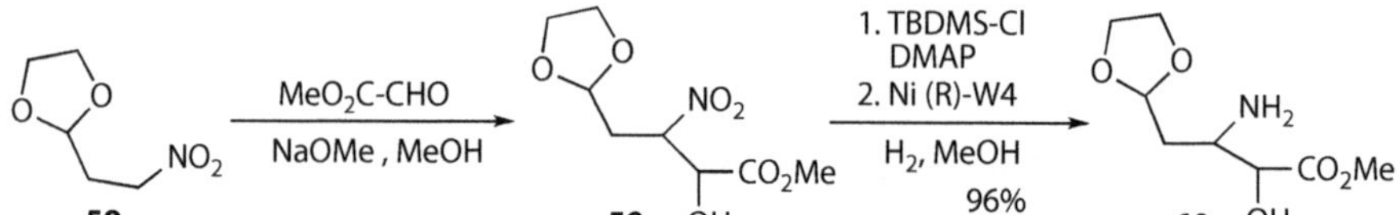

58 59 60

A nitro aldol reaction (also known as the Henry reaction)[41] of glyoxylate ***61*** with 1-nitro-1-phenylmethane in the presence of activated alumina, followed by hydrogenation and conversion to the *N*-Boc derivative, gave ***62*** with high stereoselectivity.[42] Removal of the chiral methyl auxiliary was followed by conversion to the Taxotere[43] side chain, ***63***. A similar reaction sequence that used 1-nitro-2-phenylethane led to (−)-bestatin hydrochloride (see Chapter 6, Section 6.1.3), an adjuvant in cancer chemotherapy.[44]

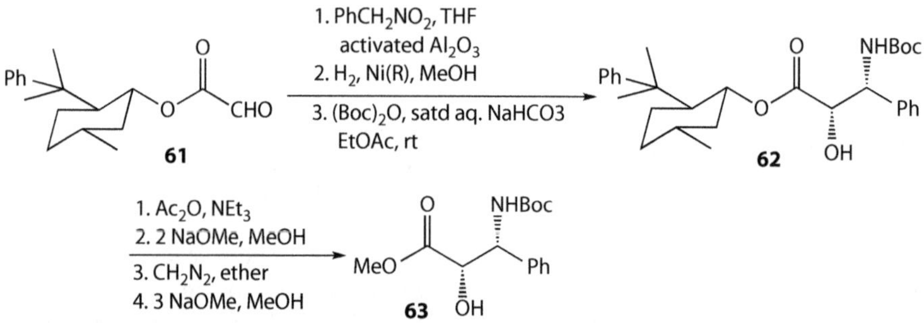

61 62 63

[39] Ram, S.; Ehrenkaufer, R.E. *Synthesis* ***1986***, 133.
[40] Williams, T.M.; Crumbie, R.; Mosher, H.S. *J. Org. Chem.* ***1985***, *50*, 91.
[41] (a) Henry, L. *Compt. Rend.* ***1895***, *120*, 1265; (b) for a review, see Luzzio, F.A. *Tetrahedron* ***2001***, *57*, 915.
[42] Kudyba, I.; Raczko, J.; Jurczak, J. *J. Org. Chem.* ***2004***, *69*, 2844.
[43] See (a) Lee, S.-H.; Yoon, J.; Chung, S.-H.; Lee, Y.-S. *Tetrahedron* ***2001***, *57*, 2139; (b) Hamamoto, H.; Mamedov, V.A.; Kitamoto, M.; Hayashi, N.; Tsuboi, S. *Tetrahedron Asym.* ***2000***, *11*, 4485; (c) Choudary, B.M.; Chowdari, N.S.; Madhi, S.; Kantam, M.L. *J. Org. Chem.* ***2003***, *68*, 1736.
[44] Ino, K.; Goto, S.; Nomura, S.; Isobe, K.-I.; Nawa, A.; Okamoto, T.; Tomoda, Y. *Anticancer Res.* ***1995***, *15*, 2081.

4.4 MUKAIYAMA TYPE CONDENSATIONS[45]

Although no examples are presented in this chapter that prepare amino acids via a direct aldol condensation, examples will be presented in connection with the synthesis of statine (see Chapter 6, Section 6.6). However, the Mukaiyama aldol reaction (or just the Mukaiyama reaction) has been used for the synthesis of amino acids, and this variation of the aldol is discussed because there are reactions with molecules that contain an amine surrogate.

Silyl enol ethers react with imines under Mukaiyama conditions[47] to give amino-esters. The reaction of ***64*** with imine ***65***, for example, gave a mixture a diastereomeric mixture of methyl 3-(*N*-phenylamino)-2,3-diphenylpropanoates ***66*** and ***67*** in a ratio of 14:86.[46] The identical reaction of CH_2=C(OMe)OTMS led to an 85% yield of the *anti*-β-amino acid with none of the *syn*-diastereomer observed. Both 2- and 3-aryl-β-alanine derivatives may be obtained by the Mukaiyama reaction[47] of imines when catalyzed by titanium tetrachloride.[48] The reaction of ketene silyl acetals with *O*-alkyl-substituted imines is another source of aryl-substituted β-alanines.[49]

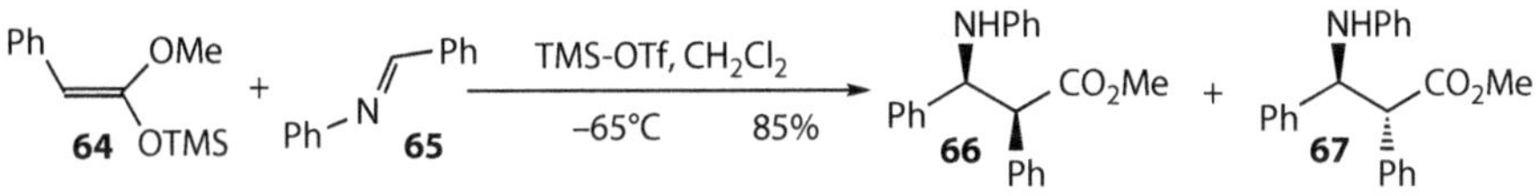

Conjugated imines such as ***69*** react with silyl enol ethers, in this case catalyzed by trimethylsilyl triflate, to give alkenyl amino acids. The reaction of ***68*** and ***69*** gave a 14:86 mixture of ***70:71***.[50] In other examples, the diastereoselectivity was higher.

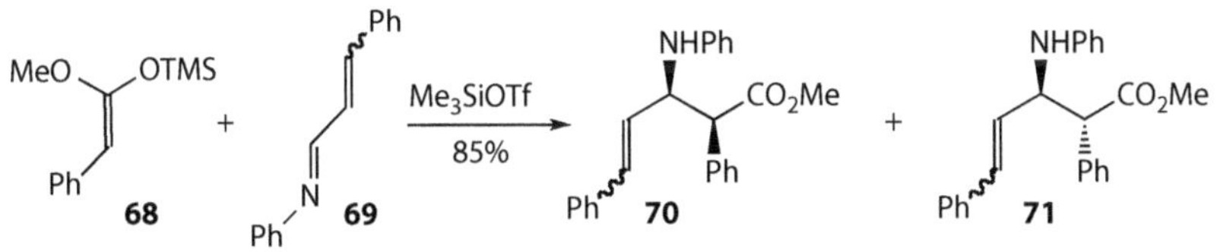

An *O*-trimethylsilyl enol ether such as ***72*** (derived from methyl 2-methylpropanoate) can react with other amine surrogates, including ***73*** to give ***74***.[51] Removal of the *N*-TMS group gave methyl-3-amino-2,2-dimethyl-propanoate, ***75***. The use

---

[45] (a) Mukaiyama, T. *Org. React.* ***1982***, *28*, 203; (b) Mukaiyama, T. *Angew. Chem. Int. Ed. Engl.* ***1977***, *16*, 817; (c) see Smith, M.B. *Organic Synthesis*, 3rd ed. Wavefunction, Inc./Elsevier, Irvine, CA/London, England, ***2010***, pp. 837–841; (d) Smith, M.B. *March's Advanced Organic Chemistry*, 7th ed. John Wiley & Sons, Hoboken, NJ, ***2013***, pp. 1151–1153.

[46] Guanti, G.; Narisano, E.; Banfi, L. *Tetrahedron Lett.* ***1987***, *28*, 4331,4335.

[47] (a) Mukaiyama, T. *Org. React.* ***1982***, *28*, 203; (b) Mukaiyama, T. *Angew. Chem. Int. Ed. Engl.* ***1977***, *16*, 817; (c) see Smith, M.B. *Organic Synthesis*, 3rd ed. Wavefunction, Inc./Elsevier, Irvine, CA/London, England, ***2010***, pp. 837–841; (d) Smith, M.B. *March's Advanced Organic Chemistry*, 7th ed. John Wiley & Sons, Hoboken, NJ, ***2013***, pp. 1151–1153.

[48] Ojima, I.; Inaba, S.; Yoshida, K. *Tetrahedron Lett.* ***1977***, 3643.

[49] Ikeda, K.; Achiwa, K.; Sekiya, M. *Tetrahedron Lett.* ***1983***, *24*, 4707.

[50] Guanti, G.; Narisand, E.; Banfi, L. *Tetrahedron Lett.* ***1987***, *28*, 4331, 4335.

[51] Shono, T.; Tsubata, K.; Okinaga, N. *J. Org. Chem.* ***1984***, *49*, 1056.

of methyl propanoate in this sequence gave 2-methyl-3-aminopropanoic acid, and the identical reaction sequence using cyclopentane carboxylic acid led to 1-amino-1-carbomethoxycyclopentane.

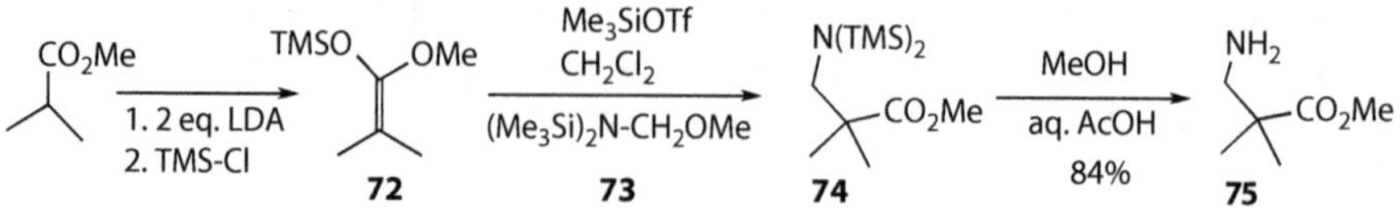

1,3,5-Triazacyclohexane derivative *76* is an unusual amine surrogate, but reaction with *72* gave methyl 2,2-dimethyl-*N*-butyl-3-aminopropanoate (*77*).[52] The analogous *N*-isopropyl, *N*-ethyl, and *N*-phenyl derivatives were also prepared, as were 2-methyl propanoate derivatives.[52]

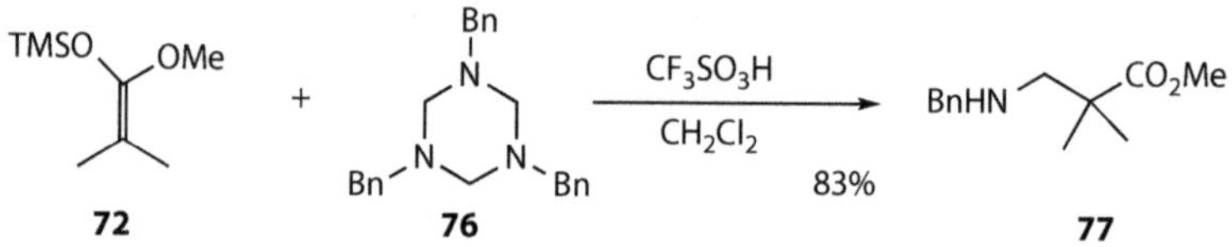

A stereoselective synthesis of β-amino acids was reported using a chiral 4-phenyloxazolidinone-controlled linear *N*-acyliminium ion reaction, in which the conversion of *78* to *79*[53] is essentially a Mukaiyama type aldol condensation.[54] Removal of the oxazolidinone moiety was followed by protection with *N*-Cbz to give ***80***.[53]

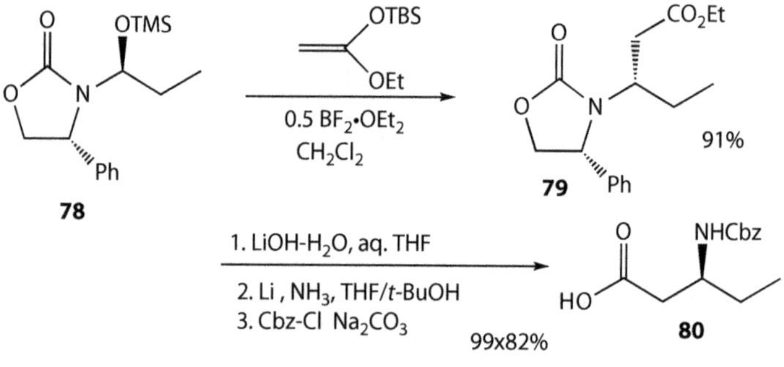

---

[52] Ikeda, K.; Achiwa, K.; Sekiya, M. *Tetrahedron Lett.* ***1983***, *24*, 913.

[53] Shin, D.-Y.; Jung, J.-K.; Seo, S.-Y.; Lee, Y.-S.; Paek, S.-M.; Chung, Y.K.; Shin, D.M.; Suh, Y.-G. *Org. Lett.* ***2003***, 3635.

[54] (a) Mukaiyama, T. *Org. React.* ***1982***, *28*, 203; (b) Mukaiyama, T. *Angew. Chem. Int. Ed. Engl.* ***1977***, *16*, 817; (c) see Smith, M.B. *Organic Synthesis*, 3rd ed. Wavefunction, Inc./Elsevier, Irvine, CA/London, England, ***2010***, pp. 837–841; (d) Smith, M.B. *March's Advanced Organic Chemistry*, 7th ed. John Wiley & Sons, Hoboken, NJ, ***2013***, pp. 1151–1153.

4.5 CONDENSATION REACTIONS INVOLVING ORGANOZINC INTERMEDIATES

There are several condensation reactions that generate an organozinc species as an intermediate rather than lithium, sodium, or magnesium enolate anions. Arguably, the two most important of these reactions are the classical Reformatsky reaction and a modified Blaise reaction.

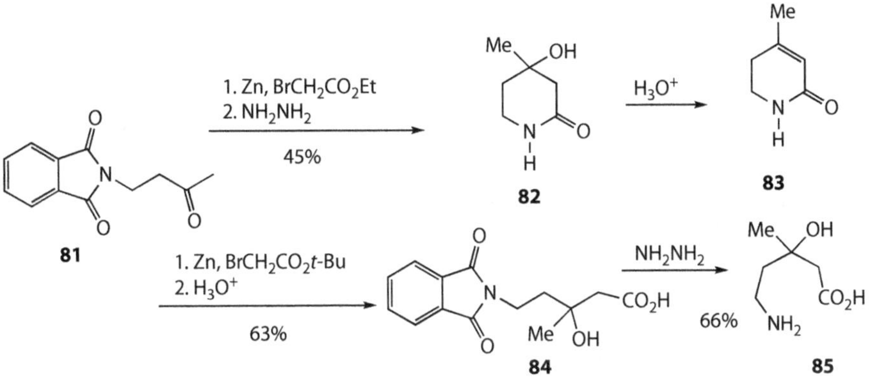

The condensation reaction of an imine derived from an aldehyde with ethyl bromoacetate under Reformatsky[55] conditions led to a mixture of amino-esters and β-lactams. An alternative approach used ketone ***81***, which was converted to an *N*-phthalimidoyl alcohol under Reformatsky conditions, but removal of the phthalimidoyl group led to formation of lactam ***82***.[56] Treatment with acid led to conjugated lactam ***83*** rather than the amino acid. If ***81*** was treated with bromo-*tert*-butyl acetate and zinc, alcohol ***84*** was obtained after hydrolysis of the ester.[56] Removal of the phthalimidoyl group with hydrazine led to 5-amino-3-hydroxy-3-methylpentanoic acid, ***85***. 2-Aryl-3-hydroxy derivatives such as 5-amino-3-hydroxyl-3-(3-chlorophenyl)pentanoic acid were also prepared by this latter approach (in 86 and 59% yield for the two steps).

An asymmetric Reformatsky type reaction[57] has been used to prepare non-α-amino acids, but the reaction is mediated by samarium iodide rather than by zinc. This methodology was used to prepare β-hydroxy amino acids.[58] The SmI_2-mediated reaction of proline with ***86*** gave ***87***. Removal of the auxiliary gave ***88***. Both statine (see Chapter 6, Section 6.6) and isostatine were prepared using this methodology.[58]

---

[55] (a) Reformatsky, S. *Ber.* ***1887***, *20*, 1210; (b) Rathke, M.W. *Org. React.* ***1975***, *22*, 423; (c) Diaper, D.G.M.; Kuksis, A. *Chem. Rev.* ***1959***, *59*, 89; (d) see Smith, M.B. *Organic Synthesis*, 3rd ed. Wavefunction, Inc./Elsevier, Irvine, CA/London, England, ***2010***, pp. 885–887; (e) Smith, M.B. *March's Advanced Organic Chemistry*, 7th ed. John Wiley & Sons, Hoboken, NJ, ***2013***, pp. 1129–1131.

[56] Carganico, G.; Cangiano, G.; Cozzi, P.; Lovisolo, P.P. *Framaco Ed. Sci.* ***1987***, *42*, 277.

[57] (a) Reformatsky, S. *Ber.* ***1887***, *20*, 1210; (b) Rathke, M.W. *Org. React.* ***1975***, *22*, 423; (c) Diaper, D.G.M.; Kuksis, A. *Chem. Rev.* ***1959***, *59*, 89; (d) see Smith, M.B. *Organic Synthesis*, 3rd ed. Wavefunction, Inc./Elsevier, Irvine, CA/London, England, ***2010***, pp. 885–887; (e) Smith, M.B. *March's Advanced Organic Chemistry*, 7th ed. John Wiley & Sons, Hoboken, NJ, ***2013***, pp. 1129–1131.

[58] Nelson, C.G.; Burke, T.R., Jr. *J. Org. Chem.* ***2012***, *77*, 733.

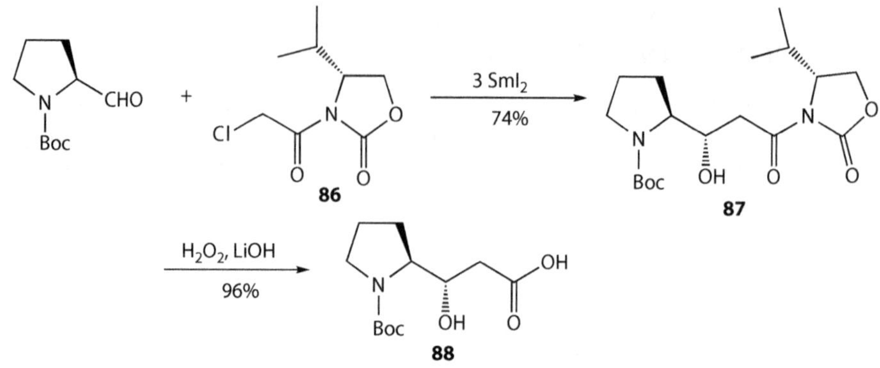

The Blaise reaction[59] involves the condensation of an α-bromoester and a nitrile, in the presence of zinc metal, to give β-keto-esters. The fundamental reaction is illustrated by the reaction of ***89*** to give ***90***, via the zinc iminium salt shown.

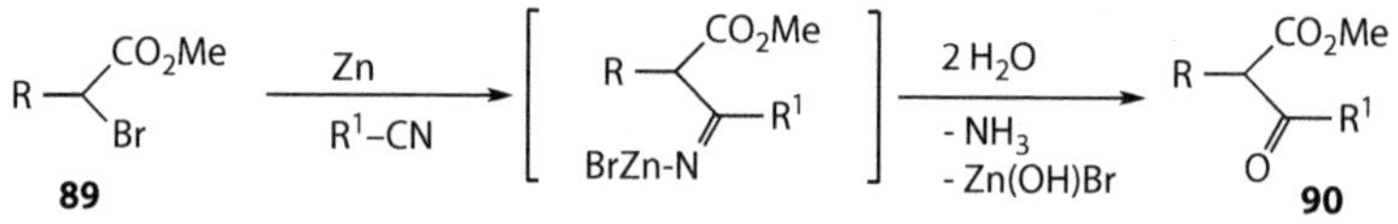

A modified Blaise reaction (the imine intermediate is converted to an enamine moiety rather than hydrolyzed to a carbonyl) can be used for the preparation of amino acids. When ***91*** reacted with benzonitrile and zinc, methyl 3-amino-2-methyl-3-phenylprop-2-enoate, ***92***, was obtained.[60] Several alkenyl amino acids were prepared, including 3-amino-7-chlorohept-2-enoate (in 95% yield) using 5-chloropentanenitrile as a precursor.[60]

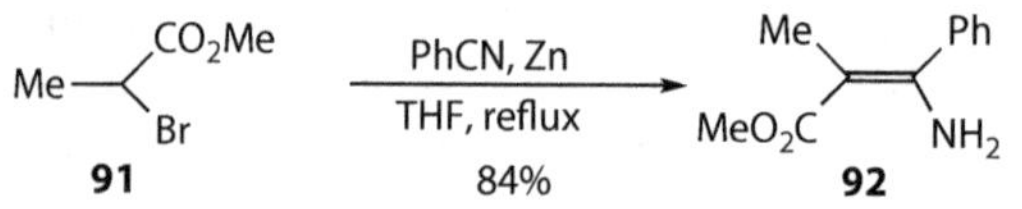

4.6 YLID REACTIONS

Ylids may be considered a class of carbanions used for the synthesis of amino acids. When thiophene-2-carbaldehyde (***93***) reacted with an appropriate ylid, Wittig olefination[61] gave the conjugated ester. Subsequent reaction with the enolate anion of nitromethane (formed by reaction of the base, Triton B (benzyltrimethylammonium

[59] (a) Blaise, E.E. *C. R. Hebd. Sci. Acad. Sci.* ***1901,*** *132*, 478; (b) Cason, J.; Rinehart, K.L.; Thornton, S.D. *J. Org. Chem.* ***1953***, *18*, 1594; (c) Smith, M.B. *March's Advanced Organic Chemistry*, 7th ed. John Wiley & Sons, Hoboken, NJ, ***2013***, p. 1131.

[60] Hannick, S.M.; Kishi, Y. *J. Org. Chem.* ***1983***, *48*, 3833.

[61] See (a) Smith, M.B. *Organic Synthesis*, 3rd ed. Wavefunction, Inc./Elsevier, Irvine, CA/London, England, ***2010***, pp. 729–739; (b) Smith, M.B. *March's Advanced Organic Chemistry*, 7th ed. John Wiley & Sons, Hoboken, NJ, ***2013***, pp. 1165–1173.

hydroxide)) via conjugate addition gave ***94***.[62] Catalytic hydrogenation of the nitro moiety gave a 1:1 mixture (19% overall yield) of ethyl 4-amino-3-(2-thiophene)butanoate, ***95***, and 4-(2-thiophene)-2-pyrrolidinone, ***96***. Basic hydrolysis of this mixture led to isolation of 4-amino-3-(2-thiophene)butanoic acid in 42% yield.[62]

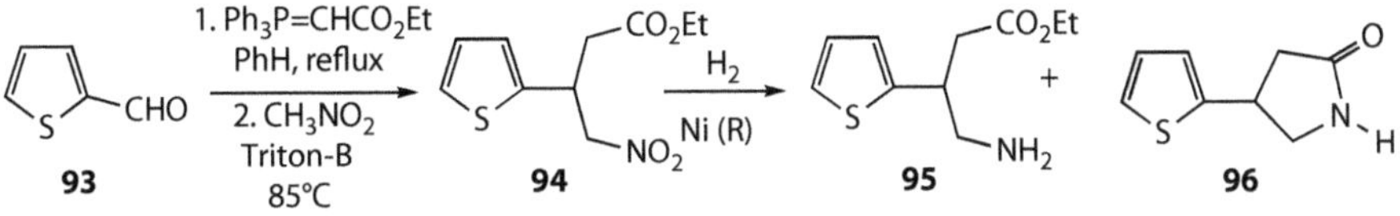

93 **94** **95** **96**

This Wittig type strategy is actually rather common. Only a single example is shown here to illustrate the idea, but other Wittig type reactions[61] are shown in the context of stereoselectivity in Chapter 5 and specific amino acid types in Chapter 6.

[62] Bethelot, P.; Vaccher, C.; Flouquet, N.; Debaert, M.; Luyckx, M.; Brunet, C. *J. Med. Chem.* ***1991***, *34*, 2557.

5 Diastereoselective and Enantioselective Syntheses

Most of the syntheses presented in Chapters 1–4 generated achiral amino acids or chiral, racemic amino acids. A few chiral, nonracemic amino acids were prepared, but the purpose was to illustrate some specific point of a synthesis. Preparation of non-α-amino acids using methods that proceed with high diastereoselectivity or enantioselectivity is very important since any biological activity usually resides in a single diastereomer, if not a single enantiomer (see Chapter 6 for biologically active amino acids). The enantioselective methods presented will include the use of chiral auxiliaries, chiral catalysts, or chiral templates (as defined by Hanessian).[1]

5.1 α-AMINO ACID TEMPLATES

Chiral, nonracemic α-amino acids will be manipulated to produce chiral, nonracemic non-α-amino acids. The fundamental premise was introduced in Chapter 1, Section 1.4.1. Indeed, several of the synthetic methods discussed in previous chapters will be used here, but the emphasis is on diastereoselectivity and enantioselectivity.

In many cases, the carboxyl portion of the amino acid starting material is converted to an aldehyde, allowing subsequent reactions that are not available to carboxylic acids. An example of this approach is the conversion of *N*-Boc alanine to *N*-Boc alanol (***1***)[2] via conversion to the ester and reduction with lithium borohydride. The next step will be a common feature of this and succeeding chapters. Oxidation of the alcohol moiety in ***1*** to an aldehyde, in this case using SO_3•pyridine in dimethyl sulfoxide (DMSO),[3] allowed condensation with sodium cyanide to give a cyanohydrin. Hydrolysis of the nitrile moiety gave the *N*-Boc-protected corresponding amino carboxylic acid, 2-hydroxy-3-aminobutanoic acid (***2***). Similarly, Boc-glycine led to ethyl 2-hydroxy-3-aminobutanoate in about 84% yield.[2]

---

[1] (a) Hanessian, S. *Total Synthesis of Natural Products: The "Chiron" Approach*. Pergamon, ***1983***; (b) Hanessian, S. *Acc. Chem. Res.* ***1979***, *12*, 159; (c) Hanessian, S.; Franco, J.; Larouche, B. *Pure Appl. Chem.* ***1990***, *62*, 1887.

[2] Iizuka, K.; Kamijo, T.; Harada, H.; Akahane, K.; Kubota, T.; Etoh, Y.; Shimaoka, I.; Tsubaki, A.; Murakami, M.; Yamaguchi, T.; Iyobe, A.; Umeyama, H.; Kiso, Y. *Chem. Pharm. Bull.* **1990**, *38*, 2487.

[3] See Parikh, J.R.; von E. Doering, W. *J. Am. Chem. Soc.* ***1967***, *89*, 5505.

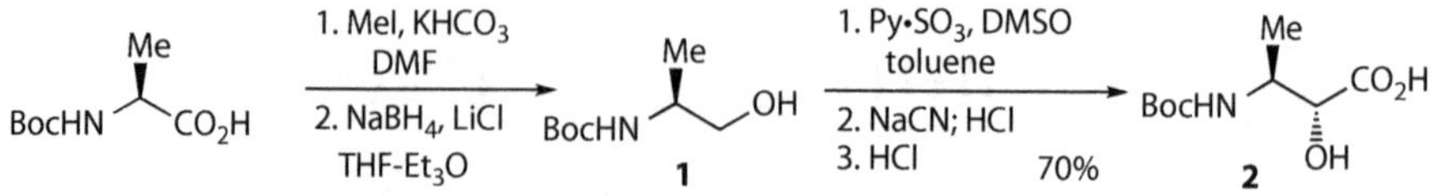

Olefination is another important reaction made possible by the presence of the aldehyde moiety, and it extends the chain. With proper choice of the ylid in a Wittig type reaction,[4,5] a new carboxyl moiety can be incorporated. A Horner-Wadsworth-Emmons olefination[6,7] of *N*-Cbz alanal (***3***), for example, gave methyl 4-(*N*-Cbz amino)pent-2-enoate (***4***) as a mixture of *cis:trans* isomers (>30:1).[8] A number of other derivatives were prepared by using different amino acid starting materials.

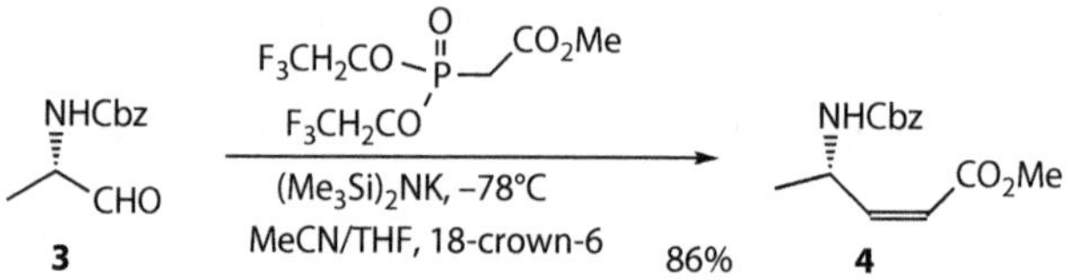

A different synthetic route converted alanine to the diastereomeric mixture of oxazolidin-2-ones, ***5*** and ***6***.[9] After separation of these diastereomers, oxidative cleavage of the alkene moiety in ***5*** (with $RuCl_3$ and $NaIO_4$) led to ***7*** and basic hydrolysis gave *D*-isothreonine (***9***).[10] *L*-Allo-threonine (***10***) was formed in a similar manner from diastereomer ***6***, via ***8***.

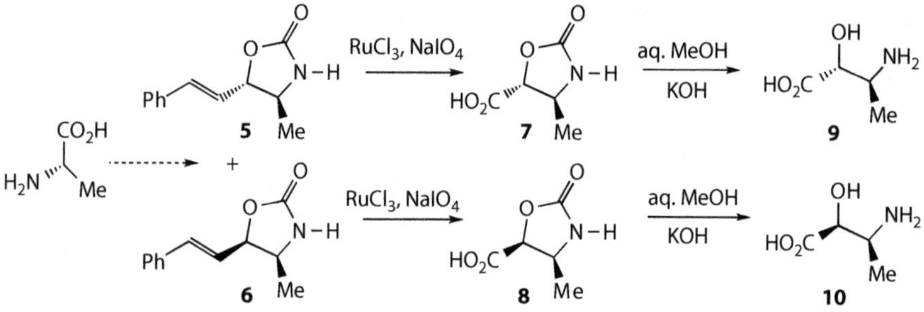

[4] (a) Wittig, G.; Rieber, M. *Annalen* ***1949***, *562*, 187; (b) Wittig, G.; Geissler, G. *Annalen* ***1953***, *580*, 44; (c) Wittig, G.; Schöllkopf, U. *Chem. Ber.* ***1954***, *87*, 1318; (d) Gensler, W.J. *Chem. Rev.* ***1957***, *57*, 191 (see p. 218).

[5] See (a) Smith, M.B. *Organic Synthesis*, 3rd ed. Wavefunction, Inc./Elsevier, Irvine, CA/London, England, ***2010***, pp. 729–739; (b) Smith, M.B. *March's Advanced Organic Chemistry*, 7th ed. John Wiley & Sons, Hoboken, NJ, ***2013***, pp. 1165–1173.

[6] (a) Horner, L.; Hoffmann, H.; Wippel, J.H.; Klahre, G. *Ber.* ***1959***, *92*, 2499; (b) Wadsworth, W.S., Jr.; Emmons, W.D. *J. Am. Chem. Soc.* ***1961***, *83*, 1733; (c) Boutagy, J.; Thomas, R. *Chem. Rev.* ***1974***, *74*, 87.

[7] See (a) Smith, M.B. *Organic Synthesis*, 3rd ed. Wavefunction, Inc./Elsevier, Irvine, CA/London, England, ***2010***, pp. 739–744; (b) Smith, M.B. *March's Advanced Organic Chemistry*, 7th ed. John Wiley & Sons, Hoboken, NJ, ***2013***, pp. 1169–1170.

[8] Kogen, H.; Nishi,T. *J. Chem. Soc. Chem. Commun.* ***1987***, 311.

[9] Wolf, J.-P.; Pfander, H. *Helv. Chim. Acta*, ***1986***, *69*, 918.

[10] Wolf, J.-P.; Pfander, H. *Helv. Chim. Acta*, ***1987***, *70*, 116.

The reaction of an α-amino acid such as *N*-benzyl alanine with (diacetoxyiodo) benzene and iodine under photolysis conditions allowed coupling with a silyl enol ether to give the chain-extended amino acid.[11] In this case, alanine was converted to β-amino acid ***11*** in 89% yield.

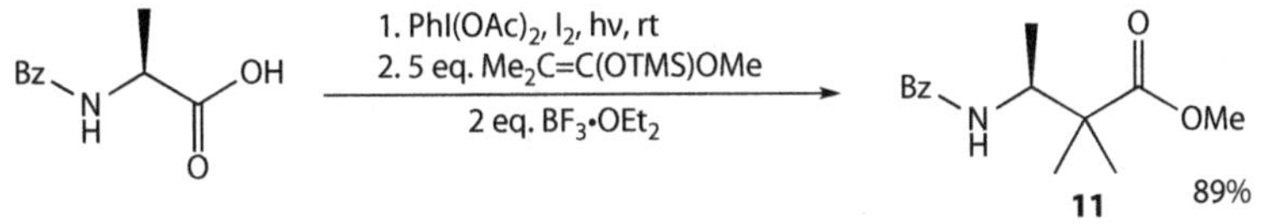

The common α-amino acids asparagine and aspartic acid are important synthetic precursors. In this approach, the α-carbonyl was converted to an alkyl group and the "side chain" carbonyl or amide was converted to a carbonyl to produce a substituted β-amino acid.

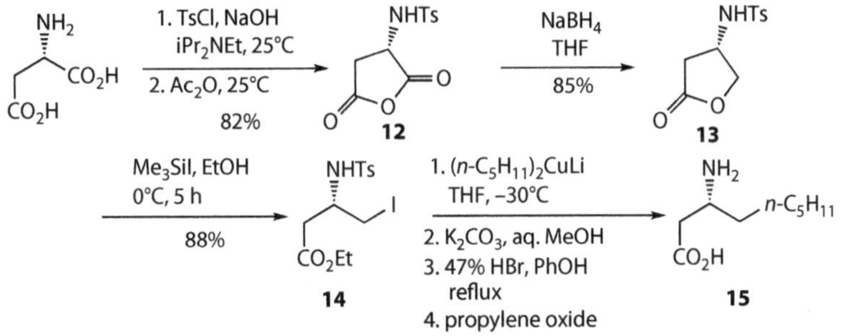

In one example, a chiral, nonracemic lactone was prepared from aspartic acid. Subsequent conversion to the *N*-tosyl derivative allowed reaction with acetic anhydride to give anhydride ***12***. Selective borohydride reduction of one carbonyl unit of the anhydride gave lactone ***13***.[12] Opening the lactone with iodotrimethylsilane gave ethyl 3-(*N*-tosylamino)4-iodobutanoate (***14***), which was converted to 3-aminononanoic acid, ***15***. The use of the appropriate organocuprate in the final sequence allowed the synthesis of 3-aminopentanoic acid, 3-aminooctanoic acid, as well as several alkyl-substituted amino acids.[12]

The amino group in *L*-aspartic acid was converted to an alcohol moiety in the mono-amide of *L*-2-hydroxysuccinic acid, ***16***.[13] Dehydration of the amide moiety led to nitrile ***17***, which was incorporated as an amine surrogate. Reduction of the nitrile and hydrolysis of the acetate gave 2-hydroxy-4-aminobutanoic acid (***18***).[13,14]

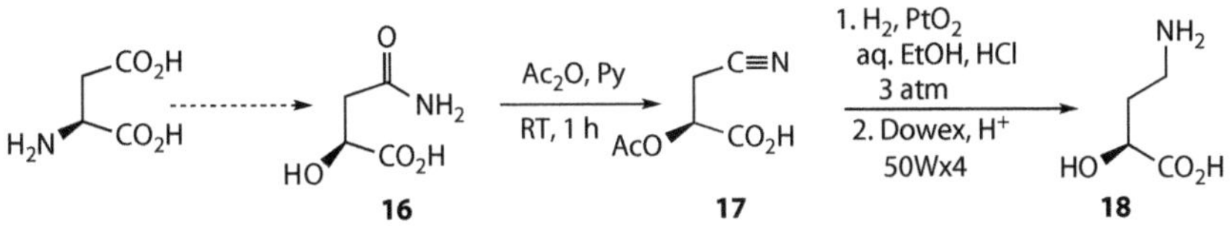

---

[11] Saavedra, C.; Hernández, R.; Boto, A.; Álvarez, E. *J. Org. Chem.* ***2009***, *74*, 4655.
[12] Jefford, C.W.; Wang, J. *Tetrahedron Lett.* ***1993***, *34*, 1111.
[13] Ohshiro, S.; Kuroda, K.; Fujita, T. *Yakugaku Zasshi* ***1967***, *87*, 1184 [*Chem. Abstr.* ***1968***, *68*: 40031w].
[14] Yoneta, T.; Shibahara, S.; Fukatsu, S.; Seki, S. *Bull. Chem. Soc. Jpn.* ***1978***, *51*, 3296.

A different approach reduced both acid moieties in aspartic acid to give an amino-diol, which was cyclized to an oxazolidinone by treatment with sodium hydride[15] Oxidation of the remaining alcohol moiety led to aldehyde ***19***. The aldehyde was protected as the dithiolane, the heterocyclic ring was hydrolyzed to the corresponding amino alcohol, and the amine group was protected to give ***20***. Reaction of the alcohol moiety in ***20*** with phosgene generated a primary chloride, which reacted with the lithium enolate of ethyl acetate to give a 5.5:1 mixture of 3*S*,4*S*:3*R*,4*S* ethyl 4-(*N*-Boc-amino)-3-hydroxypentanoate 5-carbaldehyde protected as the dithiolane (***21***).[15]

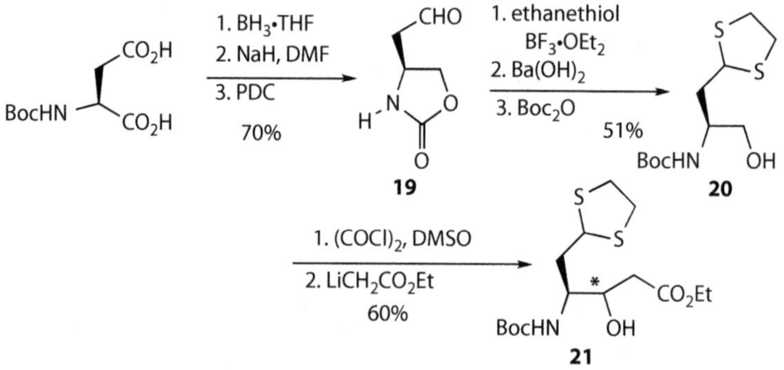

Asparagine is obviously related to aspartic acid, since it is the mono-amide of aspartic acid. Asparagine was first converted to the *N*,*N*-dibenzyl derivative[16] and then to the benzyl ester (***22***). Subsequent treatment with tosyl chloride and pyridine gave the corresponding nitrile (***23***) in 65% overall yield.[17] Reduction of the ester to the corresponding alcohol with $LiAlH_4$ and conversion to the mesylate (91% from ***24***) allowed displacement by organocuprates to give

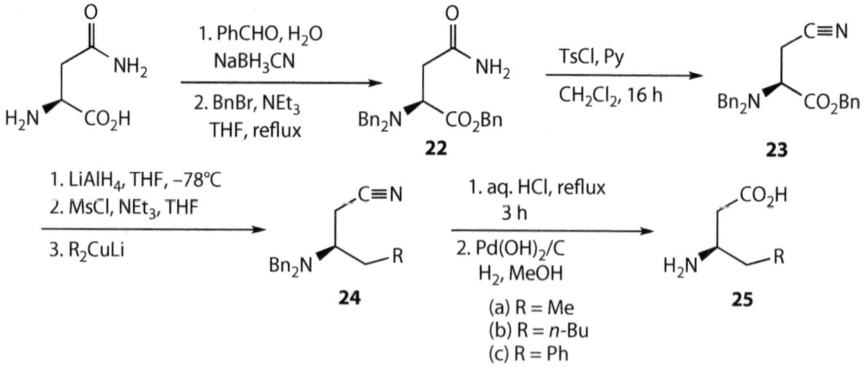

[15] Sham, H.L.; Rempel, C.A.; Stein, H.; Cohen, J. *J. Chem. Soc. Chem. Commun.* ***1987***, 683.
[16] Gmeiner, P.; Lerche, H. *Heterocycles* ***1990***, *31*, 9.
[17] Gmeiner, P. *Tetrahedron Lett.* ***1990***, *31*, 5717.

24 (**a**, R = Me, 65%; **b**, R = *n*-Bu, 72%; **c**, R = Ph, 48% from the mesylate).[17] Hydrolysis with aqueous hydrochloric acid gave the protected amino acid in about 90% yield and removal of the *N*-benzyl groups gave the free amino acid (3-aminopentanoic acid, ***25a***, R = Me[18]; 3-aminooctanoic acid, ***25b***, R = *n*-Bu; 3-amino-4-phenylbutanoic acid, ***25c***, R = Ph) in 90–95% yield.[18] In a similar manner, aspartic acid was converted to asparagine and then to ***25b*** and also to 3-amino-5-phenylpentanoic acid.[19] In a different experiment, Juaristi used *S*-asparagine as a starting material to prepare 2-methyl-3-aminopropanoic acid.[20]

L-Glutamic acid is a readily available starting material, and treatment with nitrous acid (via $NaNO_2$ and HCl) led to formation of the chiral, nonracemic γ-butyrolactone 4-carboxylic acid (***26***),[21] via spontaneous cyclization of the initially formed hydroxy acid to the five-membered ring lactone. Selective reduction of the carboxyl group in ***26*** with borane, followed by mesylation of the resulting alcohol, was followed by a displacement reaction with azide to give azido-lactone ***27*** in 27% overall yield. Catalytic hydrogenation of the azide generated the amine, which attacked the lactone to form lactam ***28***.[21] Basic hydrolysis of the lactam gave 4-hydroxy-5-aminopentanoic acid, ***29***. It is noted that lactone ***26*** has been used in several synthetic transformations.[22]

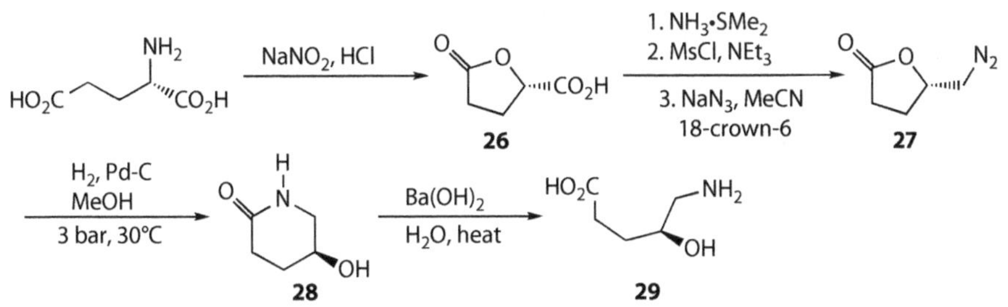

Treatment of glutamic acid with thionyl chloride and ethanol gave an 84% yield of 5*S*-carboethoxy-2-pyrrolidinone (ethyl pyroglutamate, ***30***).[23] Reduction of the ester moiety with lithium borohydride gave ***31***. Subsequent acid hydrolysis and isolation via passage through a Dowex ion exchange column gave 4-amino-5-hydroxypentanoic acid (***32***).[23] It is important to note that in this sequence, the stereogenic center in ***30*** did not racemize during this synthetic sequence. Amino acids such as ***32***

[18] Gmeiner, P. *Liebigs Ann. Chem.* ***1991***, 501.

[19] (a) Mohan, R.; Chou, Y.-L.; Bihovsky, R.; Lumma, W.C., Jr.; Erhardt, P.W.; Shaw, K.J. *J. Med. Chem.* ***1991***, *34*, 2402; (b) Nishizawa, R.; Saino, T.; Takita, T.; Suda, H.; Aoyagi, T.; Umezawa, H. *J. Med. Chem.* ***1977***, *20*, 510.

[20] (a) Juaristi, E.; Quintana, D. *Tetrahedron Asym.* ***1992***, *3*, 723; (b) Juaristi, E.; Escalante, J. *J. Org. Chem.* ***1993***, *58*, 2282.

[21] (a) Herdeis, C. *Synthesis* ***1986***, 232; (b) Herdeis, C.; Dimmerling, A. *Arch. Pharm.* (Weinheim) ***1984***, *317*, 304.

[22] Coppola, G.M.; Shuster, H.F. *Asymmetric Synthesis*. Wiley, New York, ***1987***, pp. 237–242.

[23] Silverman, R.B.; Levy, M.A. *J. Org. Chem.* ***1980***, *45*, 815. Also see Barlos, K.; Mamos, P.; Papaioannou, D.; Patrianakou, S. *J. Chem. Soc. Chem. Commun.* ***1987***, 1583.

were prepared as potential irreversible inactivators of GABA transferase[24] (also see Chapter 6, Section 6.2).

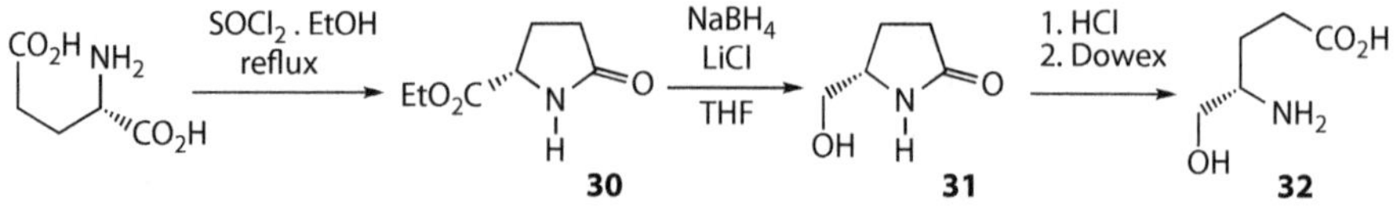

Silverman prepared 5*S*-hydroxymethyl-2-pyrrolidinone (***31***)[25] by essentially the same route as shown above. Conversion of the hydroxymethyl group to a mesylate was followed by its reduction to a methyl group with zinc and iodomethane, giving 5*R*-methyl-2-pyrrolidinone (***33***).[26] The 5*S*-methyl derivative was prepared in an identical manner from *D*-glutamic acid. When ***33*** was heated with 6N HCl, ethyl 4*R*-aminopentanoic acid (***34***) was obtained.[26] Other alkylated amino acids have been prepared from ***31***,[27] and pyroglutamate (***30***) has been used to prepare many derivatives.[19,20] It is noted that Smith and coworkers prepared ***33*** by a slightly different route, and subsequent reaction with formaldehyde and chlorotrimethylsilane led to the synthesis of 5*R*-methyl-*N*-(chloromethyl)-2-pyrrolidinone.[28] Interestingly, it was further shown that ***33*** reacts with alcohols and the resulting *N*-alkoxymethyl derivative could be analyzed by proton nuclear magnetic resonance (NMR) to determine the enantiomeric excess if the alcohol reactant contained a stereogenic center.[28] Further work described the basis for this chiral recognition,[29] and the reaction was extended to show that ***33*** reacted with amines to give a derivative that allowed detection of the enantiomeric excess via proton NMR.[30]

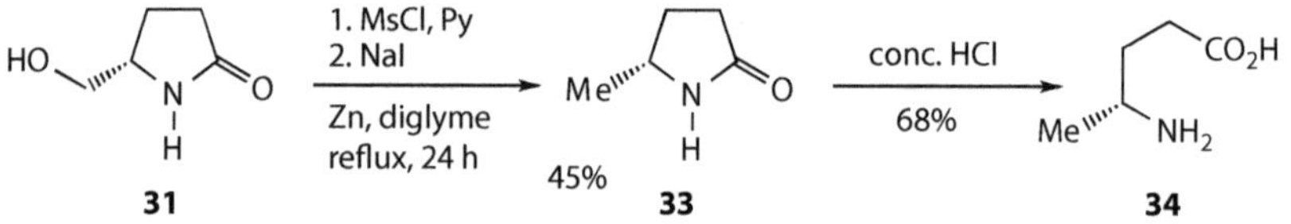

Leucine is an important synthetic starting material for the preparation of other amino acids. Reduction of the ethyl ester of *N*-Cbz leucine (***35***) with diisobutylaluminum hydride gave the corresponding aldehyde (***36***).[31] Aldehyde ***36*** reacted with

[24] Silverman, R.B.; Levy, M.A. *J. Biol. Chem.* ***1981***, *256*, 11565.

[25] Silverman, R.B.; Levy, M.A. *J. Org. Chem.* ***1980***, *45*, 815.

[26] Anduszkiewicz, R.; Silverman, R.B. *J. Biol. Chem.* ***1990***, *265*, 22288.

[27] (a) Smith, A.L.; Williams, S.F.; Holmes, A.B.; Hughes, L.R.; Lidert, Z.; Swithenbank, C. *J. Am. Chem. Soc.* ***1988***, *110*, 8696; (b) Holmes, A.B.; Smith, A.L.; Williams, S.F.; Hughes, L.R.; Lidert, Z.; Swithenbank, C. *J. Org. Chem.* ***1991***, *56*, 1393.

[28] Smith, M.B.; Dembofsky, B.T.; Son, Y.C. *J. Org. Chem.* ***1994***, *59*, 1719.

[29] Latypov, S.; Riguera, R.; Smith, M.B.; Polivkova, J. *J. Org. Chem.* ***1998***, *63*, 8682.

[30] Chen, P.; Tao, S.; Smith, M.B. *Synth. Commun.* ***1998***, *28*, 1641.

[31] (a) Rich, D.H.; Moon, B.J.; Boparai, A.S. *J. Org. Chem.* ***1980***, *45*, 2288; (b) Peet, N.P.; Burkhart, J.P.; Angelastro, M.R.; Giroux, E.L.; Mehdi, S.; Bey, P.; Kolb, M.; Neises, B.; Schirlin, D. *J. Med. Chem.* ***1990***, *33*, 394. Also see Izuka, K.; Kamijo, T.; Harada, H.; Akahane, K.; Kubota, T.; Etoh, Y.; Shimaoka, I.; Tsubaki, A.; Murakami, M.; Yamasuchi, T.; Iyobe, A.; Umeyama, H.; Kiso, Y. *Chem Pharm. Bull.* ***1990***, *38*, 2487.

potassium cyanide and sodium bisulfite to give cyanohydrin ***37***. Hydrolysis and deprotection gave 3-amino-2-hydroxy-5-methylhexanoic acid ***38***, after chromatographic separation of the diastereomeric mixture.

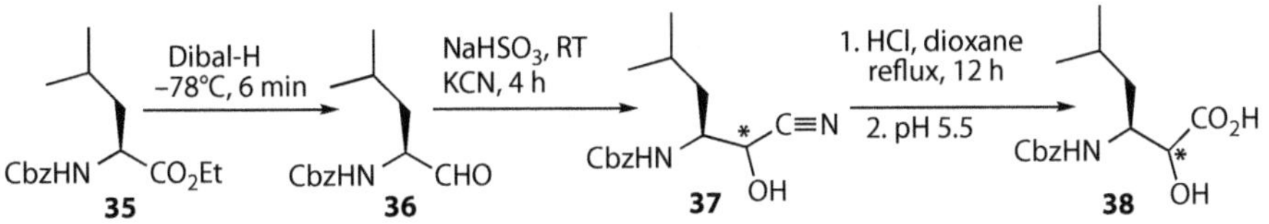

Once an aldehyde such as ***36*** is formed, other reactions are possible. In one example a Wittig reaction[32,33] with the *N*-Boc ethyl ester of leucine (***39***), gave ***40***, and this was converted to dipeptide ***41***.[34] Subsequent reaction with di-*tert*-butylcuprate gave an 80:20 mixture of ***42:43***.

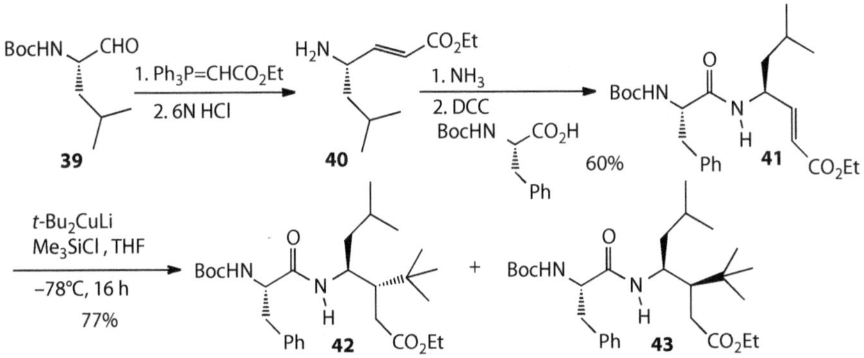

The methyl ester of *L*-valine was converted to *N*-Boc-valinal (***44***) in 82% yield.[35] Subsequent treatment with sodium bisulfite and potassium cyanide,[36] followed by acid hydrolysis, gave 3-amino-2-hydroxy-4-methylpentanoic acid (***45***), as a mixture of diastereomers.[37] Another readily available α-amino acid is *L*-ornithine, which reacted with $NaNO_2$ in sulfuric acid (forming HONO in situ). This reagent converted

[32] (a) Wittig, G.; Rieber, M. *Ann.* ***1949***, *562*, 187; (b) Wittig, G.; Geissler, G. *Ann.* ***1953***, *580*, 44; (c) Wittig, G.; Schöllkopf, U. *Chem. Ber.* ***1954***, *87*, 1318; (d) Gensler, W.J. *Chem. Rev.* ***1957***, *57*, 191 (see p. 218). For the use of a Horner-Wadsworth-Emmons reaction, see Kogen, H.; Nishi, T. *J. Chem. Soc. Chem. Commun.* ***1987***, 311.

[33] See (a) Smith, M.B. *Organic Synthesis*, 3rd ed. Wavefunction, Inc./Elsevier, Irvine, CA/London, England, ***2010***, pp. 729–739; (b) Smith, M.B. *March's Advanced Organic Chemistry*, 7th ed. John Wiley & Sons, Hoboken, NJ, ***2013***, pp. 1165–1173.

[34] Reetz, M.T.; Kanand, J.; Griebenow, N.; Harms, K. *Angew. Chem. Int. Ed. Engl.* ***1992***, *31*, 1626.

[35] (a) Rich, D.H.; Sun, E.T.; Boparai, A.S. *J. Org. Chem.* ***1978***, *43*, 3624; (b) Stanfield, C.F.; Parker, J.E.; Kanellis, P. *J. Org. Chem.* ***1981***, *46*, 4797.

[36] See Nishizawa, R.; Saino,T.; Takita, T.; Suda, H.; Aoyagi, T.; Umezawa, H. *J. Med. Chem.* ***1977***, *20*, 510.

[37] (a) Mohan, R.; Chou, Y.-L.; Bihovsky, R.; Lumma, W.C., Jr.; Erhardt, P.W.; Shaw, K.J. *J. Med. Chem.* ***1991***, *34*, 2402; (b) Peet, N.P.; Burkhart, J.P.; Angelastro, M.R.; Giroux, E.L.; Mehdi, S.; Bey, P.; Kolb, M.; Newses, B.; Schirlin, D. *J. Med. Chem.* ***1990***, *33*, 394.

the α-amino group to a hydroxyl moiety, and hydrolysis generated 2-hydroxy-5-aminopentanoic acid (***46***) in good yield.[38]

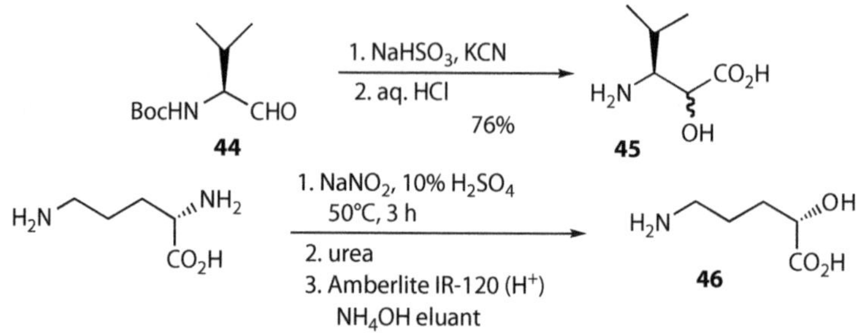

Phenylalanine is another important amino acid precursor. In one example, the isopropyl carbamate of *R*-phenylalanine was converted to aldehyde ***47*** in four steps,[39] and reacted first with sodium cyanide and then with aqueous acid to give ***48***, quantitatively, as a 81:19 mixture of *anti:syn* diastereomers ***48a:48b***.[40]

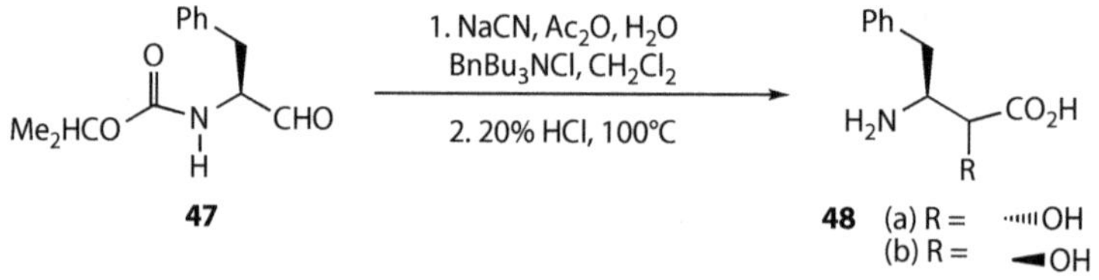

There are other synthetic transformations of phenylalanine, including conversion of the acid moiety to a halide (see ***50***), allowing a chain extension reaction by coupling with the anion of dimethyl malonate.[41] In this case, the amino group was incorporated in the starting material rather than using an amine surrogate by conversion of phenylalanine to ***49***, by a known route.[42] Protection of the amine and conversion of the alcohol moiety to chloride ***50*** via reaction with thionyl chloride was followed by reaction with the enolate anion of dimethyl malonate to give methyl 4-(*N*-benzoylamino)-2-carbomethoxy-5-phenylpentanoate, ***51***. Treatment of this malonate derivative with 47% HBr led to hydrolysis and decarboxylation, giving 4-amino-5-phenylpentanoic acid (***52***) as the final product.[41] Phenylalanal derivatives

[38] Ohshiro, S.; Kuroda, K.; Fujita, T. *Yakugaku Zasshi* ***1967***, *87*, 1184 [*Chem. Abstr.* ***1968***, *68*: 40031w].

[39] Kamijo, T.; Harada, H.; Tsubaki, A.; Yamaguchi, T.; Iyobe, A.; Iizuka, K.; Kiso, Y. *Jpn. Kokai Tokkyo Koho*, JP 1-172365, ***1989***.

[40] Matsuda, F.; Matsumoto, T.; Ohsaki, M.; Ito, Y.; Terashima, S. *Chem. Lett.* ***1990***, 723. Also see Nishi, T.; Saito, F.; Nagahori, H.; Kataoka, M.; Morisawa, Y.; Yabe, Y.; Sakurai, M.; Higashida, S.; Shoji, M.; Matsushita, Y.; Iijima, Y.; Ohizumi, K.; Koike, H. *Chem. Pharm. Bull.* ***1990***, *38*, 103.

[41] Tseng, C.C.; Terashima, S.; Yamada, S. *Chem. Pharm. Bull.* ***1977***, *25*, 29.

[42] Karrer, P.; Portmann, P.; Suter, M. *Helv. Chim. Acta* ***1948***, *31*, 1617.

can also be reacted with potassium cyanide to generate hydroxy amino acids after further manipulation of the product.[43,44,45]

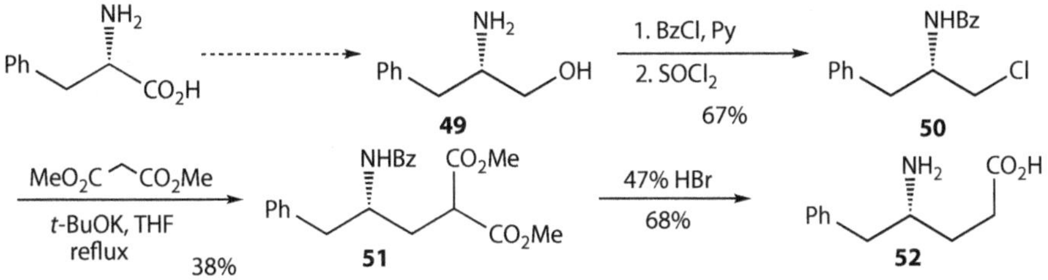

Ester ***53*** contains a chiral auxiliary in the form of the alcohol portion of the ester. The enolate anion derived from ***53*** reacted with *N*-Boc phenylalanal (derived by reduction of the phenylalanine derivative) to give a 9:1 mixture of ***54*** and ***55***.[46] Treatment of this mixture with sodium methoxide led to a 9:1 mixture of ***56*** and ***57*** in 60% overall yield.

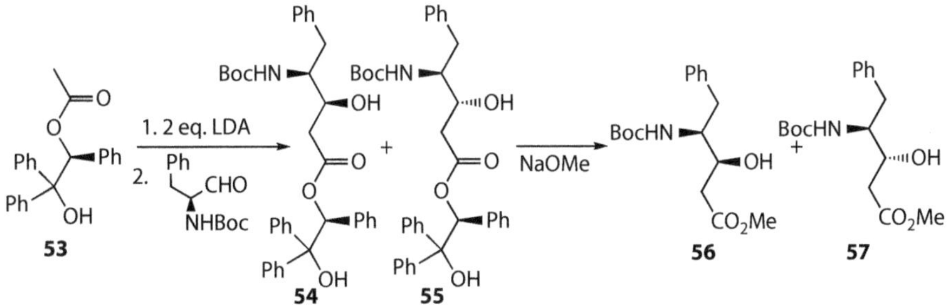

N-Phenyl phenylalanine, ***58***, was converted to Weinreb amide[47] ***59***. Subsequent reduction to the aldehyde and coupling with phosphonium ylid ***60*** gave ***61***, extending the carbon chain of the original amino acid.[48] The use of a chiral, carbene-oxazoline catalyst allowed asymmetric hydrogenation to give α-methyl amino acid ***62***.

[43] Matsuda, F.; Matsumoto, T.; Ohsaki, M.; Ito, Y.; Terashima, S. *Chem. Lett.* ***1990***, 723.

[44] (a) Metcalf, B.W.; Casara, P. *Tetrahedron Lett.* ***1975***, 3337; (b) Danzin, C.; Claverie, N.; Jung, M.J. *Biochem. Pharmacol.* ***1984***, *33*, 1741.

[45] DeCamp, A.E.; Kawaguchi, A.T.; Volante, R.P.; Shinkai, I. *Tetrehedron Lett.* ***1991***, *32*, 1867.

[46] Devant, R.M.; Radunz, H.-E. *Tetrahedron Lett.* ***1988***, *29*, 2307.

[47] (a) Nahm, S.; Weinreb, S.M. *Tetrahedron Lett.* ***1981***, *22*, 3815; (b) Mundy, B.P.; Ellerd, M.G.; Favaloro, F.G., Jr. *Name Reactions and Reagents in Organic Synthesis*, 2nd ed. Wiley-Interscience, Hoboken, NJ, ***2005***, p. 866.

[48] Zhu, Y.; Khumsubdee, S.; Schaefer, A.; Burgess, K. *J. Org. Chem.* ***2011***, *76*, 7449.

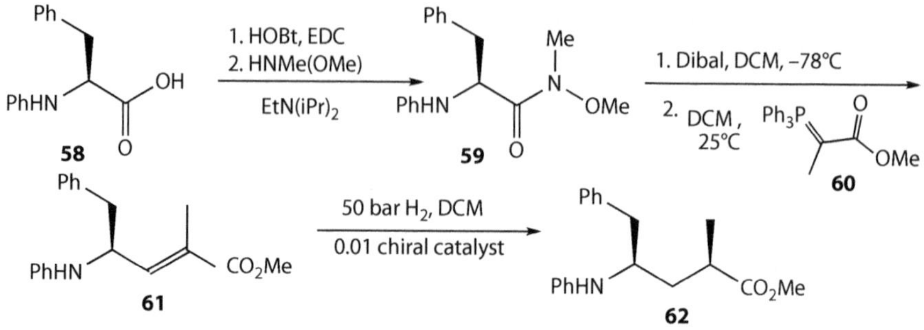

Sulfur ylids such as dimethylsulfonium methylid react with aldehydes derived from amino acids, in this case with *N*-Boc phenylalanal, to give an epoxide. In the example shown, subsequent opening of the epoxide ring by *R*-phenethylamine gave 2*R*,3*S*,5*R* ***63***,[49] where the phenethyl group is both a nitrogen-protecting group and a chiral auxiliary. Condensation of ***63*** with the sodium enolate of ethyl 2-benzyl malonate led to lactone ***64***, which was saponified and decarboxylated to give *N*-Boc 5-amino-4-hydroxy-6-phenyl-2-benzylhexanoic acid (***65***).[49]

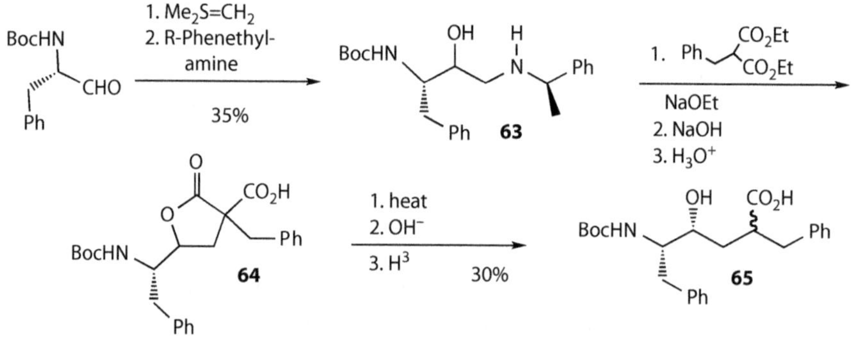

In addition to reducing the acid moiety to an aldehyde, phenylalanine is an important precursor to amino acids that contain a cyclohexyl ring. Clearly, a phenyl ring can be reduced to a cyclohexane derivative, usually by catalytic hydrogenation or by treatment with an alkali metal in liquid ammonia (Birch reduction).[50] Treatment of ***66*** with hydrogen and a rhodium catalyst, for example, reduced the phenyl ring to a cyclohexane. Subsequent oxidation with sulfur trioxide•pyridine and DMSO[3] gave the corresponding aldehyde, ***67***.[51] This aldehyde was then treated with HCN to give

[49] Evans, B.E.; Rittle, K.E.; Homnick, C.F.; Springer, J.P.; Hirshfield, J.; Veber, D.F. *J. Org. Chem.* ***1985***, *50*, 4615.

[50] For a brief review of the Birch reduction and its application to the synthesis of natural products, see (a) Schultz, A.G. *Chem. Commun.* ***1999***, 1263. Also see (b) Smith, M.B. *Organic Synthesis*, 3rd ed. Wavefunction, Inc./Elsevier, Irvine, CA/London, England, ***2010***, pp. 459–461; (c) Smith, M.B. *March's Advanced Organic Chemistry*, 7th ed. John Wiley & Sons, Hoboken, NJ, ***2013***, pp. 912–916.

[51] Iizuka, K.; Kamijo, T.; Harada, H.; Akahane, K.; Kuboto, T.; Umeyama, H.; Kiso,Y. *J. Chem. Soc. Chem. Commun.* ***1989***, 1678. For preparation of 2*S*,3*S*-***68***, see Iizuka, K.; Kamijo, T.; Harada, H.; Akahane, K.; Kubota, T.; Umeyama, H.; Ishida, T.; Kiso, Y. *J. Med. Chem.* ***1990***, *33*, 2707.

3-amino-4-cyclohexyl-2*R*-hydroxybutanoic acid, ***68***, as a mixture of diastereomers (7:3 2*R*,3*S*:2*S*,3*S*) in 60% overall yield.[51]

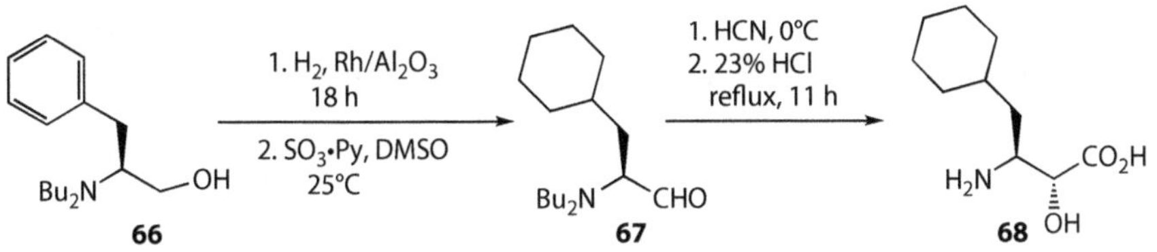

The use of *L*-serine as a template incorporates a hydroxyl group that may be manipulated to incorporate other functional groups and used to generate new amino acids. The amino group of serine was converted to a bromine, for example, with complete retention of configuration by reaction with NOBr to give *S*-***69***.[52] Upon treatment with ammonium hydroxide, epoxide ***70*** was generated *in situ* and then opened by ammonia present in the reaction to give *D*-isoserine (***71***).[52] The overall result is that the CH_2OH moiety of serine was transformed into a $-CH_2NH_2$ moiety and the $-NH_2$ was converted to the $-OH$ moiety in *S*-***71***. Similarly, *L*-threonine was converted to 2*R*,3*S*-2-hydroxy-3-aminobutanoic acid (isothreonine).[53]

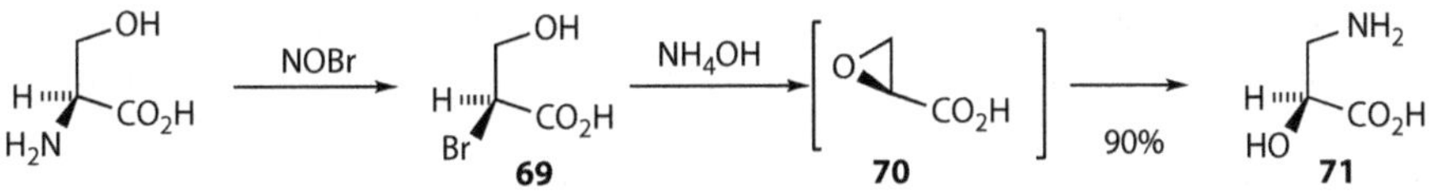

A somewhat different synthetic approach began with lactone ***72*** (derived from homoserine). Reduction with diisobutylaluminum hydride gave hemiacetal derivatives ***73***:***74*** as a 1:5 mixture.[54] Subsequent Wittig olefination[32,33] gave ***75*** in 80% overall yield from ***72***. Catalytic hydrogenation of the alkene moiety and cleavage of the trityl group gave 6-hydroxy-4-aminohexanoic acid, ***76***.[54] In another sequence, the methyl ester of *N*-Boc-*D*-serine was

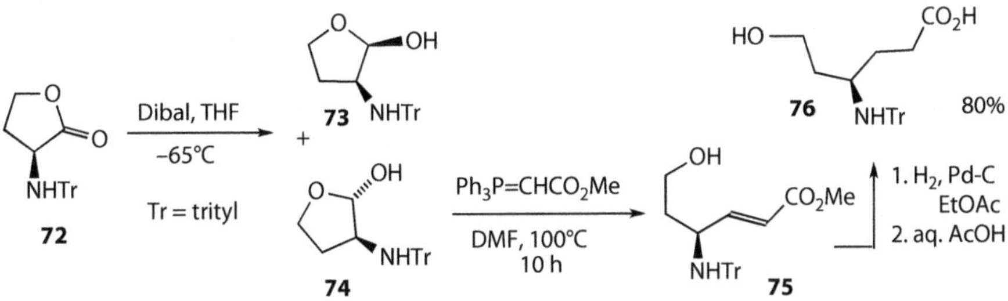

[52] Shimohigashi, Y.; Waki, M.; Izumiya, N. *Mem. Fac. Sci. Kyushu Univ. Ser. C* ***1979***, *11*, 217 [*Chem. Abstr.* ***1979***, *91*: 39804u]. Also see Pons, D.; Savignac, M.; Genet, J.-P. *Tetrahedron Lett.* ***1990***, *31*, 5023.

[53] Shimohigashi, Y.; Waki, M.; Izumiya, N. *Bull. Chem. Soc. Jpn.* ***1979***, *52*, 949.

[54] Papaioannou, D.; Stavropoulos, G.; Sivas, M.; Barlos, K.; Francis, G.W.; Aksnes, D.W.; Maartmann-Moe, K. *Acta Chem. Scand.* ***1991***, *45*, 99.

converted to isoxazolidine carbaldehyde *77* (known as Garner's aldehyde).[55] Garner used a Diels-Alder reaction[56] (see Chapter 2, Section 2.5) between the aldehyde moiety in *77* and Danishefsky's diene[57] to give *78*, as a 3:1 β:α-hydrogen mixture. The pyranose moiety was converted to a hydroxyester and deprotection of the isoxazolidine group gave methyl 4-(*N*-Boc amino)-3,5-dihydroxypentanoate, *79*.[55]

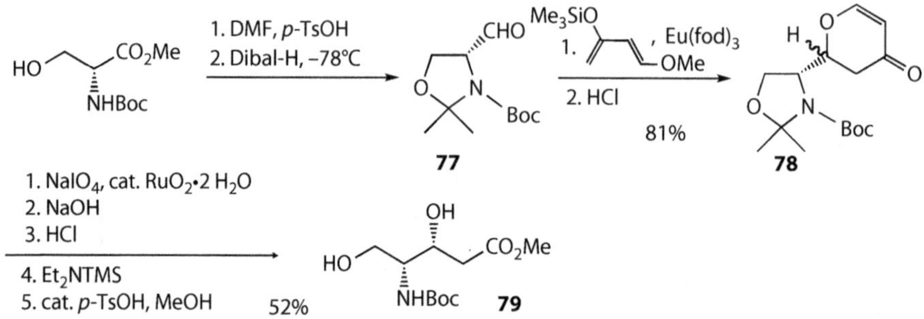

5.2 NON-α-AMINO ACID TEMPLATES

Some non-α-amino acids are either commercially available or readily prepared, and so serve as useful templates for the preparation of new non-α-amino acids, as introduced in Chapter 1, Section 1.4.2. This section will show selected syntheses of this type to illustrate the utility of this approach as a complement to modification of α-amino acids in Section 5.1.

A relatively straightforward method for converting one amino acid to another uses an enolate alkylation reaction. Beginning with a non-α-amino acid starting material, alkylation of the enolate anion can generate a new amino acid, sometimes with high diastereoselectivity. Seebach used an ester enolate strategy to produce alkyl-substituted amino alkanoic acids, but used β-amino-ester derivatives as precursors. When *80* was treated with two molar equivalents of lithium diisopropylamide and then iodoethane, racemic methyl 3-(*N*-benzoylamino)-2-ethylbutanoate was formed as a 16:1 mixture of diastereomers *81:82*.[58] A similar reaction with iodomethane gave a 50% yield of methyl *N*-benzyl-3-amino-2-methylbutanoate (4:1 *anti:syn*). Reaction with allyl bromide gave a 90% yield of methyl *N*-benzyl-2-allyl-3-amino-butanoate (31:1 *anti:syn*), and benzyl bromide led to a 90% yield of methyl *N*-benzoyl-3-amino-2-benzylbutanoate (36:1 *anti:syn*).[58] Seebach also used amino-esters such as *80*[59] where the stereogenic center at C3 (the amino-bearing

[55] Garner, P. *Tetrahedron Lett.* ***1984***, *25*, 5855.

[56] (a) Desimoni, G.; Tacconi, G.; Barco, A.; Pollini, G.P. *Natural Product Synthesis through Pericyclic Reactions*. American Chemical Society, Washington, DC, ***1983***; (b) see Smith, M.B. *Organic Synthesis*, 3rd ed. Wavefunction, Inc./Elsevier, Irvine, CA/London, England, ***2010***, pp. 1013–1075; (c) Smith, M.B. *March's Advanced Organic Chemistry*, 7th ed. John Wiley & Sons, Hoboken, NJ, ***2013***, pp. 1020–1039.

[57] (a) Danishefsky, S. *Accts. Chem. Res.* ***1981***, *14*, 400; (b) Danishefsky, S. *Aldrichim. Acta* ***1986***, *19*, 59.

[58] Seebach, D.; Estermann, H. *Tetrahedron Lett.* ***1987***, *28*, 3103.

[59] Estermann, H.; Seebach, D. *Helv. Chim. Acta* ***1988***, *71*, 1824.

carbon) makes this amino-ester a chiral template. In subsequent reactions, that stereogenic center will induce asymmetry into the new products. Ethyl *N*-benzyl-3-aminobutanoate (***83***), for example, was treated with 5.2 molar equivalents of lithium diisopropylamide to give dianion ***84***.[59] Quenching with iodomethane gave ***85a*** (80%de, where de refers to diastereomeric excess) and reaction with iodoethane gave ***85b*** (93%de). The alkylation reaction was highly diastereoselective, and each diastereomer was produced with good enantiopurity. The methyl ester gave similar results, although the yield of the alkylation product was observed to vary with the nature of the halide that was used. Dianion ***84*** was also condensed with aldehydes to give hydroxy amino acid derivatives.[59]

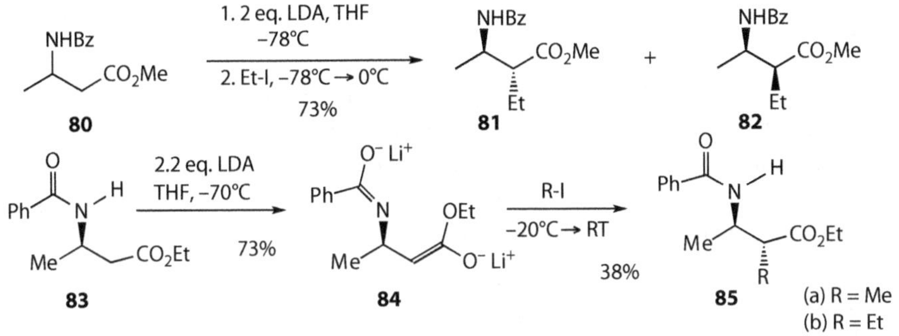

Addition of three equivalents of lithium chloride followed by formation of the lithium enolate anion of ***86*** led to the *anti*-diastereomer [methyl 3-(*N*-benzoylamino)-2-methylbutanoate, ***87***], after reaction with iodomethane, with excellent diastereoselectivity.[59]

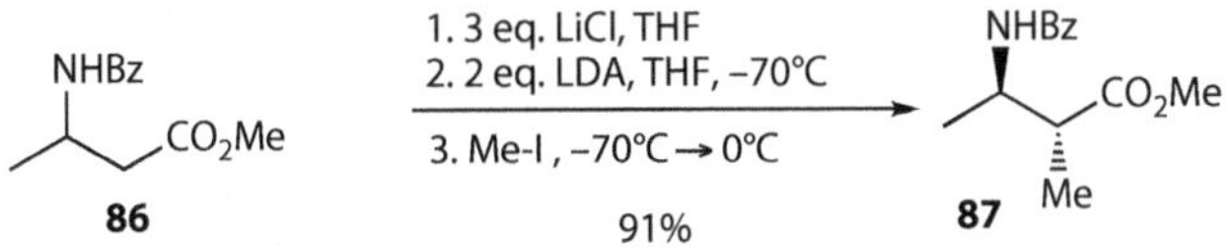

A more elaborate synthesis involved condensation of ***88*** and ***89*** to give ***90***[60] via the boron enolate anion of ***89***.[61] Removal of the acyl-protecting group and reduction gave ***91***. Treatment with camphorsulfonic acid manipulated the "protecting group" (converted an oxazolidine ring to an acetonide) to give ***92***. Oxidation of the primary alcohol moiety to a carboxylic acid gave the acetonide of 5-(*N*-Boc amino)-6-cyclohexyl-2-isopropyl-3,4-dihydroxyhexanoic acid, ***93***.

[60] Thaisrivongs, S.; Tomasselli, A.G.; Moon, J.B.; Hui, J.; McQuade, T.J.; Turner, S.R.; Strohbach, J.W.; Howe, W.J.; Tarpley, W.G.; Heinrikson, R.L. *J. Med. Chem.* ***1991***, *34*, 2344.

[61] See, for example, Evans, D.A.; Nelson, J.V.; Vogel, E.; Taber, T.R. *J. Am. Chem. Soc.* ***1981***, *103*, 3099.

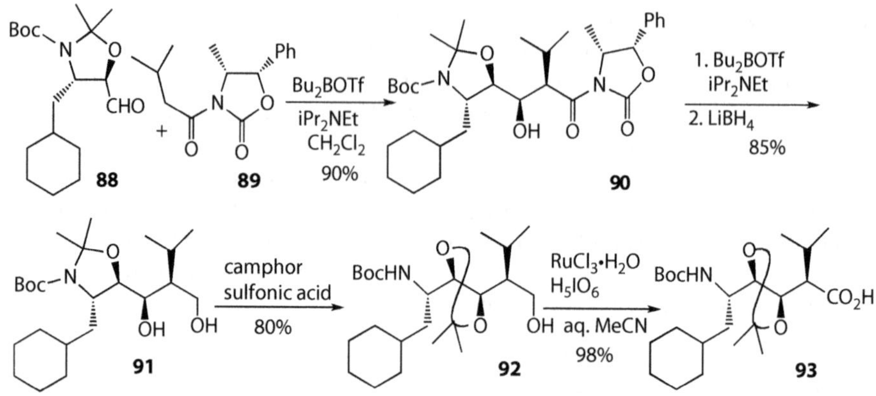

Aldehydes can also be generated by oxidative cleavage of alkenes. Acetylation of ***94*** was followed by oxidative cleavage of the alkene, and subsequent reaction of the resultant aldehyde with Wittig reagent[wittig] ***95*** gave ***96***.[62] Other amino acids were prepared by using an analog of ***94***, from carboalkoxymethyl to alkyl, by changing the alkyl substituents in the Wittig reagent.

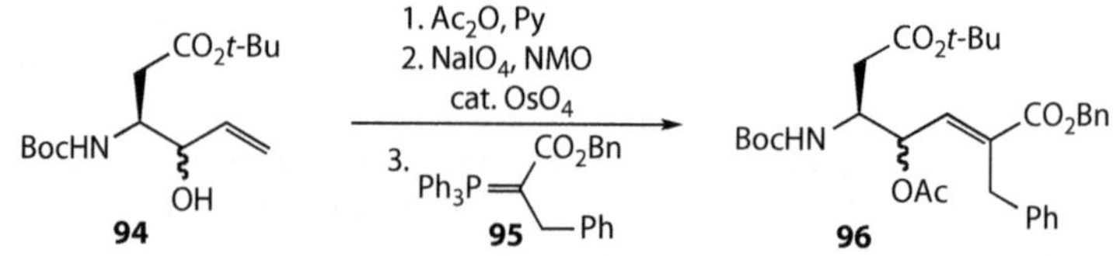

L-β-Malamidic acid (***97***) is a useful chiral template, and it was converted to ***98*** using a hypervalent iodine species.[63] Acid hydrolysis gave *S*-isoserine (3-amino-2-hydroxypropanoic acid, ***R-71***).

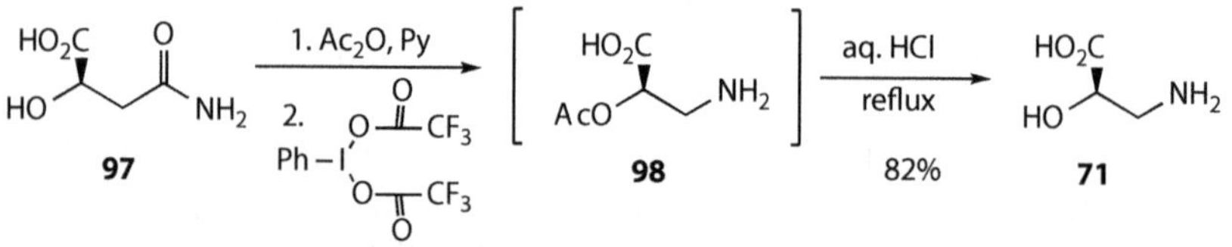

Hydrogenation of enamino-ester derivatives leads to β-amino acid derivatives, and the use of a chiral hydrogenation catalyst gives an enantioselective preparation of chiral amino acids. A chiral catalyst (***100***) developed by Achiwa facilitated reduction of ***99*** to give methyl *N*-acetyl-3-aminobutanoate (***101***), quantitatively (53%ee, *R*).[64] Similarly, methyl *N*-acetyl-3-amino-3-phenyl-propanoate was formed in 90% yield (47%ee, *R*). Similarly, Noyori used the BINAP-Ru(II) catalyst [***103***, where BINAP is 2,2'-*bis*-(diphenylphosphino)-1,1'-binaphthyl] to reduce β-aminopropenoic acid as

---

[62] Shue, Y.-K.; Carrera, G.M., Jr.; Nadzan, A.M. *Tetrahedron Lett.* ***1987***, *28*, 3225.
[63] Andruskiewicz, R.; Czeruinski, A.; Grzybowska, J. *Synthesis* ***1983***, 31.
[64] Achiwa, K.; Soga, T. *Tetrahedron Lett.* ***1978***, 1119.

well as several derivatives.[65] The reaction of ***99*** with hydrogenation and rhodium *R*-BINAP gave a good yield of ***101***, but in only 5%ee, *R*. For several examples, the *E*-isomer showed greater selectivity during hydrogenation. Reduction of *E*-***102***, for example, gave a quantitative yield of ***101*** (96%ee, *S*).

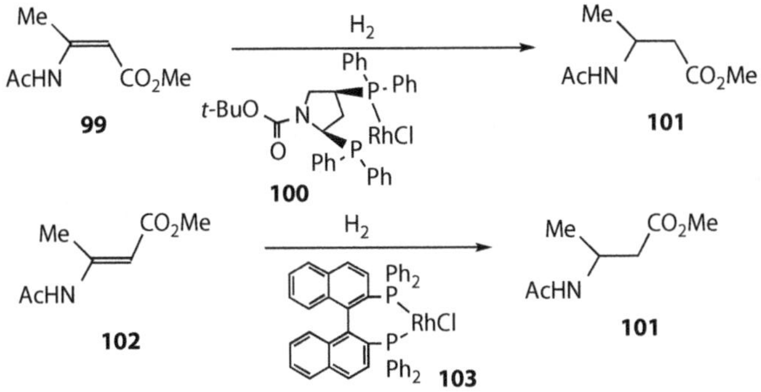

BINAP derivatives such as ***104*** (chiral *ortho*-substituted BINAPO ligands; *o*-BINAPO) have been developed for the enantioselective reduction of systems similar to ***99*** or ***102***.[66] bis-Ferrocenyl diphosphines such as ***105*** (TRAP; known as EtTRAP when R = ethyl, BuTRAP when R = butyl) have been used as chiral hydrogenation ligands with rhodium catalyst.[67] Unsymmetrical ferrocenylethylamine-derived mono-phosphoramidites were used for Rh-catalyzed enantioselective hydrogenation of enamino-esters or enamides.[68]

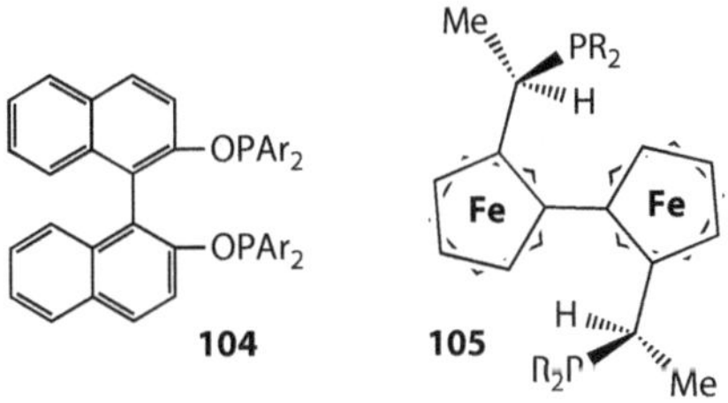

5.3 SYNTHESIS FROM OTHER STARTING MATERIALS

Two important methods used to induce asymmetry in a synthesis are incorporation of a chiral auxiliary and using a chiral, nonracemic starting material (a chiral template), as shown in previous sections for synthesis using amino acids. A variety of starting materials are known.

[65] Lubell, W.D.; Kitamura, M.; Noyori, R. *Tetrahedron Asym.* **1991**, *2*, 543. Also see Qiu, L.; Kwong, F.Y.; Wu, J.; Lam, W.H.; Chan, S.; Yu, W.-Y.; Li, Y.-M.; Guo, R.; Zhou, Z.; Chan, A.S.C. *J. Am. Chem. Soc.* **2006**, *128*, 5955.

[66] Zhou, Y.-G.; Tang, W.; Wang, W.-B.; Li, W.; Zhang, X. *J. Am. Chem. Soc.* **2002**, *124*, 4952.

[67] Sawamura, M.; Kuwano, R.; Ito, Y. *J. Am. Chem. Soc.* **1995**, *117*, 9602.

[68] Zeng, Q.-H.; Hu, X.-P.; Duan, Z.-C.; Liang, X.-M.; Zheng, Z. *J. Org. Chem.* **2006**, *71*, 393.

A chiral, nonracemic heterocyclic intermediate (oxazolidine ***107***) was prepared from 2*R*-phenyl-2-aminoethanol (***106***). A Reformatsky type reaction[69] (also see Chapter 4, Section 4.4) proceeded with ring opening and led to ***108***. Catalytic hydrogenation cleaved the benzylic auxiliary to give ***109*** (ethyl 3-aminopentanoate) in 55% overall yield (72%ee, *R*).[70] Both ethyl 3-aminobutanoate and ethyl 3-aminoheptanoate were prepared by this method.

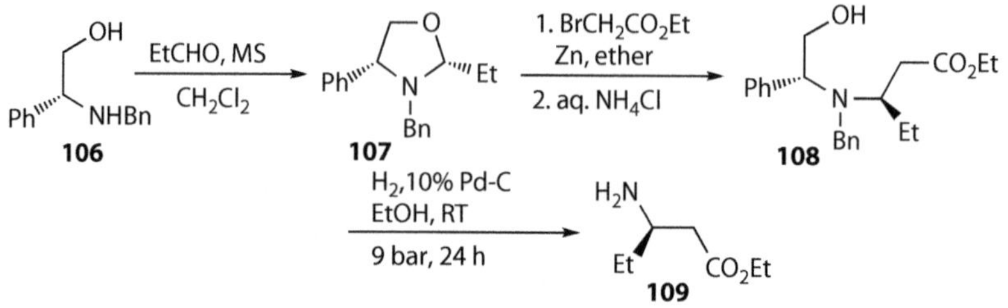

Conjugate addition of the lithium salt of a chiral amine (a homochiral lithium amide) to a β-substituted α,β-unsaturated ester leads to formation of a chiral amino acid. Routes to amino acids that relied on relations with conjugated systems were introduced in Chapter 3. A representative example of a reaction protocol that utilized conjugate addition is addition of the chiral, nonracemic lithium amide ***111*** (which has a phenethyl auxiliary) to ***110*** to give the amino-ester.[71] Catalytic hydrogenation removed both benzylic groups (the auxiliary and the benzyl group), and acid hydrolysis of the ester moiety led to 3-amino-3-(4-benzyloxyphenyl)-propanoic acid, ***112***.[71] Amino acid ***112*** was formed with >99%dr (dr is diastereomeric ratio), and with high enantioselectivity.

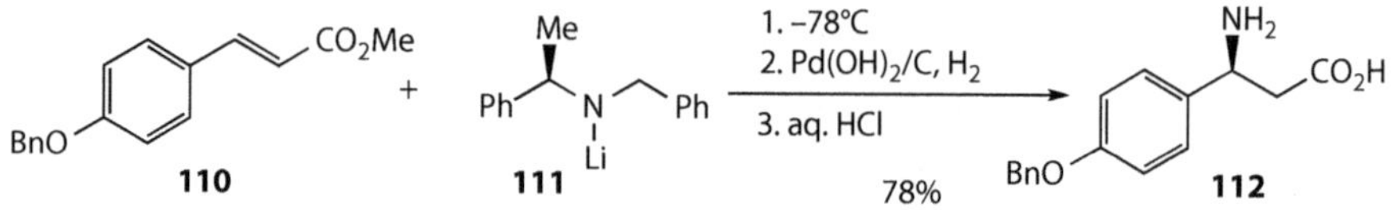

A different approach attached a chiral auxiliary to a conjugated ester, and the chiral substrate was subsequently reacted with an achiral amine. This approach was used by d'Angelo when diphenylmethylamine was added to 2-(1-β-naphthyl-1-methylethyl)-5-methylcyclohexyl, but-2*E*-enolate (***113***), a menthol derivative, under

[69] (a) Reformatsky, S. *Ber.* ***1887***, *20*, 1210; (b) Rathke, M.W. *Org. React.* ***1975***, *22*, 423; (c) Diaper, D.G.M.; Kuksis, A. *Chem. Rev.* ***1959***, *59*, 89; (d) see Smith, M.B. *Organic Synthesis*, 3rd ed. Wavefunction, Inc./Elsevier, Irvine, CA/London, England, ***2010***, pp. 885–887; (e) Smith, M.B. *March's Advanced Organic Chemistry*, 7th ed. John Wiley & Sons, Hoboken, NJ, ***2013***, pp. 1129–1131.

[70] Andrés, C.; González, A.; Pedrosa, R.; Pérez-Encabo, A. *Tetrahedron Lett.* ***1992***, *33*, 2895.

[71] Davies, S.G.; Ichihara, O. *Tetrahedron Asym.* ***1991***, *2*, 183. Also see Davies, S.G.; Garrido, N.M.; Kruchinin, D.; Ichihara, O.; Kotchie, L.J.; Price, P.D.; Mortimer, A.J.P.; Russell, A.J.; Smith, A.D. *Tetrahedron Asym.* ***2006***, *17*, 1793.

high-pressure conditions[72] (15 kbar) to give 3*S*-(*N*-diphenylmethylamino)butanoic acid (***114***) in >99%ee.[73]

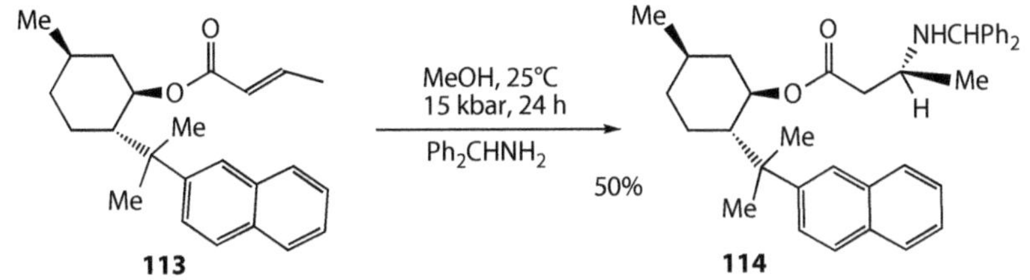

113 **114**

A variation involved "conjugate" addition to α,β-epoxy esters using amine nucleophiles. A problem with this approach is that the amine may attack epoxy carbons at both the α- and the β-position relative to the ester moiety. An example is the reaction at the β-carbon of epoxy acid ***115***[74] with diethylamine, which opened the epoxide ring to give a 1:6 mixture of ***116:117.*** The diastereoselectivity of this ring opening was improved to >20:1 (***116:117***) by addition of *tetrakis*-isopropoxy titanium(IV) to the initial reaction with diethylamine.[74]

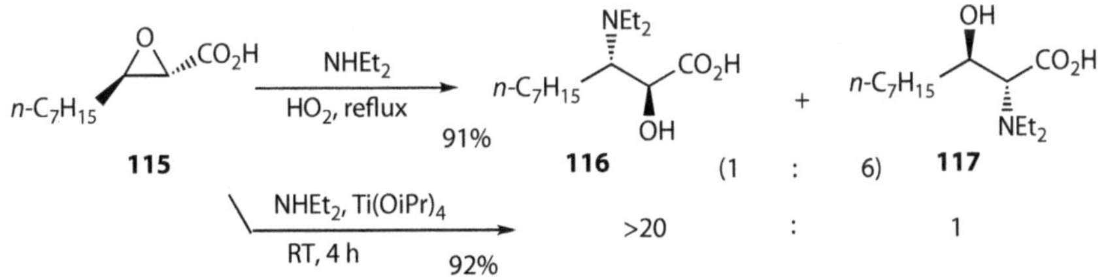

A different conjugate addition involved the reaction of a conjugated ester with an isocyanate. α-Fluoro-ester ***118***, for example, was converted to aldehyde ***119*** and then Horner-Wadsworth-Emmons olefination[75] gave ***120***.[76] Reaction of ***120*** with chlorosulfonyl isocyanate, followed by treatment with base, gave ***121.*** Subsequent basic hydrolysis gave 3-amino-4-fluoro-5-hydroxyl-4-methylhexanoic acid (***122***).

---

[72] (a) Matsumoto, K.; Morris, A.R. *Organic Synthesis at High Pressure.* Wiley, New York, ***1991***; (b) Matsumoto, K.; Sera, A.; Uchida, T. *Synthesis* ***1985***, 1; (c) Matsumoto, K.; Sera, A. *Synthesis* ***1985***, 999.

[73] d'Angelo, J.; Maddaluno, J. *J. Am. Chem. Soc.* ***1986***, *108*, 8112.

[74] Chong, J.M.; Sharpless, K.B. *J. Org. Chem.* ***1985***, *50*, 1560.

[75] See (a) Smith, M.B. *Organic Synthesis*, 3rd ed. Wavefunction, Inc./Elsevier, Irvine, CA/London, England, ***2010***, pp. 739–744; (b) Smith, M.B. *March's Advanced Organic Chemistry*, 7th ed. John Wiley & Sons, Hoboken, NJ, ***2013***, pp. 1169–1170.

[76] Kitazume, T.; Yamamoto,T.; Ohnogi, T.; Yamazaki, T. *Chem. Express* ***1989***, *4*, 657.

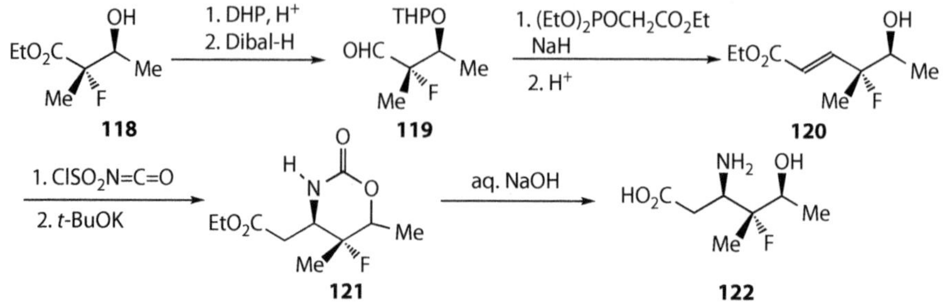

An internal conjugate addition reaction was reported in which a pedant carbamoyl nitrogen was added to a conjugated amide moiety to give the heterocycle. Reaction of ***123*** with mercuric triflate, for example, was followed by reduction with sodium borohydride to give a 2:1 mixture of ***124:125***.[77] Acid hydrolysis of this mixture gave 3-phenyl-3-aminopropanoic acid (***126***, in 33%ee).

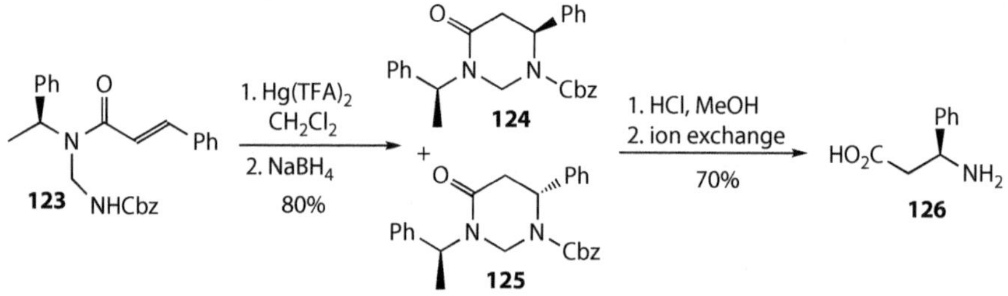

Malic acid can be obtained in optically pure form, and it serves as a useful chiral template for the synthesis of amino acids. *S*-Malic acid (***127***) was easily converted to ***128*** via reaction with formaldehyde.[78] This conversion allowed the preparation of diazoketone ***129***, and a Curtius rearrangement[79,80] led to 2-hydroxy-3-aminopropanoic acid (isoserine, ***71***) after hydrolysis. In this work, isoserine was cited as having important biological activity[81] as well as being a constituent of the antibiotic edeine.[82] *S*-Malic acid was also converted to a chiral, nonracemic

[77] Amoroso, R.; Cardillo, G.; Tomasini, C.; Tortoreto, P. *J. Org. Chem.* ***1992***, *57*, 1082.

[78] Milewska, M.J.; Polonski, T. *Synthesis* ***1988***, 475.

[79] (a) Curtius, T. *J. Prakt. Chem.* ***1894***, *50*, 275; (b) Smith, P.A.S. *Org. React.* ***1946***, *3*, 337; (c) Saunders, J.H.; Slocombe, R.J. *Chem. Rev.* ***1948***, *43*, 203.

[80] See (a) Smith, M.B. *Organic Synthesis*, 3rd ed. Wavefunction, Inc./Elsevier, Irvine, CA/London, England, ***2010***, pp. 192, 1051; (b) Smith, M.B. *March's Advanced Organic Chemistry*, 7th ed. John Wiley & Sons, Hoboken, NJ, ***2013***, pp. 1361–1362.

[81] Liebman, K.C.; Fellner, S.K. *J. Org. Chem.* ***1962***, *27*, 438.

[82] Roncari, G.; Kurylo-Borowska, Z.; Craig, L.C. *Biochemistry* ***1963***, *5*, 2153.

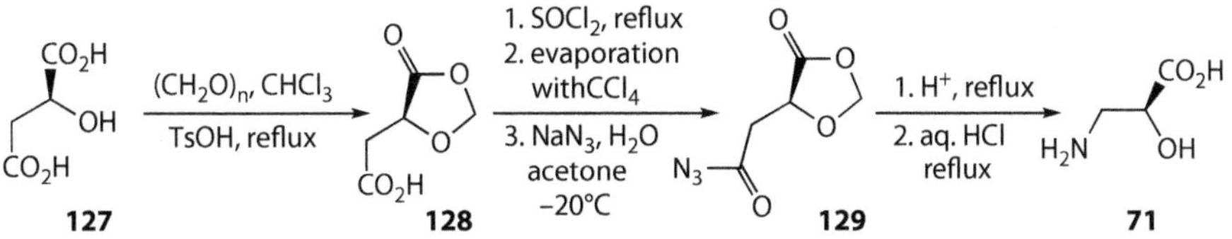

imide that was an important synthetic intermediate. Reaction of malic acid with benzylamine gave the chiral, nonracemic imide, ***130***.[83] Reduction of one imide carbonyl with sodium borohydride, protection of the resulting alcohol, and coupling with tri-*n*-butylallyltin led to selective formation of ***131*** (as a 3:1 *trans:cis* mixture). Removal of both the *O*- and *N*-benzylic groups gave ***132***. A new set of protecting groups (see ***133***) was necessary to allow separation of diastereomers before final ring opening of the lactam[84] led to 4*S*-amino-3*S*-hydroxyheptanoic acid, ***134***.[83]

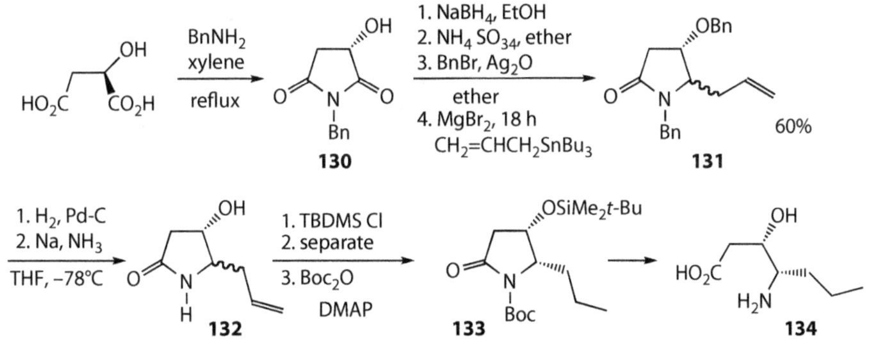

Alkyne ***135*** was prepared from succinic anhydride in seven steps (19% overall yield). The auxiliary on the alcohol moiety allowed the preparation of chiral, nonracemic ***136*** after removal of the auxiliary.[85] Mitsunobu reaction[86,87] with phthalimide gave the phthalimidoyl group, with inversion of configuration (see ***13***). Subsequent steps deprotected the alcohol, allowing oxidation to the acid, and removal of the protecting group from the alkyne and conversion of the phthalimidoyl group to an amine gave 4-aminohex-5-ynoic acid, ***138***.[85]

---

[83] Bernardi, A.; Micheli, F.; Potenza, D.; Scolastico, C.; Villa, R. *Tetrahedron Lett.* ***1990***, *31*, 4949.

[84] (a) Katsuki, T.; Yamaguchi, M. *Bull. Chem. Soc., Jpn.* ***1976***, *49*, 3287; (b) Jouin, P.; Castro, B. *J. Chem. Soc. Perkin Trans. I* ***1987***, 1177; (c) Klutchko, S.; O'Brien, P.; Hodges, J.C. *Synth. Commun.* ***1989***, *19*, 2573; (d) Andrew, R.G.; Conrow, R.E.; Elliott, J.D.; Johnson, W.S.; Ramezani, S. *Tetrahedron Lett.* ***1987***, 6535.

[85] (a) Tabor, A.B.; Holmes, A.B.; Baker, R. *J. Chem. Soc. Chem. Commun.* ***1989***, 1025; (b) Holmes, A.B.; Tabor, A.B.; Baker, R. *J. Chem. Soc. Perkin Trans. I* ***1991***, 3301.

[86] Mitsunobu, O. *Synthesis* ***1981***, 1.

[87] See (a) Smith, M.B. *Organic Synthesis*, 3rd ed. Wavefunction, Inc./Elsevier, Irvine, CA/London, England, ***2010***, pp. 1213–126; (b) Smith, M.B. *March's Advanced Organic Chemistry*, 7th ed. John Wiley & Sons, Hoboken, NJ, ***2013***, pp. 469–470.

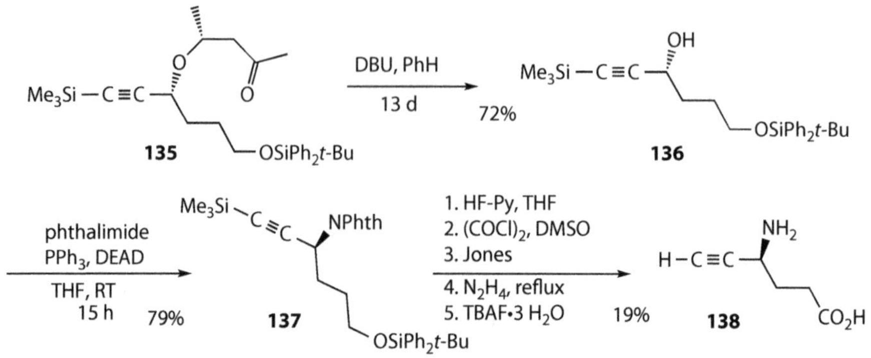

Seebach reported an asymmetric synthesis of a 3-aminobutanoic acid derivative using ***139*** as a chiral, nonracemic precursor.[88] Reaction of ***139*** with benzylamine gave 3*S*-(*N*-benzylamino)butanoic acid, *S*-***140***, with asymmetry induced by the presence of the auxiliary. The stereochemistry of the final amino acid could be "reversed" by initial conversion of ***139*** to the β-lactone (***141***) followed by treatment with *n*-butyllithium and benzylamine, giving 3*R*-(*N*-benzylamino)butanoic acid, *R*-***140***.[88]

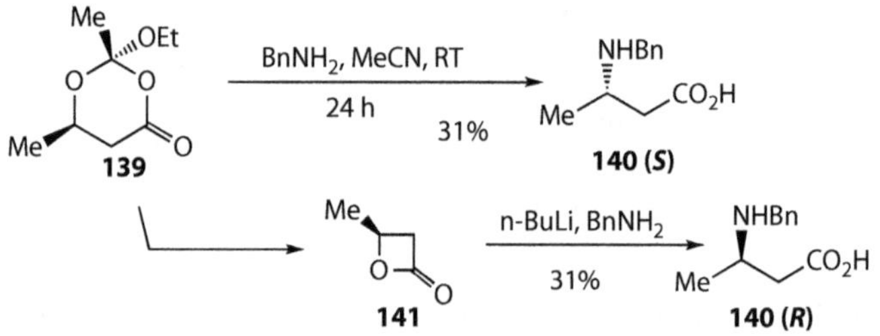

A benzene ring can be oxidized to a carboxylic acid moiety, making aryl-substituted alkyl amines putative sources of amino acids. In one interesting example, chiral, nonracemic amine ***142*** was the chiral template, and the amine group was protected as its *N*-Boc derivative. This was followed by Birch reduction[50] to give ***143***.[89] Ozonolysis cleaved the vinyl ether moieties in the ring to esters, and catalytic hydrogenation gave ***144***. *R*-3-aminobutanoic acid (***145***) was the final product after deprotection, decarboxylation, and hydrolysis.[89]

[88] Griesbeck, A.; Seebach, D. *Helv. Chim. Acta* ***1987***, *70*, 1326.
[89] Bringmann, G.; Geuder, T. *Synthesis* ***1991***, 829.

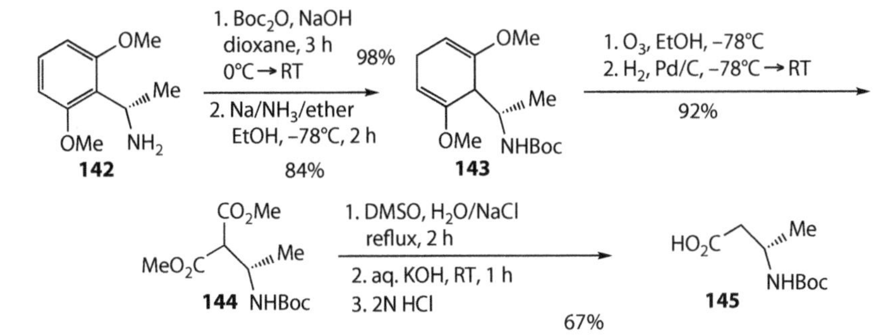

An interesting transformation used citronellic acid (***146***, derived from pulegone)[90] as a template. The acid moiety was converted to an amine and the alkene moiety was converted to a nitrile and thereby to an acid.[91] This transformation was accomplished by reaction of ***146*** with urea to give ***147***, and reduction followed by acylation and ozonolysis (with a reductive workup) gave ***148***.[91] Conversion to the chloride and displacement with cyanide gave ***149***, and acid hydrolysis led to 5-methyl-7-aminoheptanoic acid, ***150***. In this lengthy sequence, the carboxylic acid moiety in ***146*** functioned as the amine precursor and the alkenyl group functioned as the eventual acid moiety.

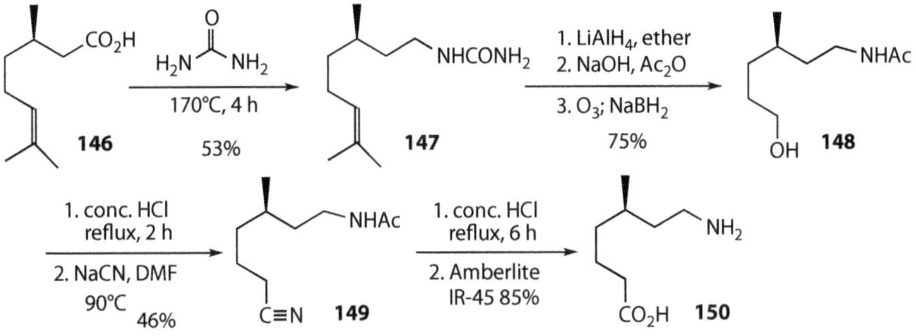

In addition to oxidative cleavage of alkenes by ozone, another cleavage method is illustrated by the conversion of the alcohol moiety in ***151*** to a mesylate and then to azide ***52***.[92] Oxidative cleavage with periodate and ruthenium trichloride was followed by hydrogenation of the azide to give 3-amino-2-methylpentanoic acid, ***153***.

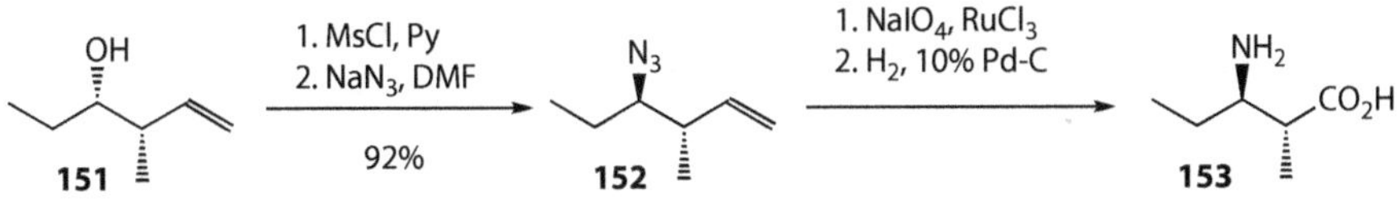

Displacement reactions in allylic systems are well known. In some cases, a nucleophile attacks the C=C moiety rather than the C–X moiety in an allylic system

[90] Lukes, R.; Zabácová, A.; Plesek, J. *Croat. Chem. Acta* ***1957***, *29*, 201 [*Chem. Abstr.* ***1959***, *53*, 17898c.
[91] Overberger, C.G.; Takekoshi, T. *Macromolecules* ***1968***, *1*, 1.
[92] Bates, R.B.; Gangwar, S. *Tetrahedron Asym.* ***1993***, *4*, 69.

in what is known as an S_N2' reaction.[93] The use of allylic acetates with malonate derivatives, catalyzed by palladium(0), is one example. Reaction of ***154*** with diethyl malonate, catalyzed by palladium, led to ethyl 6-(*N*-Boc amino)-2-carboethoxy-8-methylnon-4*E*-enoate, ***155***.[94] It is noted that enantiopure ***154*** must be prepared for its use as a template in this reaction. Similarly prepared was the 6-amino-6-(4-benzyloxyphenyl) derivative.

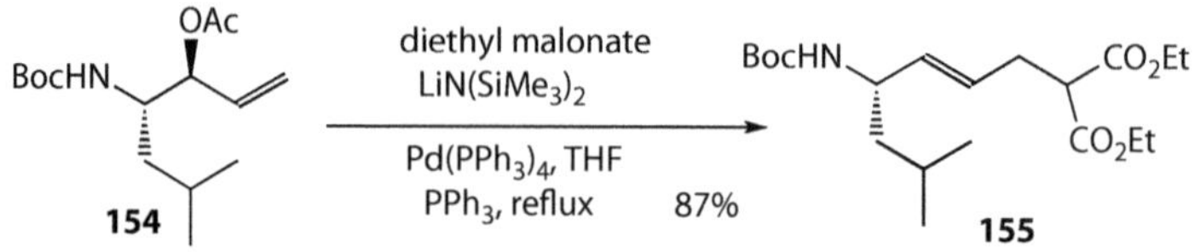

A lengthy synthetic route converted α-bromoacetic acid to isoserine. Chiral, nonracemic alcohol ***156*** (which is a menthol-like derivative), reacted with bromoacetic acid to give the ester, ***157***. The "alcohol" becomes the chiral auxiliary in the chiral, nonracemic ester.[95] Reaction of the bromide with silver nitrate and then sodium acetate and nitromethane/potassium fluroide (KF) led to azido-hydroxyester ***158***, with >95%de. Protection of the alcohol was followed by reduction of the azide. Hydrolysis and deprotection gave *S*-(–)-isoserine (***71***).[95] This example demonstrates that even simple achiral molecules may require lengthy syntheses in order to produce a chiral, nonracemic product from an achiral starting material.

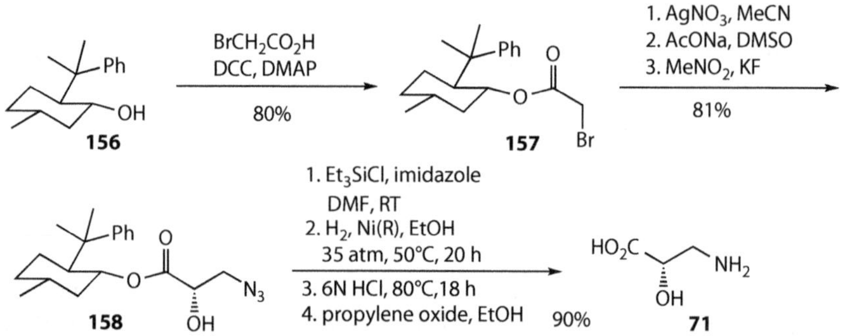

Several γ-heteroatom-substituted β-keto-esters were reduced using a ruthenium catalyst with a chiral ligand, and the substrates included γ-ether, sulfone, and carbamate-protected amino moieties. The example shown reduced the oxo moiety in ***159*** to give ***160*** in quantitative yield, and with 98%ee.[96]

---

[93] (a) Magid, R.M. *Tetrahedron* ***1980***, *36*, 1901; (b) Bordwell, F.G.; Pagani, G.A. *J. Am. Chem. Soc.* ***1975***, *97*, 118; (c) Bordwell, F.G.; Mecca, T.G. *J. Am. Chem. Soc.* ***1975***, *97*, 123, 127; (d) Bordwell, F.G.; Wiley, P.F.; Mecca, T.G. *J. Am. Chem. Soc.* ***1975***, *97*, 132.

[94] Thompson, W.J.; Tucker, T.J.; Schwering, J.E.; Barnes, J.L. *Tetrahedron Lett.* ***1990***, *31*, 6819.

[95] Solladie-Cavallo, A.; Khiar, N. *Tetrahedron Lett.* ***1988***, *29*, 2189.

[96] Fan, W.; Li, W.; Ma, X.; Tao, X.; Li, X.; Yao, Y.; Xie, X.; Zhang, Z. *J. Org. Chem.* ***2011***, *76*, 9444.

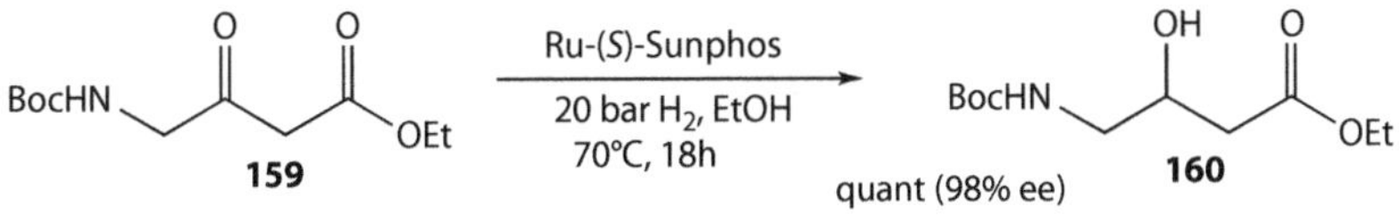

5.4 TRANSFORMATIONS USING CARBOHYDRATES AS CHIRAL TEMPLATES

In Hanessian's monograph,[1] carbohydrates are shown to be important chiral templates in many total syntheses. The chirality and functionality inherent to carbohydrates make them ideal for this purpose. Carbohydrates can also be used as chiral templates to prepare amino acids. Furanose ***161*** was converted to ***162*** in seven steps, and ***161*** was also converted to ***163*** in seven steps.

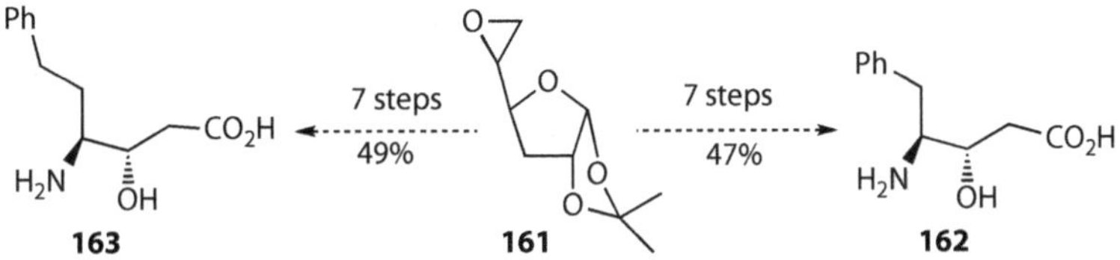

Tartaric acid is certainly related to carbohydrates since it is a glycaric acid (an aldonic acid), and an asymmetric synthesis of an amino acid was reported using tartrate derivative ***164***, which was converted to ***165***.[97] Reduction of the bromomethyl moiety, removal of one acetate-protecting group, and conversion of the other to a tosylate gave ***166***. Treatment of this mono-tosylate with methoxide generated an alkoxide that displaced the tosyl group to give an epoxide, and this was converted to azide ***167***. Hydrogenation of the azide gave methyl 3-amino-2-hydroxybutanoate (***168***). In another synthesis, *L*-(+)-tartaric acid was converted to an amide-ester, and then acetylated to give ***169***.[98] Dehydration of the amide gave nitrile ***170***, and

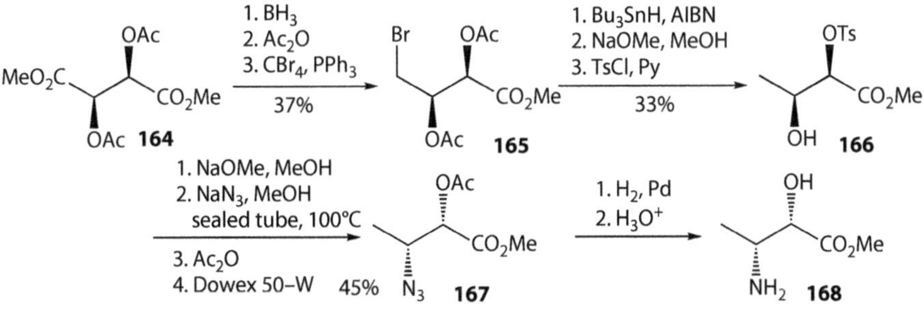

[97] Umemura, E.; Tsuchiya, T.; Umezawa, S. *J. Antibiot.* ***1988***, *41*, 530.
[98] Yokoo, A.; Akutagawa, S. *Bull. Chem. Soc. Jpn.* ***1962***, *35*, 644.

subsequent acid hydrolysis gave a mixture of 14% of ***171*** (4-amino-2,3-dihydroxybutanoic acid) and 19% of *S-**18*** (4-amino-2-hydroxybutanoic acid). The poor yields in the final steps and the mixture of hydroxy amino acids make this approach less useful.

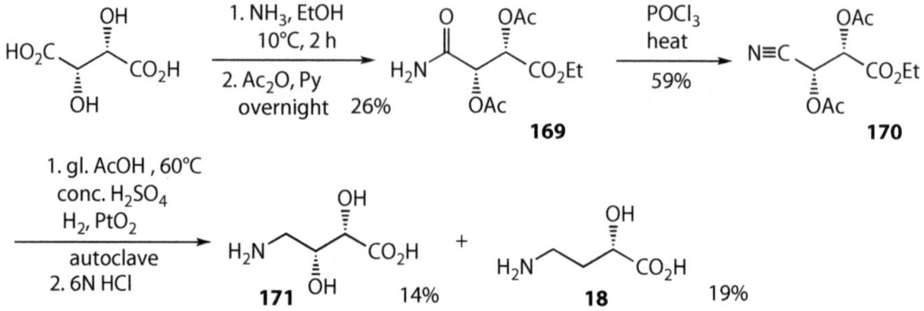

In another lengthy transformation using a carbohydrate template, ***172*** was converted to azide ***173*** and the pyran ring was opened to give ***174***.[99] Unmasking the aldehyde and oxidation to an acid allowed its conversion to ***175***. Protection and an internal Mitsunobu reaction[86,87] led to aziridine ***176***. This was converted to 5-(*N*-Boc-amino)-4-benzyloxy-6-phenylhexanoic acid (***177***) by opening the aziridine ring and protecting the functional groups.[99]

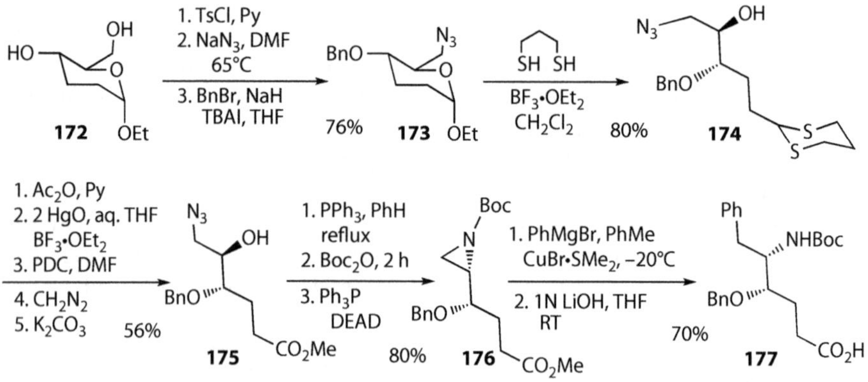

5.5 ENZYMATIC TRANSFORMATIONS

There are many enzymatic processes that may be used for the preparation of amino acids. Enzymatic kinetic resolution and the action of lipases are quite important for preparation of enantiopure β-amino acids, for example.[100]

The carbonyl moiety of oxo amino acids can be reduced to give hydroxy amino acids. Many common reducing agents do not give highly diastereoselective reductions in the absence of steric effects or neighboring group effects. Enzymatic reductions,

[99] Chakraborty, T.K.; Gangakhedkar, K.K. *Tetrahedron Lett.* ***1991***, *32*, 1897.
[100] For a review, see Liljeblad, A.; Kanerva, L.T. *Tetrahedron* ***2006***, *62*, 5831.

however, often proceed with excellent enantioselectivity and good to excellent diastereoselectivity. In one example, exposure to the organism *Saccharomyces carlsbergensis* ATCC 2345 led to reduction of the ketone moiety selectively in methyl 2-oxo-4-aminobutanoate (***178***), giving methyl 4-amino-2*S*-hydroxybutanoate, ***179***.[101]

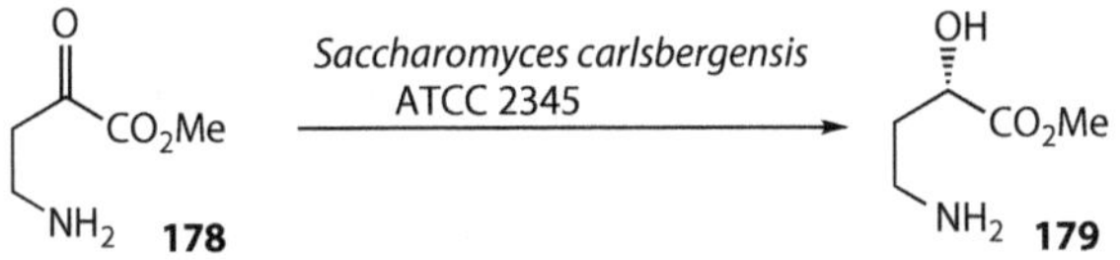

178 **179**

Specific enzymatic processes can introduce functionality in an enantioselective process. *N*-Benzyl-2-pyrrolidinone (***180***) was hydroxylated directly using an available enzymatic system, *Beauveria sulfurescens* (ATCC 7159).[102] Hydroxy-lactam ***181*** was subsequently hydrolyzed and debenzylated to give 2*S*-hydroxy-4-aminobutanoic acid, ***18***.

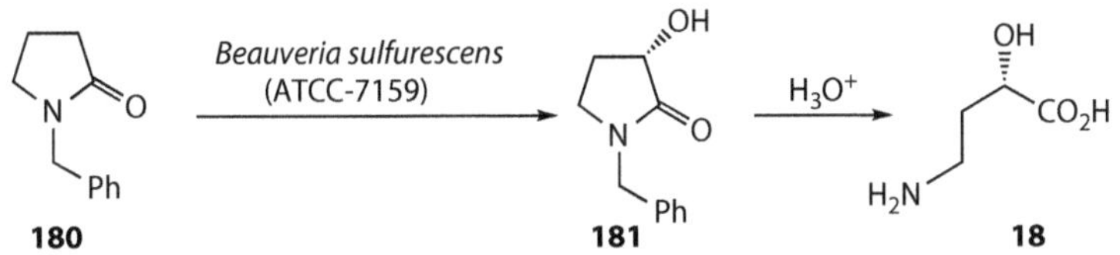

180 **181** **18**

One method for preparing chiral, nonracemic molecules is by kinetic resolution. If a reaction occurs with one enantiomer faster than with its antipode, that enantiomer forms a product and its antipode remains unchanged. Since these different diastereomers can be separated, resolution can be achieved. When diester ***182*** was treated with pig liver esterase, a 46% yield of half-ester (***183***) was obtained (since only 50% of the ester could react, 50% represents 100% reaction and a 46% yield represents a 92% yield based on reactivity).[103] The unreacted diester was separated from the half-ester product of the enzymatic hydrolysis. Reaction of the acid moiety in ***183*** with diphenylphosphoryl azide (DPPA) followed by conversion of the resulting amine to its carbamate (with methanol) resulted in ***184***. Acid hydrolysis cleaved the carbamoyl group to give ***185*** (3*S*-methyl 4-aminobutanoic acid) with high overall enantioselectivity.[103] It is noted that when ***183*** reacted with ammonia and an iodoso compound, a 60% yield of 3*R*-methyl 4-aminobutanoic acid was obtained.[103]

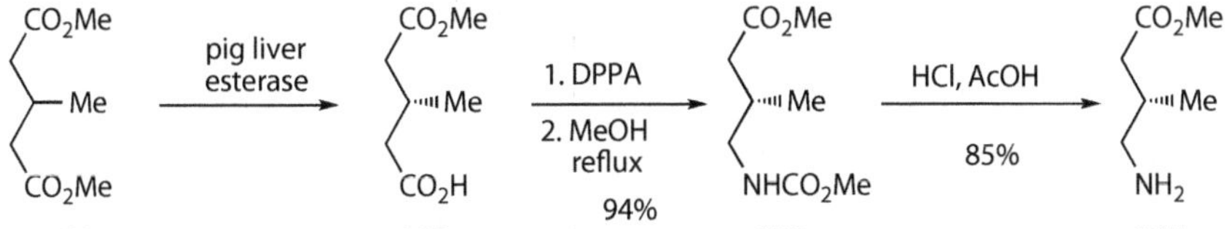

182 **183** **184** **185**

[101] Harris, K.J.; Sih, C.J. *Biocatalysis* ***1992***, *5*, 195.
[102] Srairi, D.; Maurey, G. *Bull. Soc. Chim. Fr.* ***1987***, 297.
[103] Andruszkiewicz, R.; Barrett, A.G.M.; Silverman, R.B. *Synth. Commun.* ***1990***, *20*, 159.

The desymmetrization of dimethyl 3-(benzylamino)glutarate (***186***) was accomplished via aminolysis using lipase B from *Candida antarctica*. Reaction with benzylamine and the lipase B gave ***187*** in 92% yield.[104] This enantiopure product was subsequently converted to 3,4-diaminobutanoic acid ***188*** in five steps, as shown. Hofmann rearrangement[105] of the amide to the amine was followed by deprotection, which was complicated by lactam formation, requiring the final acid hydrolysis step. Other lipases resolved racemic methyl 3-*tert*-butoxycarbonylaminobutanoate and allowed the isolation of (+)-methyl 3-*tert*-butoxycarbonylaminobutanoate.[106] Enzymatic desymmetrization of 3-alkyl- and 3-arylglutaronitriles was reported using *Rhodococcus* sp. AJ270 cells to give (*S*)-3-substituted 4-cyanobutanoic acids with good enantioselectivity in the presence of acetone.[107]

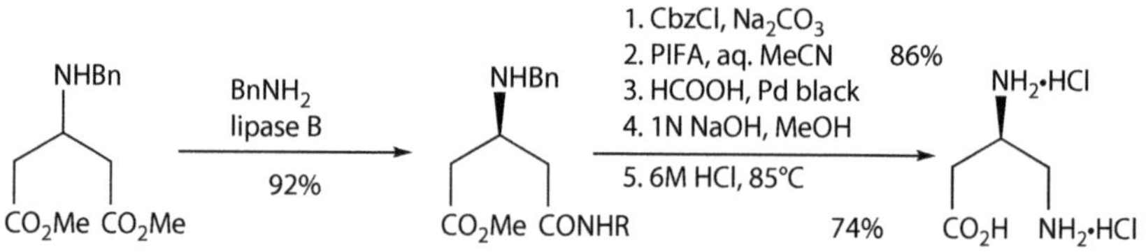

186 **187** **188**

The enzymatic hydrolysis of racemic *N*-phenylacetyl α-alkyl β-amino acids with penicillin G acylase (PGA) in water, with $CaCO_3$ at pH 11, led to preferential hydrolysis of the *S*-isomer. The resulting products were enantio-enriched *S*-α-alkyl β-amino acid and the *R*-*N*-phenylacetyl starting material that was not hydrolyzed under these conditions.[108] The yields for this kinetic resolution process were greater than 90%.

The penicillin acylase-catalyzed hydrolysis of *N*-phenylacetyl β-aryl-β-amino acid derivatives led to rapid hydrolysis of the (*R*)-enantiomer to the corresponding β-aryl-β-amino acids.[109] The less reactive (*S*)-enantiomer was recovered as the *N*-phenylacetyl derivative. Kinetic resolution of the phenylacetamides of β-amino acids is also known, using penicillin G amidase.[110]

Isomerases are an important class of enzymes. One from a *Taxus* species, phenylalanine aminomutase, isomerizes α-amino acids bearing an aryl group to give chiral, β-amino acids. An example is the conversion of 4-methylphenylalanine (***189***) to ***190***.[111]

---

[104] López-García, M.; Alfonso, I.; Gotor, V. *J. Org. Chem.* ***2003***, *68*, 648.

[105] (a) Hofmann, A.W. *Ber.* ***1881***, *14*, 2725; (b) Wallis, E.S.; Lane, J.F. *Org. React.* ***1949***, *3*, 267; (c) see Smith, M.B. *Organic Synthesis*, 3rd ed. Wavefunction, Inc./Elsevier, Irvine, CA/London, England, ***2010***, pp. 190–192; (d) Smith, M.B. *March's Advanced Organic Chemistry*, 7th ed. John Wiley & Sons, Hoboken, NJ, ***2013***, pp. 1360–1361.

[106] Chisso Corp. *U.S. Pat.* 5 518 903, *Process Biochem.* ***1997***, *32*, 72.

[107] Wang, M.-X.; Liu, C.-S.; Li, J.-S. *Tetrahedron Asym.* ***2002***, *12*, 3367.

[108] Cardillo, G.; Tolomelli, A.; Tomasini, C. *J. Org. Chem.* ***1996***, *61*, 8651.

[109] Soloshonok, V.A.; Fokina, N.A.; Rybakova, A.V.; Shishkina, I.P.; Galushko, S.V.; Sorochinsky, A.E.; Kukhar, V.P.; Savchenko, M.V.; Švedas, V.K. *Tetrahedron Asym.* ***1995***, *6*, 1601.

[110] (a) Soloshonok, V.A.; Svedas, V.K.; Kukhlar, V.P.; Kirilenko, A.G.; Rybakova, A.V.; Solodenko, V.A.; Fokina, N.A.; Kogut, O.V.; Galaev, I.Y.; Kozlova, E.V.; Shishkina, I.P.; Galushko, S.V. *Synlett* ***1993***, 339; (b) Soloshonok, V.A.; Fokina, N.A.; Rybakova, A.V.; Shishkina, I.P.; Galushko, S.V.; Sorochinsky, A.E.; Kukhlar, V.P.; Savchenko, M.V.; Svedas, K.V. *Tetrahedron Asym.* ***1995***, *7*, 1601.

[111] Klettke K.L.; Sanyal, S.; Mutatu, W.; Walker, K.D. *J. Am. Chem. Soc.* ***2007***, *129*, 6988.

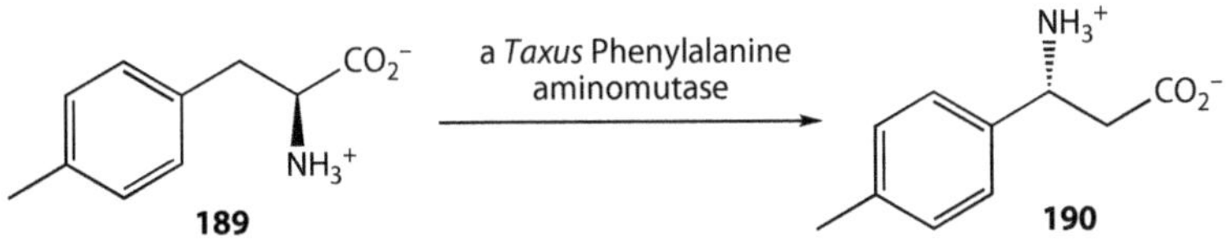

3-Aryl conjugated acids are converted to a mixture of α- and β-amino acids with the enzyme phenylalanine aminomutase. In the example shown, ***191*** gave mostly ***193***, with small amounts of ***192***.[112] In addition, ***193*** was formed in >99%ee.[112] With other substituents on the benzene ring or different reaction conditions, a higher percentage of ***192*** is obtained. Indeed, in some cases, the α-amino acid is the major product.

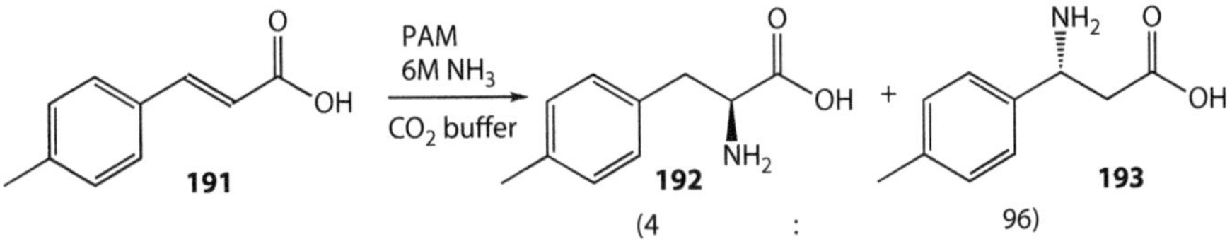

[112] Szymanski, W.; Wu, B.; Weiner, B.; de Wildeman, S.; Feringa, B.L.; Janssen, D.B. *J. Org. Chem.* ***2009***, *74*, 9152.

6 Biologically Important Amino Acids

Non-α-amino acids are known to play a prominent role in nature and in important biological compounds. Indeed, there are many examples of non-α-amino acids that possess important biological properties. This chapter discusses several types of these amino acids in order to illustrate both the impetus for synthesizing amino acids and the various structural types that are found. Those synthetic approaches presented in this chapter span the range of methods presented in the first five chapters, and include some new strategies as well.

6.1 PEPTIDES AND PROTEINS

6.1.1 As Residues in Peptides and Proteins

Natural peptides are obviously important biomolecules, and the structure is usually a poly(α-amino acid) structure such as *1*. Non-α-amino acids are found in many proteins, or they are the target of important enzymes. This section will describe examples in which these amino acids are important components of peptides or proteins. The examples are presented in a somewhat random order in one sense, but they are presented in this manner to illustrate that incorporation of non-α-amino acids is rather widespread.

There is considerable research on the preparation and study of peptides consisting of homologated proteinogenic amino acids.[1] There are unnatural β-peptides that contain β-amino acids (see ***2*** or ***3***). Conformational studies have been reported for β-amino acids and the so-called β-peptides[2] generated from such amino acids.[3] Peptides such as ***2*** are known as β^3-peptides,[4] and ***3*** are known as β^2-peptides.[5] γ-Peptides such as ***4*** (γ^4-peptides) are known that contain γ-amino acids.[6] As stated by Voss and Ramos, "unlike natural α-peptide chains, which fold and unfold in a

---

[1] Voss, G.; Ramos, G. *Chemistry of Crop Protection: Progress and Prospects in Science and Regulation*. Wiley-VCH Verlag GmbH & Co. KgaA, Weinheim, FRG, ***2004***, pp. 17–29.

[2] Cheng, R.P.; Gellman, S.H.; DeGrado, W.F. *Chem. Rev.* ***2001***, *101*, 3219.

[3] Beke, T.; Somlai, C.; Magyarfalvi, G.; Perczel, A.; Tarczay, G. *J. Phys. Chem. B* **2009**, *113*, 7918.

[4] Matthews, J.L.; Braun, C.; Guibourdenche, C.; Overhand, M.; Seebach, D., in *Enantioselective Synthesis of b-Amino Acids*, Juaristi, E., ed. Wiley-VCH, New York, ***1997***, pp. 105–126.

[5] (a) Seebach, D.; Sifferlen, T.; Mathieu, P.A.; Häne, A.M.; Krell, C.M.; Bierbaum, D.J.; Abele, S. *Helv. Chim. Acta* ***2000***, *83*, 2849; (b) Micuch, P.; Seebach, D. *Helv. Chim. Acta* ***2002***, *85*, 1567; (c) Seebach, D.; Abele, S.; Gademann, K.; Guichard, G.; Hintermann, T.; Jaun, B.; Matthews, J.L.; Schreiber, J.V.; Oberer, L.; Hommel, U.; Widmer, H. *Helv. Chim. Acta* ***1998***, *81*, 932.

[6] Hintermann, T.; Gademann, K.; Jaun, B.; Seebach, D. *Helv. Chim. Acta* ***1998***, *81*, 983.

cooperative way, β-peptide folding is non-cooperative.[7] The shape, the handedness, the resulting helices are all different[8] as we go from α- to β- to γ-peptides, with increasing stability of the secondary structures in this order.[9]”[1]

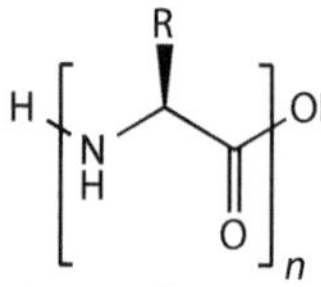

1

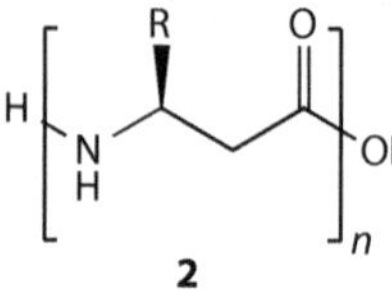

2

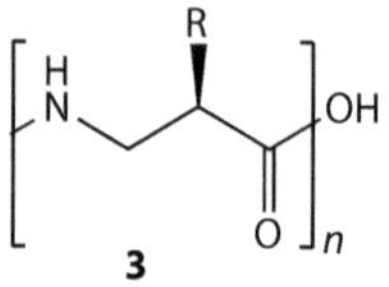

3

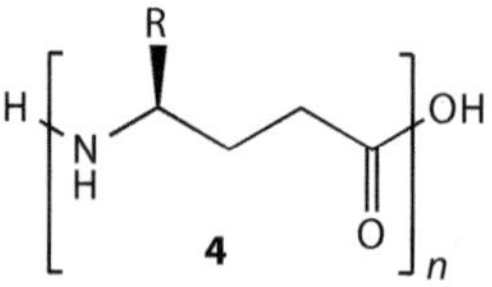

4

A more general way to view such peptides is that β, γ, . . . , ω peptides contain homologated residues. In such peptides, incorporation of the additional atoms between peptide units, when compared to α-amino acids, will increase the degrees of torsional freedom, and expand the number of conformations that are available to the polypeptide.[10,11] For β-peptides, new classes of folded polyamide structures (foldamers) are generated.[12] It has been observed that incorporation of non-α-amino acids into the strand segments of a peptide hairpin does not disrupt the overall fold of the peptide.[13]

Considerable work has been done that relates to peptides that either contain non-α-amino acid residues or are composed entirely of non-α-amino acid residues. Synthetic peptides have been prepared that contain non-α-amino residues. One example was the preparation of ω-amino acid oligomers, containing 6-aminohexanoic acid residues, which were shown to form helical structures.[14] Another example is the class of synthetic peptides that are conjugated to 9-aminoacridine (**5**). Some of the peptides contain non-α-amino acids, including 6-aminohexanoic acid units.[15] These conjugates were studied for their DNA-binding capability and were shown to be weak DNA binders.[15] These conjugates may modulate DNA interaction with strong bis(acridine) binders.

---

[7] (a) Gademann, K.; Jaun, B.; Seebach, D.; Perozzo, R.; Scapozza, L.; Folkers, G. *Helv. Chim. Acta* ***1999***, *82*, 1; (b) Seebach, D.; Beck, A.K.; Brenner, M.; Gaul, C.; Heckel, A. *Chimia* ***2001***, *55*, 831.

[8] Seebach, D.; Beck, A.K.; Rueping, M.; Schreiber, J.V.; Seilner, H. *Chimia* ***2001***, *55*, 98.

[9] Seebach, D.; Beck, A.K.; Brenner, M.; Gaul, C.; Heckel, A. *Chimia* ***2001***, *55*, 831.

[10] Vasudev, P.G.; Chatterjee, S.; Shamala, N.; Balaram, P. *Chem. Rev.* ***2011***, *111*, 657.

[11] (a) Seebach, D.; Matthews, J.L. *Chem. Commun.* ***1997***, *21*, 2015; (b) Seebach, D., Beck, A.K.; Bierbaum, D.J. *Chem. Biodiversity* ***2004***, *1*, 1111; (c) Kimmerlin, T.; Seebach, D. *J. Pept. Res.* ***2005***, *65*, 229; (d) Seebach, D.; Hook, D.F.; Glattli, A. *Biopolymers* ***2006***, *84*, 23; (e) Lelais, G.; Seebach, D. *Biopolymers* ***2004***, *76*, 20.

[12] (a) Gellman, S.H. *Acc. Chem. Res.* ***1998***, *31*, 173; (b) Cheng, R.P.; Gellman, S.H.; DeGrado, W.F. *Chem. Rev.* ***2001***, *101*, 3219.

[13] Roy, R.S; Gopi H.N.; Raghothama, S.; Karle, I.L; Balaram, P. *Chem. Eur. J.* ***2006***, *12*, 3295.

[14] Schramm, P.; Hofmann, H.-J. *J. Mol. Struct. THEOCHEM* ***2009***, *907*, 109.

[15] Šebestík, J.; Stibor, I; Hlaváček, J. *J. Peptide Sci.* ***2006***, *12*, 472.

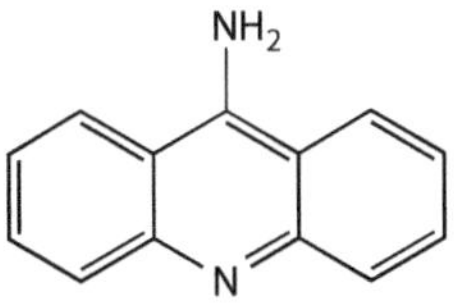

5

The growth factor receptor-bound protein (Grb2) is an SH2 domain-containing docking module, and several synthetic Grb2-SH2 domain inhibitors have been prepared for their anticancer properties. A 5-aminopentanoic acid linker was incorporated into a nonphosphorylated cyclic pentapeptide scaffold to generate several new inhibitors.[16] Four protected blocked tripeptides that contained the 5-aminopentanoic acid (δ-aminovaleric acid) residue were shown to self-assemble to form supramolecular β-sheet structures.[17] This self-aggregation led to formation of fibrillar structures that resemble neurodegenerative disease causing amyloid fibrils.

Elastase (a serine protease that hydrolyzes amides and esters, produced in the pancreas as an inactive zymogen, and activated in the duodenum by trypsin) substrates were prepared with peptide sequences that included 4-aminobutyric acid or 5-aminovaleric acid spacer units. These prodrugs containing an Ala-Ala-Ala sequence were effective substrates for both porcine pancreatic elastase and human polymorphonuclear elastase, but hydrolytic cleavage occurred exclusively at the ester bond, not at the peptide-spacer bond.[18] Further work was done to address this issue.

"Human melanin-concentrating hormone (hMCH) and many of its analogues are potent but nonspecific ligands for human melanin-concentrating hormone receptors 1 and 2 (hMCH-1R and hMCH-2R)."[19] Analogs of the nonselective agonist Ac-Arg6-cyclo(S–S)(Cys7-Met8-Leu9-Gly10-Arg11-Val12-Tyr13-Arg14-Pro15-Cys16)-NH_2 were prepared by incorporation of 5-aminovaleric acid (Ava) in place of the Leu9-Gly10 or Arg14-Pro15 segments, and they were "high affinity antagonists selective for hMCH-1R."[19]

Neuropeptide B (NPB) is an endogenous ligand for GPR7 (NPBW1; G-protein-coupled receptor 7) and GPR8 (NPBW2; G-protein-coupled receptor 7), with relatively high selectivity for GPR7.[20] Several NPB analogs were generated by replacing the consecutive amino acids from Pro4 to Val13 with several units of 5-aminopentanoic acid, and they exhibited potent agonistic activities for GPR7 expressed in HEK293 cells.[21]

Relaxin analogs have been prepared that have a modified A chain loop, and one of the new analogs contained a ω-aminooctanoic acid, but this led to a significant

[16] (a) Song, Y.-L.; Peach, M.L.; Roller, P.P.; Qiu, S.; Wang, S.; Long, Y.-Q. *J. Med. Chem.* ***2006***, *49*, 1585; (b) Song, Y.-Li; Tan, J.; Luo, X.-M.; Long, Y.-Q. *Org. Biomol. Chem.* ***2006***, *4*, 659.

[17] Banerjee, A.; Das, A.K.; Drew, Michael G.B.; Banerjee, A. *Tetrahedron* ***2005***, *61*, 5906.

[18] Achilles, K. *Archiv. Pharmazie* (Weinheim), ***2002***, *335*, 325.

[19] Bednarek, M.A.; Hreniuk, D.L.; Tan, C.; Palyha, O.C.; MacNeil, D.J.; Van der Ploeg, L.H.Y.; Howard, A.D.; Feighner, S.D. *Biochemistry* ***2002***, *41*, 6383.

[20] Singh, G.; Davenport, A.P. *Br. J. Pharmacol.* ***2006***, *148*, 1033.

[21] Maki, K.; Masao, M.; Atsushi, H.; Kenichi, T.; Akio, K.; Tokita Shigeru, T. *J. Peptide Sci.* ***2007***, *13*, 379.

loss of biological activity.[22] Relaxin is a hormone produced by the ovaries during pregnancy that causes pelvic and cervical expansion and relaxation.[23]

New analogs of the *N*-formyl and *N*-Boc derivatives of the tripeptide Met-Leu-Phe-OMe have been prepared in which the hydrophobic leucine residue was replaced with 5-aminopentanoic acid, 4-aminobutanoic acid, or 6-aminohexanoic acid.[24] Two of these analogs show good and selective antagonist activity on superoxide anion production. The tripeptide Met-Leu-Phe-OMe "peptides are involved in the defense mechanism against bacterial infections through binding with specific receptors located on the neutrophil membranes. In addition to the cell-directed migration (chemotaxis), the peptide receptor interaction gives rise to a series of biochemical events including production of superoxide anions and lysosomal enzyme release."[24]

A family of enzymes that includes PLA2 (phospholipase A2) catalyze the hydrolysis of the *sn*-2 fatty acyl bond of phospholipids to liberate free fatty acid and lysophospholipid.[25] This enzyme plays a role in brain pathology in patients suffering from Alzheimer's disease.[26] An indomethacin prodrug known as DP-155 (***6***) has been prepared in which indomethacin (***7***; a nonsteroidal anti-inflammatory drug (NSAID) is connected to lecithin via a 6-aminohexanoic acid linker, at the *sn*-2 position. Another way to say this is that DP-155 is a phospholipid prodrug of ***7***. In vivo, DP-155 is cleaved to indomethacin by PLA2 in the brain and in the GI tract.[27]

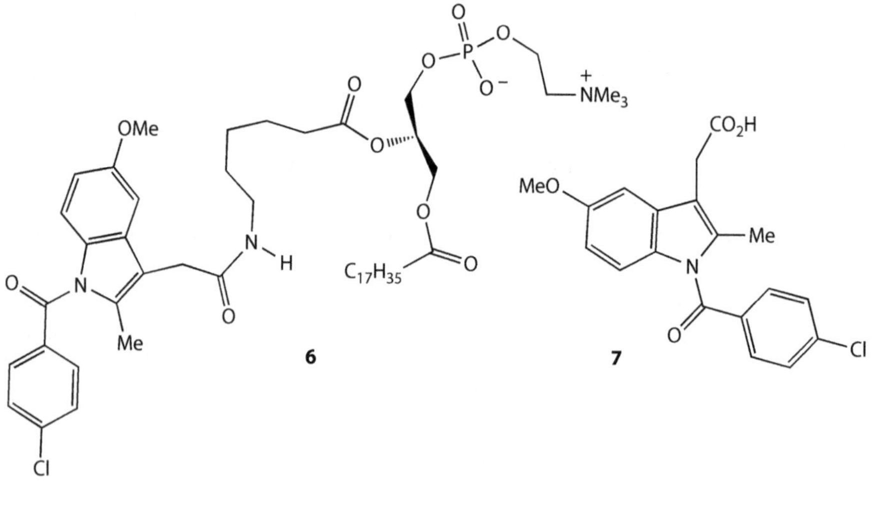

6 **7**

[22] Büllesbach, E.E.; Schwabe, C. *Int. J. Peptide Protein Res.* ***1995****, 46*, *238.*

[23] Conrad, K.P. *Am. J. Physiol. Regul. Integr. Comp. Physiol.* ***2011****, 301*, R267.

[24] Pagani Zecchini, G.; Morera, E.; Nalli, M.; Paglialunga Paradisi, M.; Lucente, G.; Spisani, S. *Farmaco* ***2001****, 56*, 851. Also see Spisani, S.; Fraulini, A.; Varani, K.; Falzarano, S.; Cavicchioni, G. *Eur. J. Pharmacol.* ***2007****, 567*, 171.

[25] Kudo, I.; Murakami, M. *Prostaglandins Other Lipid Mediat.* ***2002****, 68–69*, 3.

[26] (a) Stephenson, D.T.; Lemere, C.A.; Selkoe, D.J.; Clemens, J.A. *Neurobiol. Dis.* ***1996****, 3*, 51; Colangelo, V.; Schurr, J.; Ball, M.J.; Pelaez, R.P.; Bazan, N.G.; Lukiw, W.J. *J. Neurosci. Res.* ***2002****, 70*, 462; (c) Zipp, F.; Aktas, O. *Trends Neurosci.* ***2006****, 29*, 518.

[27] Dvir, E.; Elman, A.; Simmons, D.; Shapiro, I.; Duvdevani, R.; Dahan, A.; Hoffman, A.; Friedman, J.E. *CNS Drug Rev.* ***2007****, 13*, 260.

Natural products that exhibit interesting biological activity relating to peptides are known. A class of neuropeptides have reactive cell-surface receptors in malignant tissues, making them useful for cancer imaging and therapy.[28] Since many malignant tumors express neuropeptide receptors, they may be useful for imaging and therapy with radiolabeled synthetic antagonists.

The amphibian peptide bombesin[29] and also gastrin-releasing peptide have shown interesting biological activity.[30] They are potent autocrine growth factors in human small cell lung carcinoma. Bombesin analogs such as ***8*** have been prepared for radiolabeling with rhenium-188, for possible use for cancer imaging and therapy.[31] Analog ***8*** was prepared with a *N*-hydroxyl 6-aminohexanoic acid moiety. The ^{19}F labeled[32] and ^{64}Cu-labeled[33] analogs have also been prepared and the *in vivo* properties studied.

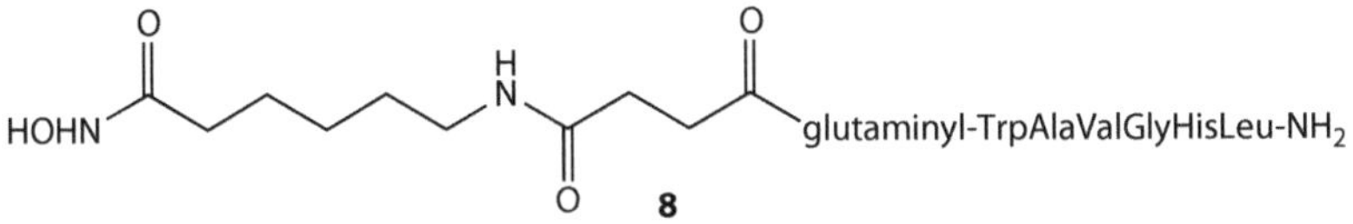

8

The last example in this section shows a slightly different use for non-α-amino acids, in the context of their incorporation into peptides or proteins. One method to probe proteins in their native context is to tag proteins with small, bright reporter groups. In one study, an engineered variant of *Escherichia coli* lipoic acid ligase (LplA) was attached covalently to a fluorescent substrate such as 7-hydroxycoumarin (***9***), with a 13-residue peptide recognition sequence genetically fused to a protein.[34] An attempt to improve the characteristics of the reporter group led to the synthesis of 7-aminocoumarin (***12***).[35] The acyl substitution reaction of ***9*** with 5-aminopentanoic acid led to ***10***. Subsequent conversion to the triflate allowed a Buchwald-Hartwig cross-coupling[36] reaction to prepare ***11***, and deprotection gave ***12***.

---

[28] Fischman, A.J., Babich, J.W., Strauss, H.W. *J. Nucl. Med.* ***1993***, *34*, 2253.

[29] Moody, T.W., Pert, C.B., Gazdar, A.F., Carney, D.N., Minna, J.D. *Science* ***1981***, *214*, 246.

[30] Coy, D.H.; Taylor, J.E.; Jiang, N.Y.; Kim, S.H.; Wang, L.H.; Huang, S.C.; Moreau, J.P.; Gardner, J.D.; Jensen, R.T. *J. Biol. Chem.* ***1989***, *264*, 14691.

[31] Safavy, A.; Khazaeli, M.B.; Qin, H.; Buchsbaum, D.J. *Cancer,* ***1997***, *80*, 2354.

[32] Höhne, A.; Mu, L.; Honer, M.; Schubiger, P.A.; Ametamey, S.M.; Graham, K.; Stellfeld, T.; Borkowski, S.; Berndorff, D.; Klar, U.; Voigtmann, U.; Cyr, J.E.; Friebe, M.; Dinkelborg, L.; Srinivasan, A. *Bioconjugate Chem.* ***2008***, *19*, 1871.

[33] Lane, S.R.; Nanda, P.; Rold, T.L.; Sieckman, G.L.; Figueroa, S.D.; Hoffman, T.J.; Jurisson, S.S.; Smith, C.J. *Nuc. Med. Biol.* ***2010***, *37*, 751.

[34] Uttamapinant, C.; White, K.A.; Baruah, H.; Thompson, S.; Fernandez-Suarez, M.; Puthenveetil, S.; Ting, A.Y. *Proc. Natl. Acad. Sci. USA* ***2010***, *107*, 10914.

[35] Jin, X.; Uttamapinant, C.; Ting, A.Y. *ChemBioChem* ***2011***, *12*, 65.

[36] See Anderson, K.W.; Tundel, R.E.; Ikawa, T.; Altman, R.A.; Buchwald, S.L. *Angew. Chem. Int. Ed.* ***2006***, *45*, 6523.

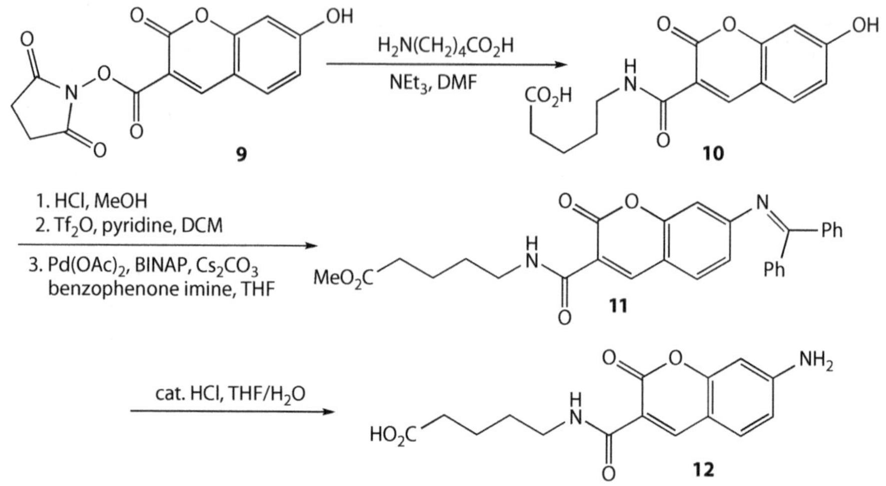

6.1.2 As Substrates for Peptides and Proteins

Non-α-amino acids not only appear as residues in the structure of peptides, but also serve as substrates for peptides and enzymes. They are incorporated into smaller molecules that often inhibit the action of proteins or enzymes. A few examples are presented to show that non-α-amino acids are important in molecules that are targeted by proteins or that target proteins.

The protein PepT1 transports small peptides,[37] for example, and is found in the wall of the small intestine of mammals. PepT1 transports certain drugs into the circulatory system,[38] especially amino acids, which allows them to be taken orally. Substrates for this protein include α-amino acids, di- and tripeptides, as well as 4-aminobutanoic acid, 5-aminopentanoic acid, and (4-aminophenyl)acetic acid.[39] The cysteine protease clostripain has been used as a biocatalyst for the synthesis of peptide isosteres, including the use of non-α-amino acids as substrates.[40]

GABA-related derivatives (4-aminobutanoic acid derivatives) have been used as human GIVA PLA_2 (group IVA cytosolic phospholipase A_2) inhibitors.[41] Several oxoamides such as ***13*** were prepared by oxidation of the alcohol precursor, followed

[37] Meredith, D.; Boyd, C.A.R. *J. Membrane Biol.* ***1995***, *145*, 1.

[38] (a) Amidon, G.L.; Lee, H.J. *Ann. Rev. Pharmacol. Toxicol.* ***1994***, *34*, 321; (b) Balimane, P.V.; Tamai, I.; Guo, A.; Nakanishi, T.; Kitada, H.; Leibach, F.H.; Tsuji, A.; Sinko, P.J. *Biochem. Biophys. Res. Commun.* ***1998***, *250*, 246; (c) Tamai, I.; Nakanishi, T.; Nakahara, H.; Sai, Y.; Ganapathy, V.; Leibach, F.H.; Tsuji, A. *J. Pharm. Sci.* ***1998***, *87*, 1542.

[39] Bailey, P.D.; Boyd, C.A.R.; Bronk, J.R.; Collier, I.D.; Meredith, D.; Morgan, K.M., Temple, C.S. *Angew. Chem.* ***2000***, *112*, 515.

[40] Guenther, R.; Stein, A.; Bordusa, F. *J. Org. Chem.* ***2000***, *65*, 1672.

[41] Kokotos, G.; Six, D.A.; Loukas, V.; Smith, T.; Constantinou-Kokotou, V.; Hadjipavlou-Litina, D.; Kotsovolou, S.; Chiou, A.; Beltzner, C.C.; Dennis, E.A. *J. Med. Chem.*, ***2004***, *47*(14), 3615.

by Wittig olefination[42] to give ***14***. Hydrogenation and conversion to amide ***15*** allowed deprotection and oxidation to give ketoamide ***16***.[41] These inhibitors block production of arachidonic acid and prostaglandin E_2 in cells, and also exhibit anti-inflammatory and analgesic activity.[41]

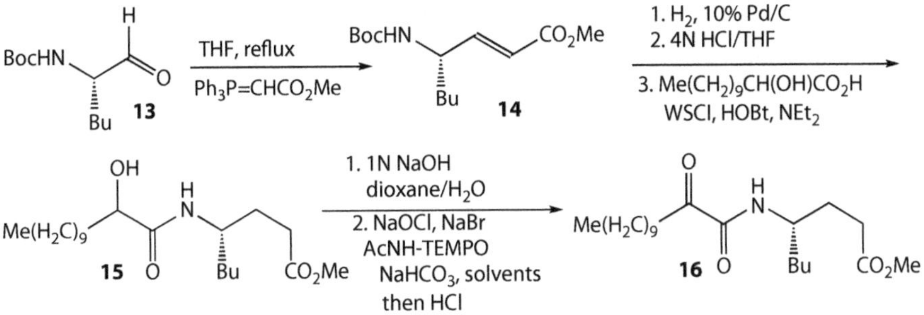

Pepstatin, ***17***,[43] is a potent inhibitor of pepsin, and a natural product that contains the statine residue, (3*S*,4*S*)-4-amino-3-hydroxy-6-methylheptanoic acid (***18***). It is known that ***18*** mimics the tetrahedral transition-state structure of peptide bond hydrolysis.[44]

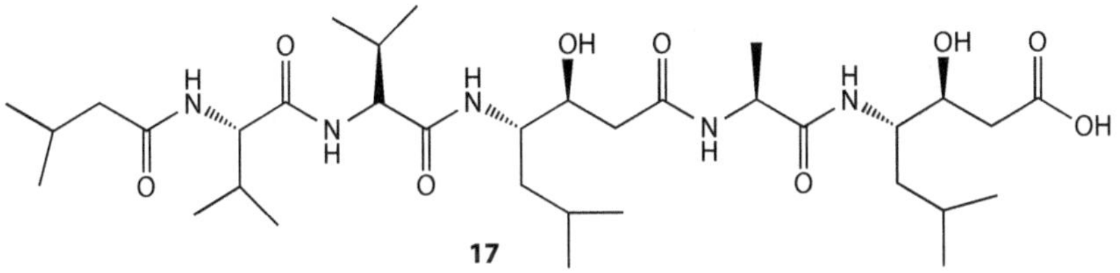

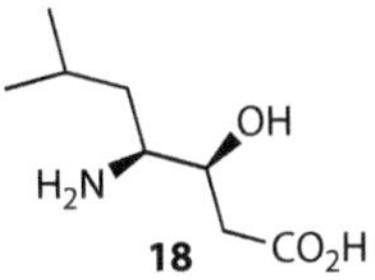

A new cell-permanent inhibitor of the β-secretase BASE1, which contributes to the production of the Aβ-peptide, has been prepared.[45] It has been suggested that the Aβ-peptide accumulates during the course of Alzheimer's disease and plays a major role in the disease.[46] This inhibitor contains a 6-aminohexanoic acid moiety as well as a statine analog (see Section 6.6), and may prevent production of Aβ-peptide

---

[42] See (a) Smith, M.B. *Organic Synthesis*, 3rd ed. Wavefunction, Inc./Elsevier, Irvine, CA/London, England, ***2010***, pp. 729–739; (b) Smith, M.B. *March's Advanced Organic Chemistry*, 7th ed. John Wiley & Sons, Hoboken, NJ, ***2013***, pp. 1165–1173.

[43] Umezawa, H.; Aoyagi, T.; Morishima, H.; Matsuzaki, M.; Hamada, M. *J. Antibiot.* ***1970***, *23*, 259.

[44] (a) Morishima, H.; Takita, T.; Aoyagi, T.; Takeuchi, T.; Umezawa, H.; *J. Antibiot.* ***1970***, *23*, 263; (b) Marks, N.; Grynbaum, A.; Lajtha, A. *Science* ***1973***, *181*, 949.

[45] Lefranc-Jullien, S.; Lisowski, V.; Hernandez, J.-F.; Martinez, J.; Checler, F. *Br. J. Pharmacol.* ***2005***, *145*, 228.

[46] Hardy, J.A.; Higgins, G.A. *Science* ***1992***, *256*, 184.

in vivo. Lipopeptides identified as the lobocyclamides were shown to incorporate the long-chain β-amino acids 3-aminooctanoic acid and 3-aminodecanoic acid.[47]

Dipeptide ***19***, known as aliskiren, is a nonpeptide peptidomimetic and the first marketed plasma renin inhibitor. Blockade of the renin angiotensin system (RAS) reduces blood pressure (BP), but a powerful counterregulatory mechanism is activated during RAS blockade. Many renin inhibitors have been prepared that have non-α-amino acid residues.[48] Renin levels are controlled by the amount of angiotensin II present in plasma and tissues, and renin inhibitors bind to the active site of renin and are currently used as a therapy for hypertension.[48]

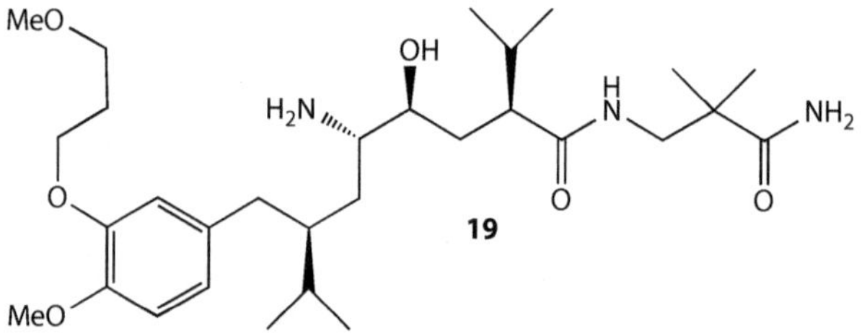

19

Aminocyclopentane derivatives such as ***20*** and ***21*** have been incorporated onto the RGD tripeptide sequence (Arg-Gly-Asp) to give integrin dual binders.[49] Integrins are heterodimeric transmembrane glycoproteins that attach cells to certain types of extracellular matrix proteins.[50] Such rigid motifs are thought to force peptides into bioactive conformations, which enhances stability toward enzymatic degradation. This work also included amino acid ***22***.[49]

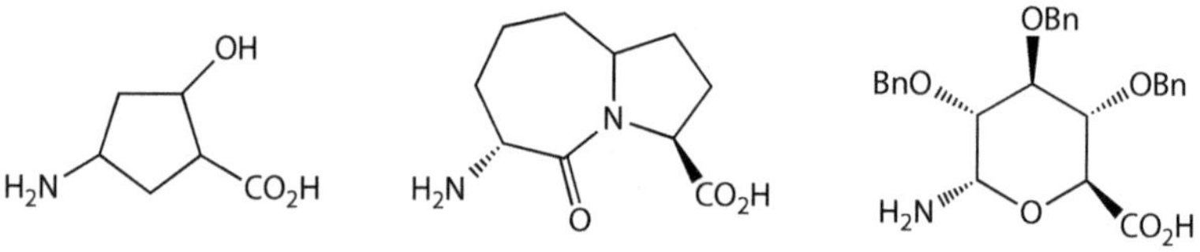

20 **21** **22**

Many small molecule kinase inhibitors are in clinical trials as anticancer drugs.[51] "It has been shown that conjugates of oligo-(*D*-arginine) peptides with adenosine-4'-dehydroxymethyl-4'-carboxylic acid (an ARC), 5-isoquinolinesulfonic acid,[52] or carbocyclic analogue of 3'-deoxyadenosine[53] are highly potent inhibitors of protein kinase A

---

[47] MacMillan, J.B.; Ernst-Russell, M.A.; de Ropp, J.S.; Molinski, T.F. *J. Org. Chem.* ***2002***, *67*, 8210.

[48] Webb, R.L.; Schiering, N.; Sedrani, R.; Maibaum, J. *J. Med. Chem.* ***2010***, *53*, 7490.

[49] Casiraghi, G.; Rassu, G.; Auzzas, L.; Burreddu, P.; Gaetani, E.; Battistini, L.; Zanardi, F.; Curti, C.; Nicastro, G.; Belvisi, L.; Motto, I.; Castorina, M.; Giannini, G.; Pisano, C. *J. Med. Chem.* ***2005***, *48*, 7675.

[50] See (a) Humphries, M.J. *Biochem. Soc. Trans.* ***2000*** *28*, 311; (b) Hynes, R. *Cell* ***2002***, *110*, 673.

[51] Gill, A.L.; Verdonk, M.; Boyle, R.G.; Taylor, R. *Curr. Top. Med. Chem.* ***2007***, *7*, 1408.

[52] Enkvist, E.; Lavogina, D.; Raidaru, G.; Vaasa, A.; Viil, I.; Lust, M.; Viht, K.; Uri, A. *J. Med. Chem.* ***2006***, *49*, 7150.

[53] Enkvist, E.; Raidaru, G.; Vaasa, A.; Pehk, T.; Lavogina, D.; Uri, A. *Bioorg. Med. Chem. Lett.* ***2007***, *17*, 5336.

(also known as cAMP-dependent protein kinase, cAPK or PKA (protein kinase A); a transferase class of enzymes)."[51] The crystal structure shows a complex of the catalytic subunit of PKA with a typical ARC type inhibitor, compound ***23*** (ARC-1034).[54] It is clear from structure ***23*** that the linking unit is 6-aminohexanoic acid.

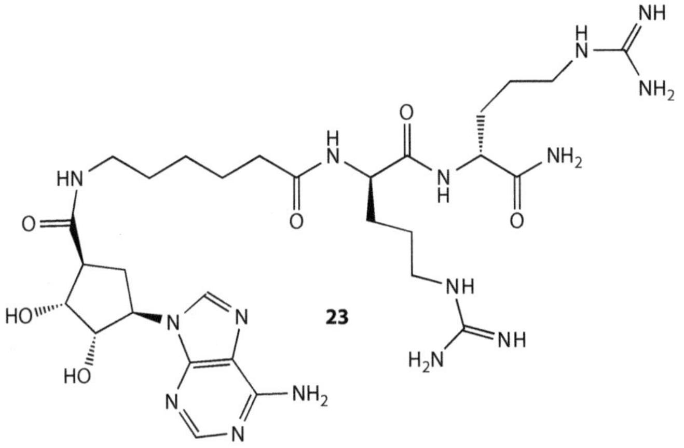

23

The preparation of oxoamide derivatives of aminobutanoic acid moieties such as ***24*** and ***25*** were shown to be inhibitors of group IVA cytosolic phospholipase A_2 (GIVA PLA_2), which is the rate-limiting provider of proinflammatory mediators.[55] These enzymes hydrolyze fatty acids from the *sn*-2 position of membrane phospholipids.[56]

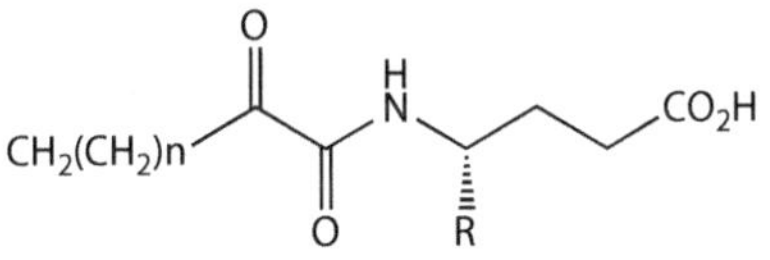

24 n = 13, (R = H)
25 n = 9, (R – C_4H_9)

Azasterols such as ***26*** have amino acids incorporated via an amide unit. These compounds were evaluated for their antiparasitic activity against organisms that cause trypanosomiasis, Chagas disease, leishmaniasis, and malaria.[57]

[54] Lavogina, D.; Lust, M.; Viil, I.; König, N.; Raidaru, G.; Rogozina, J.; Enkvist, E.; Uri, A.; Dirk, D. *J. Med. Chem.* ***2009***, *52*, 308.
[55] Kokotos, G.; Six, D.A.; Loukas, V.; Smith, T.; Constantinou-Kokotou, V.; Hadjipavlou-Litina, D.; Kotsovolou, S.; Chiou, A.; Beltzner, C.C.; Dennis, E.A. *J. Med. Chem.* ***2004***, *47*, 3615.
[56] See Burke, J.E.; Dennis, E.A. *J. Lipid. Res.* ***2009***, *50*, s237.
[57] Gros, L.; Lorente, S.O.; Jimenez, C.J.; Yardley, V.; Rattray, L.; Wharton, H.; Little, S.; Croft, S.L.; Ruiz-Perez, L.M.; Gonzalez-Pacanowska, D.; Gilbert, I.H. *J. Med. Chem.* ***2006***, *49*, 6094.

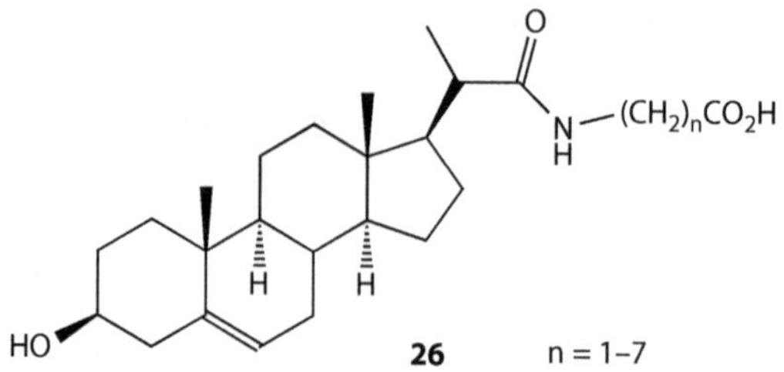

26

Amino acid AHMOA (***30***) is a key component of a renin inhibitory angiotensin peptide fragment.[58] *N*-Boc leucine was converted to an acid chloride by treatment with isobutyl chloroformate. Subsequent treatment with diazomethane gave the diazoketone (***27***), which was treated with silver benzoate. This led to a Wolff rearrangement[59] and formation of ***28***.[58] Reduction of the ester moiety to an aldehyde with diisobutylaluminum hydride allowed condensation with the lithium enolate of ethyl acetate to give ***29***. Saponification led to *N*-Boc-5-amino-3-hydroxy-7-methyloctanoic acid (***30***, given the abbreviation AHMOA).[58]

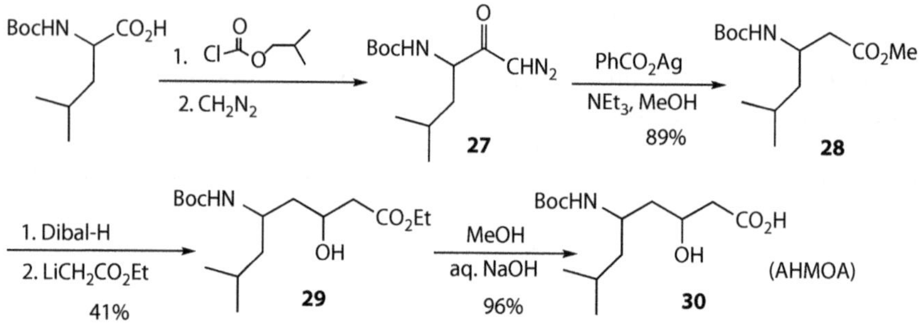

Non-α-amino acids exhibit activity with biomolecules other than proteins and peptides. The endoplasmic reticulum and Golgi network contain complex enzymatic machinery that will produce *N*-glycans.[60] A key regulatory step in *N*-glycan processing is the addition of a β1–4-linked GlcNAc residue to the central β-mannose unit of the core pentasaccharide. Chemoenzymatic synthesis produced glycans that are linked to bovine serum albumin (BSA) to give complex type biantennary (the chain splits into two branches) *N*-glycans.[61] The linking units were 6-aminohexanoic acid moieties.

---

[58] Johnson, R.L.; Verschoor, K. *J. Med. Chem.* ***1983***, *26*, 1457.

[59] (a) Wolff, L. *Ann.* ***1912***, *394*, 25; (b) Kirmse, W. *Carbene Chemistry*, 2nd ed. Academic Press, New York, ***1971***, pp. 475–492; (c) Mundy, B.P.; Ellerd, M.G. *Name Reactions and Reagents in Organic Synthesis*. Wiley, New York, ***1988***, pp. 232–233.

[60] (a) Brockhausen, I.; Schachter, H., in *Glycosciences: Status and Perspectives*, Gabius, H.-J., Gabius, S., eds. Chapman & Hall, Weinheim, ***1997***, pp. 79–113; (b) Reuter, G.; Gabius, H.-J. *Cell. Mol. Life Sci.* ***1999***, *55*, 368.

[61] André, S.; Unverzagt, C.; Kojima, S.; Frank, M.; Seifert, J.; Fink, C.; Kayser, K.; von der Lieth, C.-W.; Gabius, H.-J. *Eur. J. Biochem.* ***2004***, *271*, 118.

6.1.3 As Residues in Cyclic Peptides

Cyclic peptides are an interesting and increasingly important class of compounds. Non-α-amino acid residues play a prominent role in many of these compounds, both natural products and synthetic analogs. 5-Aminopentanoic acid was used as a linker for the cyclization of β-dipeptides, for example. Coupling of dipeptide ***31***, prepared from β-alanine (3-aminopropanoic acid), with the *tert*-butyl ester of 5-aminopentanoic acid gave an 81% yield of ***32***.[62] Subsequent deprotection and cyclization gave ***33*** in 35% overall yield from ***32***.

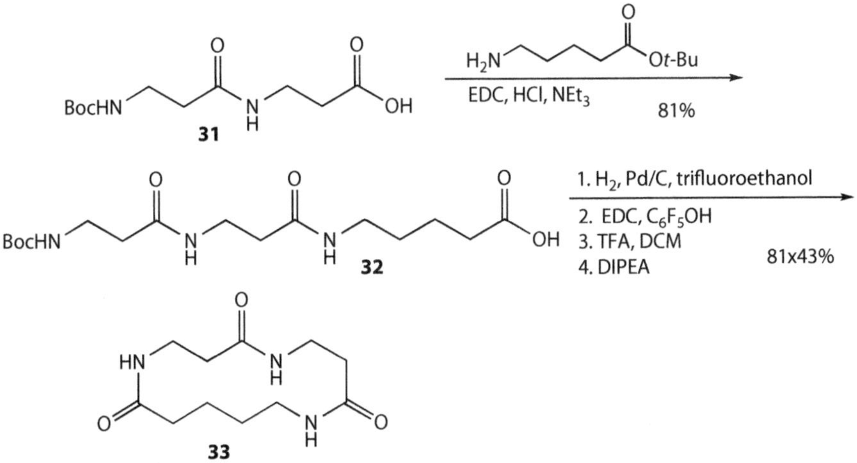

New natural products known as guineamides are cyclic depsipeptides isolated from a Papua New Guinea collection of *Lyngbya majuscula*. Several of the new compounds have β-amino acid residues, including ***34***.[63] Other isolates have the 2-methyl-3-aminobutanoic acid residue (***35***) and the 2-methyl-3-aminohexanoic acid residue, ***36***.[63]

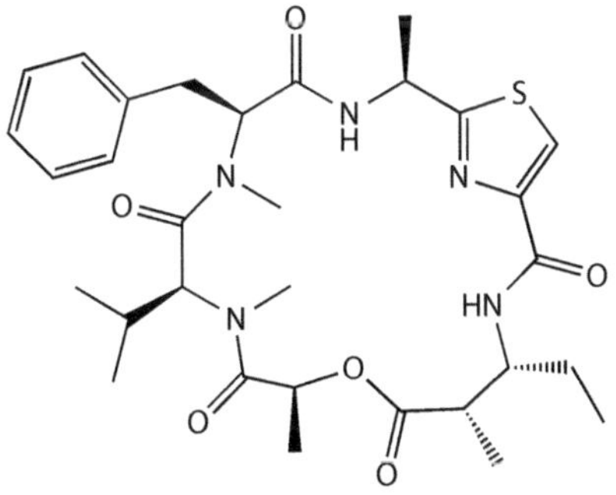

34

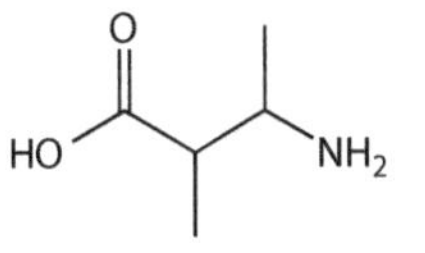

35

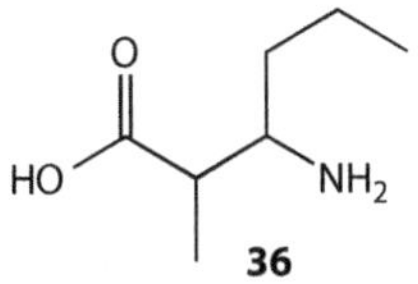

36

[62] Mengel, A.; Reiser, O.; Aube, J. *Bioorg. Med. Chem. Lett.* **2008**, *18*, 5975.
[63] Tan, L.T.; Sitachitta, N.; Gerwick, W.H. *J. Nat. Prod.* **2003**, *66*, 764.

The cyclic heptapeptide unguisin A (***37***), isolated from the marine fungus *Emericella ungis*, contains GABA as one residue.[64] This cyclic peptide shows modest antibacterial activity, and it has been synthesized using an Fmoc strategy (Fmoc is 9-fluorenylmethyl chloroformate), solid-phase peptide synthesis to prepare the linear peptide, which was then cyclized in the presence of 4-(4,6-dimethoxy-1,3,5-triazin-2-yl)-4-methylmorpholinium tetrafluoroborate.[65]

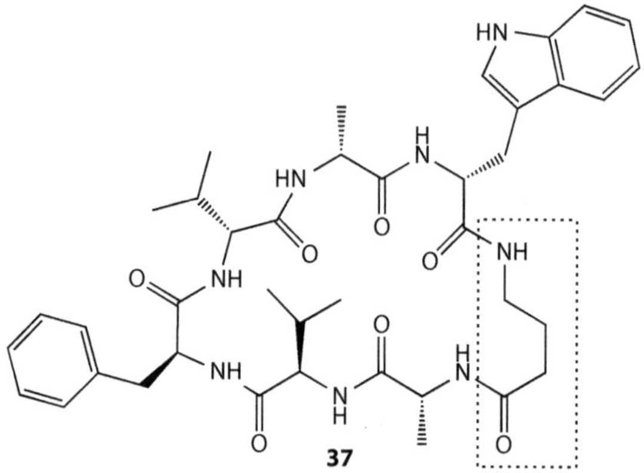

37

Other natural products are known to contain a γ-amino acid, including spiruchostatin A (***38***),[66] which was isolated from *Pseudomonas* sp. Macrocycle ***38*** contains a valine-derived (3*S*,4*R*)-statine moiety (see Section 6.6) and shows bioactivity as a histone deacetylase inhibitor.[67] The importance of such a γ-amino acid containing natural products led to work that reprogrammed the genetic code to ribosomally express backbone-cyclized peptides containing γ-amino acids.[68] In this work, peptides were expressed that bear an *N*-terminal γ-amino acid moiety by reprogramming the genetic code of initiation using various dipeptide initiators.[68] Application includes the ribosomal synthesis of backbone-macrocyclic peptides containing diverse nonstandard amino acids, allowing the preparation of exotic macrocyclic peptide libraries. This work could lead to more diverse libraries of γ-amino acid-containing peptides that will allow further exploration of biological activity.

---

[64] See (a) Malmstrøm, J. *J. Nat. Prod.* ***1999***, *62*, 787; (b) Malmstrøm, J.; Ryager, A.; Anthoni, U.; Nielsen, P.H. *Phytochem.* ***2002***, *60*, 869.

[65] Hunter, L.; Chung, J.H. *J. Org. Chem.* ***2011***, *76*, 5502.

[66] See Masuoka, Y.; Nagai, A.; Shin-ya, K.; Furihata, K.; Nagai, K.; Suzuki, K.; Hayakawa, Y.; Seto, H. *Tetrahedron Lett.* ***2001***, *42*, 41.

[67] Crabb, S.J.; Howell, M.; Rogers, H.; Ishfaq, M.; Yurek-George, A.; Carey, K.; Pickering, B.M.; East, P.; Mitter, R.; Maeda, S.; Johnson, P.W.; Townsend, P.; Shin-ya, K.; Yoshida, M.; Ganesan, A.; Packham, G. *Biochem. Pharmacol.* ***2008***, *76*, 463.

[68] Ohshiro, Y.; Nakajima, E.; Goto, Y.; Fuse, S.; Takahashi, ET.; Doi, T.; Suga, H. *ChemBioChem* ***2011***, *12*, 1183.

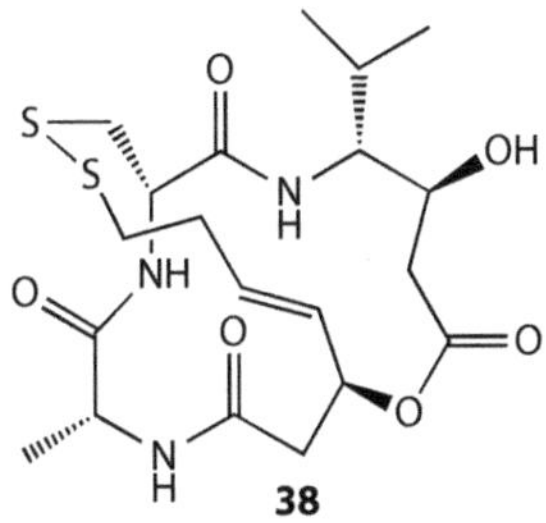

38

6.2 β-ALANINE AND AMINOPROPANOIC ACID COMPOUNDS

β-Amino acids are a very interesting class of compounds,[69] especially α-methyl derivatives.[70] These non-α-amino acids exhibit a variety of biological activity, ranging from antifungal to anticancer properties. β-Amino acids are also incorporated into peptides, and those peptides show greater resistance to protease enzymes.[69,70] This section will present a few of the myriad natural products that have a β-amino acid in their structure, and in many cases important biological applications will also be discussed. There is an emphasis on the synthetic routes to these important molecules in the discussions in this section, usually using one of the strategies or reaction sequences discussed in Chapters 1–5 of this book.

The synthesis of baccatin III derivatives[71] required amino acid ***40***. *S*-(+)-Phenylglycine (***39***) was reduced to the aldehyde, protected, and the *N*-Boc derivative condensed with vinylmagnesium bromide. Subsequent *O*-alkylation and oxidative cleavage of the alkenyl moiety gave ***40***[71] (also see Chapter 1, Section 1.4.1). Baccatin III is a natural product that has been used for the semisynthesis of taxol.[72]

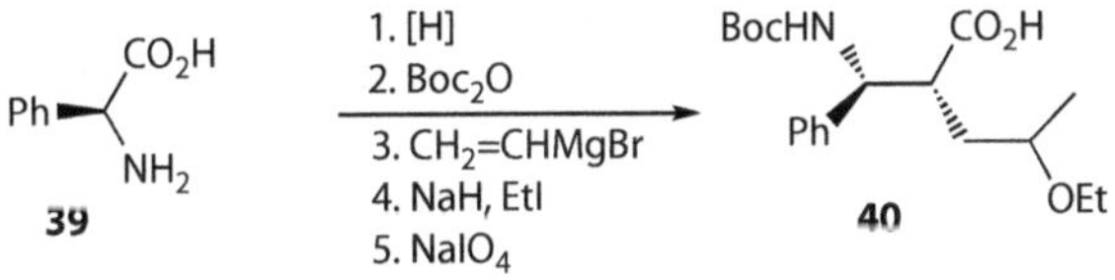

---

[69] (a) Cole, D. *Tetrahedron* ***1994***, *50*, 9517; (b) Juaristi, E.; Quintna, D.; Escalante, J. *Aldrichim. Acta* ***1994***, *27*, 3; (c) Cardillo, G.; Tomasini, C. *Chem. Soc. Rev.* ***1996***, 117; (d) *Enantioselective Synthesis of β-Amino Acids*, Juaristi, E., ed. Wiley-VCH, New York, ***1997***.

[70] See (a) Pettit, G.R.; Kamano, Y. Kizu, H.; Dufresne, C.; Herald, C.; Bontems, R.J.; Schmidt, J.M.; Boettner, F.E.; Nieman, R.A. *Heterocycles* ***1989***, *28*, 553; (b) Carter, D.C.; Moore, E.R.; Mynderwe, J.S.; Niemczura, W.P.; Todd, J.S. *J. Org. Chem.* ***1984***, *49*, 236; (d) Hu, T.; Panek, J.J. *J. Org. Chem.* ***1999***, *64*, 3000.

[71] Correa, A.; Denis, J.N.; Greene, A.E.; Grierson, D.S. *PCT Int. Appl.* WO 91,17,976 [*Chem. Abstr.* ***1992***, *16*: P129620v].

[72] See, for example, Danishefsky, S.J.; Masters, J.J.; Young, W.B.; Link, J.T.; Snyder, L.B.; Magee, T.V.; Jung, D.K.; Isaacs, R.C.A.; Bornmann, W.G.; Alaimo, C.A.; Coburn, C.A.; Di Grandi, M.J. *J. Am. Chem. Soc.* ***1996***, *118*, 2843.

β-Amino-α-hydroxy acids are known as norstatines,[73] and amastatin (***41***) is one example. Amastatin is a tetrapeptide containing three α-amino acid residues plus the 2-hydroxy-3-aminoheptanoic acid residue, and it is an inhibitor of aminopeptidases.[74] Norstatine type compounds[75] are active against malaria proteases expressed by the parasite *Plasmodium falcifarum*.[76]

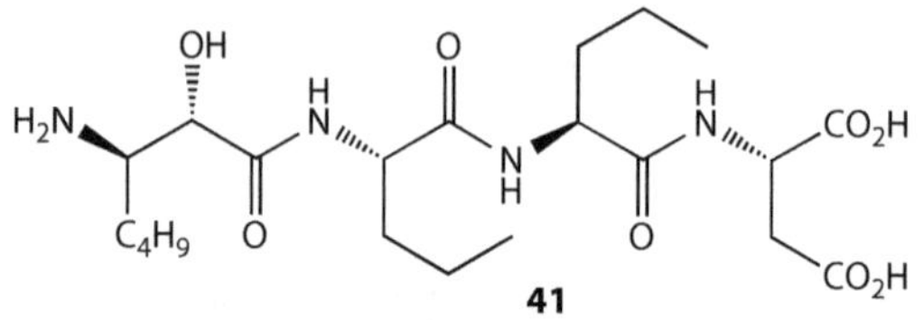

41

A synthesis of norstatines was reported using chiral Boc-aminoaldehydes ***42***, where R = Me, $PhCH_2$.[77] Initial condensation of the aldehyde ***42*** with ethynylmagnesium bromide gave the alcohol ***43*** in up to 97% yield. Subsequent protection as the tetrahydropyranyl ether was followed by oxidative cleavage of the alkyne unit with periodate-ruthenium to gave the acid, ***44***. Another approach to optically active norstatines reacted *N*-Boc aldimines with chiral enolate anions of 1,3-dioxolan-4-ones.[78]

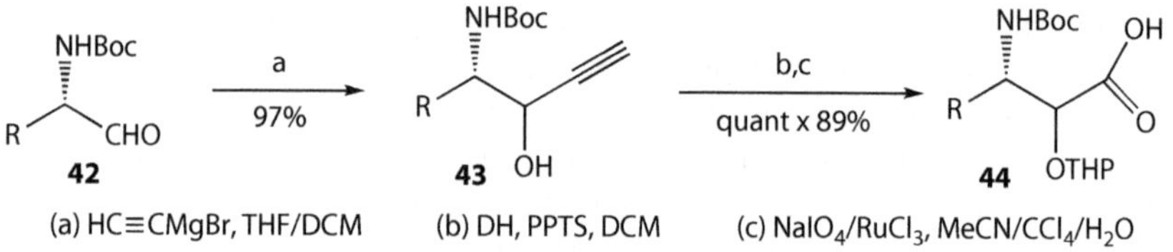

N-Benzoyl β-amino acids such as ***45*** were cyclized to give an oxazoline ***46***.[79] Subsequent ring opening with dilute aqueous acid gave the hydroxylated amino acid ***47*** in good yield with high diastereoselectivity. Several derivatives of this type were prepared by this method. Aminopeptidase N/CD13 (APN) is an important target against cancer metastasis and angiogenesis. A series of compounds targeting APN were synthesized and evaluated for their antimetastasis and antiangiogenesis potency.[80] Bestatin may be the best-studied example, but many non-α-amino acid derivatives have been studied.[80]

---

[73] Umezawa, H.; Aayagi, T.; Morishima, H.; Matzuzaki, M.; Hamada, M.; Takeuchi, T. *J. Antiobiot.* ***1970***, *23*, 259.

[74] Rich, D.H.; Moon, B.J.; Harbeson, S. *J. Med. Chem.* ***1984***, *27*, 417.

[75] Nezami, A.; Luque, I.; Rimura, T.K.; Y. Freire, E. *Biochemistry* **2002**, *41*, 2273.

[76] (a) Francis, S.E.; Gluzman, I.Y.; Oksman, A.; Knickerbocker, A.; Mueller, R.; Bryanta, M.L.; Shermana, D.R.; Russel, D.G.; Goldberg, D.E. *EMBO J.* ***1994***, *13*, 306; (b) Wyatt, D.M.; Berry, C. *FEBS Lett.* **2002**, *513*, 159.

[77] Kourtal, S.; Paris, J. *Lett. Peptide Sci* ***1996***, *3*, 73.

[78] Battaglia, A.; Guerrini, A.; Bertucci, C. *J. Org. Chem.* **2004**, *69*, 9055.

[79] Nocioni, A.M.; Papa, C.; Tomasini, C. *Tetrahedron Lett.* ***1999***, *40*, 8453.

[80] Su, L.; Cao, J.; Jia, Y.; Zhang, X.; Fang, H.; Xu, W. *ACS Med. Chem. Lett.* **2012**, *3*, 959.

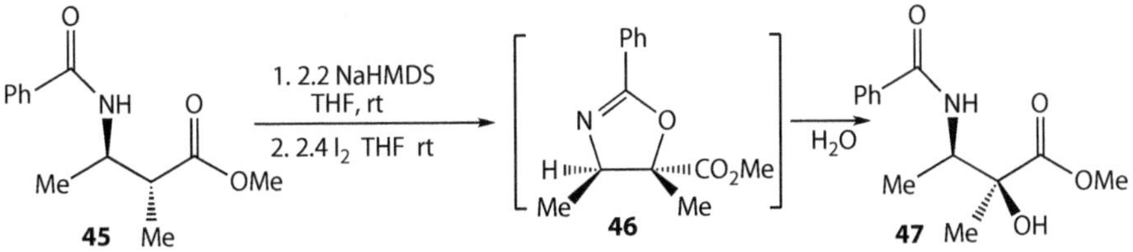

Amino acid ***51*** was used in a synthesis of the antibiotic kasugamycin.[81] This component was prepared by addition of NOCl to ***48***,[82] giving ***49*** as a reactive intermediate. Basic hydrolysis gave the keto-oxime (***50***) and catalytic hydrogenation gave *erythro*-5-hydroxy-4-aminohexanoic acid (***51***).

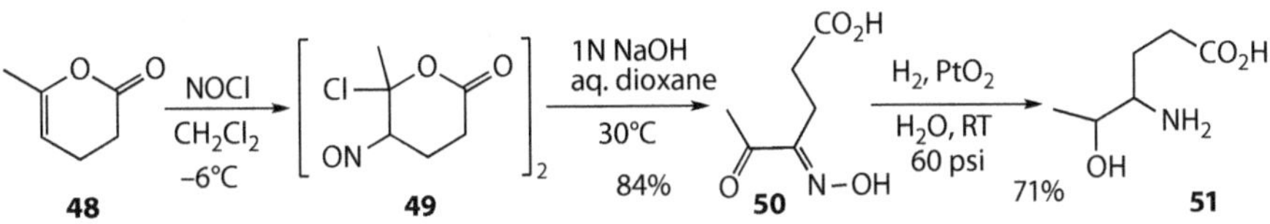

Ring opening reactions of epoxides provide another route to hydroxy amino acids. Alkenyl acid ***52*** was epoxidized and gave ***53***. The epoxide moiety reacted with benzylamine to give ***54***, quantitatively.[83] Removal of the benzyl group by catalytic hydrogenation gave a quantitative yield of ***55***. 3-Amino-2-hydroxybutanoic acid (***55***) is a component of the antibiotic kanamycin.[83b]

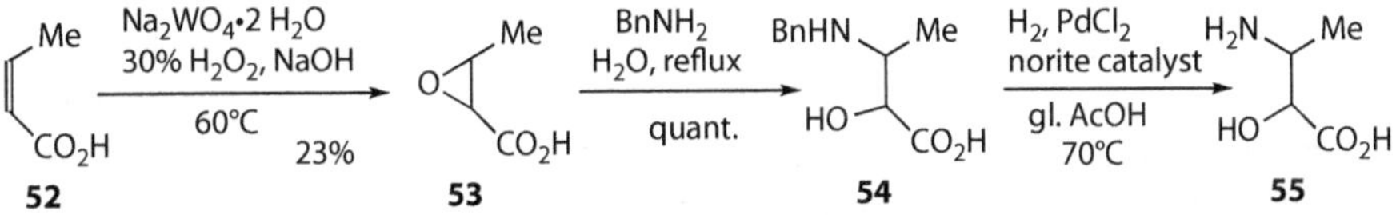

A new class of endomorphin-1 analogs (2-methylene-3-aminopropanoic acids, or MAPs; see ***61***) were prepared that contained α-methylene-β-amino acids. Binding and functional activity, metabolic stability, and antinociceptive activity were examined. Note that POM is the protecting group pivaloyl methyl. The synthesis of the α-methylene-β-amino acid began with aryl-sulfonyl amines such as ***56***.[84] Different derivatives were prepared by modification of the aryl group, and both hydrocarbon aryl and heteroaryl derivatives were prepared. Coupling of ***56*** with phosphonate ester ***57***, in the presence of a thiourea catalyst, gave ***58***. Subsequent Horner-Wadsworth-

---

[81] Suhara, Y.; Sasaki, F.; Koyama, G.; Maeda, K.; Umezawa, H.; Ohno, M. *J. Am. Chem. Soc.* ***1972***, *94*, 6501.

[82] Vorlander, D.; Knotzsch, A. *Ann. Chem.* ***1897***, *294*, 319.

[83] (a) Liwschitz, Y.; Singerman, A.; Luwish, M. *Israel J. Chem.* ***1963***, *1*, 441; (b) also see Umemura, E.; Tsuchiya, T.; Umezawa, S. *J. Antibiot.* ***1988***, *41*, 530.

[84] Wang, Y.; Xing, Y.; Liu, X.; Ji, H.; Kai, M.; Chen, Z.; Yu, J.; Zhao, D.; Ren, H.; Wang, R. *J. Med. Chem.* ***2012***, *55*, 6224.

Emmons reaction[85] with paraformaldehyde led to formation of the alkene moiety in ***59***. Deprotection gave ***60***, which was utilized in the preparation of ***61***.

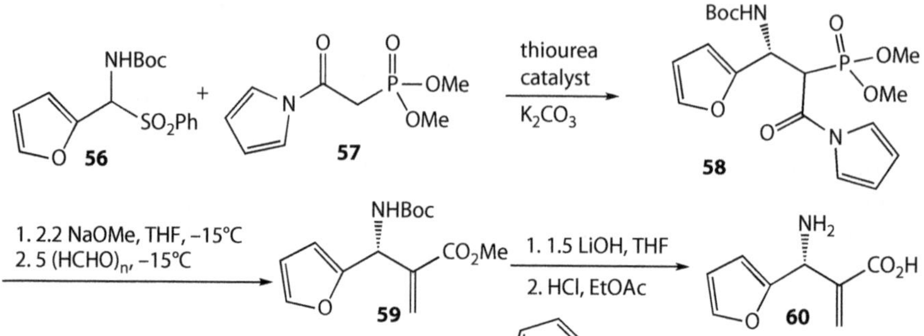

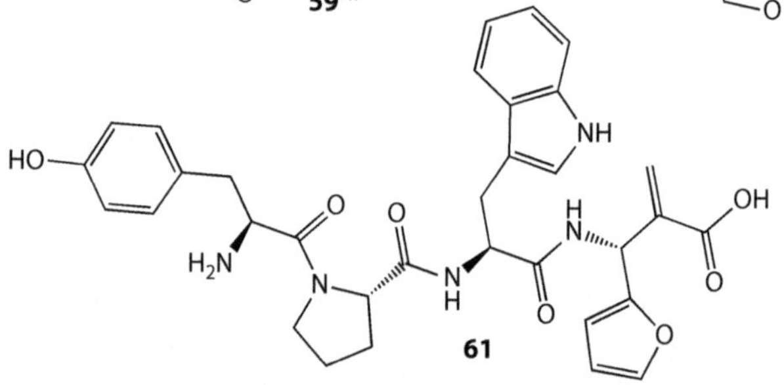

Several ω-*N*-quinone amino acids have been prepared from β-alanine, 4-aminobutanoic acid, 5-aminopentanoic, and 6-aminohexanoic acid.[86] These amino acids were used for post-chain assembly modifications of bioactive peptides known to target cancer-damaged areas. These quinone-peptide conjugates may act as site-directed antitumor drugs. The synthesis simply heats 1,4-naphthoquinone with the amino acid of interest in aqueous ethanol, in this case, 4-aminobutanoic acid, to give ***62***.[86]

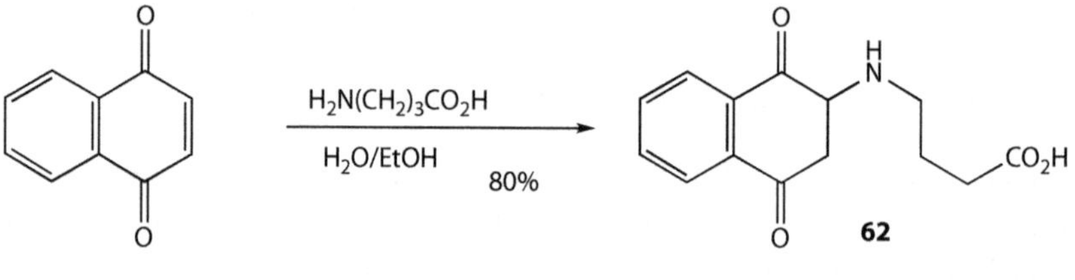

The β-amino acid residue is part of the structure of several spider toxins. An example is the Joro spider toxin, ***63***.[87] Such polyamine toxins are open-channel blockers

[85] See (a) Smith, M.B. *Organic Synthesis*, 3rd ed. Wavefunction, Inc./Elsevier, Irvine, CA/London, England, ***2010***, pp. 739–744; (b) Smith, M.B. *March's Advanced Organic Chemistry*, 7th ed. John Wiley & Sons, Hoboken, NJ, ***2013***, pp. 1169–1170.

[86] Bittner, S.; Gorohovsky, S.; Paz-Tal, O.; Becker, J.Y. *Amino Acids* **2002**, *22*, 71.

[87] Olsen, C.A.; Kristensen, A.S.; Strømgaard, K. *Angew. Chem. Int. Ed.* ***2011***, *50*, 11296

of ionotropic glutamate (iGlu) receptors.[88] These receptors mediate many excitatory synaptic transmissions in vertebrates.[89]

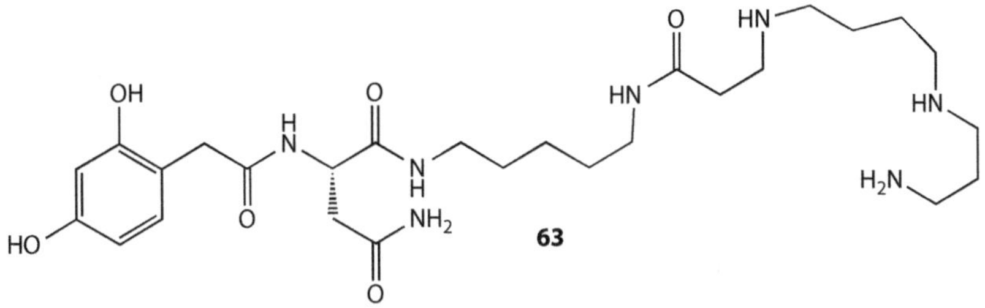

63

The functionalized amino acid 3-guanidinopropionic acid (*64*, a β-amino acid derivative) was shown to improve insulin sensitivity and promote weight loss selectively from adipose tissue in animal models of non-insulin-dependent diabetes mellitus.[90]

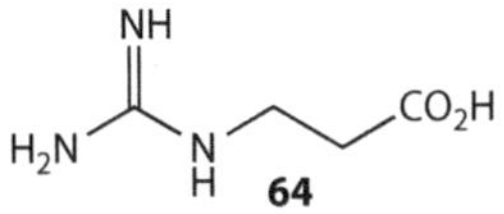

64

6.3 GABA AND RELATED COMPOUNDS

4-Aminobutanoic acid (γ-aminobutyric acid, known as GABA) is an important mammalian neurotransmitter.[91] Deficiencies in GABA are associated with diseases that exhibit neuromuscular dysfunction such as epilepsy,[92] Huntington's disease, and Parkinson's disease.[91] It has been shown that dysfunction in the neurotransmission properties of GABA can lead to cerebral ischemia-induced neuronal death, but such death may be prevented by GABAergic drugs.[93] GABA receptors are also potential targets to treat Alzheimer's disease.[94] Alcoholism is an addiction, and it is known that GABA, opioid peptides, serotonin, acetylcholine, the endocannabinoids, and glutamate systems play a role in the initial addictive process.[95] GABA is one of many compounds known to have paracrine activity (a form of cell signaling in which the target

[88] Lucas, S.; Poulsen, M.H.; Nørager, N.G.; Barslund, A.F.; Bach, T.B.; Kristensen, A.S.; Strømgaard, K. *J. Med. Chem.* ***2012***, *55*, 10297.

[89] (a) Strømgaard, K.; Mellor, I.R. *Med. Res. Rev.* ***2004***, *24*, 589; (b) Fleming, J.J.; England, P.M. *Nat. Chem. Biol.* ***2010***, *6*, 89.

[90] Larsen, S.D.; Connell, M.A.; Cudahy, M.M.; Evans, B.R.; May, P.D.; Meglasson, M.D.; O'Sullivan, T.J.; Schostarez, H.J.; Sih, J.C.; Stevens, F.C.; Tanis, S.P.; Tegley, C.M.; Tucker, J.A.; Vaillancourt, V.A.; Vidmar, T.J.; Watt, W.; Yu, J.H. *J. Med. Chem.* ***2001***, *44*, 1217.

[91] Enna, S.J. *The GABA Receptors*. Humana Press, Clifton, NJ, ***1983***. Also see Osborne, R.H. *Pharmacol. Ther.* ***1996***, *69*, 117.

[92] See (a) Mintzer, M.A.; Simanek, E.E. *Chem. Rev.* ***2009***, *109*, 259; (b) Steinlein, O.K. *Chem. Rev.* ***2012***, *112*, 6334.

[93] Schwartz-Bloom, R.D.; Sah, R. *J. Neurochem.* ***2001***, *77*, 353.

[94] Hamley, W. *Chem. Rev.* ***2012***, *112*, 5147.

[95] Ross, S.; Peselow, E. *Clin. Neuropharmacol.* ***2009***, *32*, 269.

cell is near the signal-releasing cell) or autocrine activity (a form of cell signaling in which a cell secretes an autocrine agent that binds to autocrine receptors on the same cell, leading to changes in the cell).[96] Endogenous toxicosis has been identified as a primary biological etiological factor in nervous and mental disease and violent behavior.[97] Toxins may accumulate in the hypothalamus, and the blood-brain barrier prevents toxins from reaching most regions of the brain. These toxins may include excess norepinephrine, dopamine, epinephrine, serotonin, GABA, amino acids, or peptides.[97] The toxins can accumulate in nervated neurons and inhibit release of norepinephrine into synapses. "Low levels of synaptic norepinephrine are associated with symptoms of depression."[97] Bifunctional hydrazine-bisphosphonates (HBPs), with spacers of various lengths, were synthesized and studied for their enhanced affinity to bone and use for drug targeting to bone. 4-Aminobutanoic acid was incorporated into the targeted hydrazine-bisphosphosphonates.[98]

There are several amino acids that have similar biological activities that are known as GABAergic compounds (possessing GABA-like activity). GABAergic compounds and their synthesis are also discussed in this section. The utility and mode of action of these GABAergic compounds are quite varied.[99] Aminomethylcyclopropanecarboxylic acid, ***65***, is a selective agonist for $GABA_C$ receptors, but it is inactive against $GABA_A$ receptors, for example.[100] Amino acid ***65*** and several other non-α-amino acids were studied for their ability to activate ligand-gated ion channels.[100] Other cyclic amino acids related to ***65*** are discussed in Chapter 7.

65

GABA derivatives also have biological activity. Semicarbazone derivatives of GABA were prepared from GABA, for example, with the goal of producing new drugs for the treatment of neurological disorders like epilepsy and neuropathic pain, and the new analogs produced anticonvulsant and antinociceptive actions in models of neuropathic pain.[101]

It is clear that GABA and related compounds are quite important, and there are many synthetic approaches to these compounds. Further, many GABA derivatives are known. Baclofen [4-amino-3-(*p*-chlorophenyl)butanoic acid, ***68***] is an important GABAergic compound used in the treatment of epilepsy.[102] A useful synthesis began with conjugate addition of the anion of nitromethane to the double bond in ***66*** to give

---

[96] Denef, C. *J. Neuroendocrinol.* ***2008***, *20*, 1. Also see Koob, G.F. *Alcoholism Clin. Exp. Res.* ***2003***, *27*, 232.

[97] Van Winkle, E. *Med. Hypotheses* ***2000***, *54*, 146; ***2000***, *55*, 356.

[98] Yewle, J.N.; Puleo, D.A.; Bachas, L.G. *Bioconjugate Chem.* ***2011***, *22*, 2496.

[99] See, for example, (a) Rekatas, G.V.; Tani, E.; Demopoulos, V.J.; Kourounakis, P.N. *Arkiv der Pharm.* ***1996***, *329*, 393; (b) Bryans, J.S.; Wustrow, D.J. *Med. Res. Rev.* ***1999***, *19*, 149. Also see Krogsgaard-Larsen, P.; Frølund, B.; Liljefors, T. *Chem. Rec.* ***2002***, *2*, 419.

[100] Chebib, M.; Johnston, G.A.R. *J. Med. Chem.* ***2000***, *43*, 1427.

[101] Yogeeswari, P.; Ragavendran, J.V.; Sriram, D.; Nageswari, Y.; Kavya, R.; Sreevatsan, N.; Vanitha, K.; Stables, J. *J. Med. Chem.* ***2007***, *50*, 2459.

[102] See reference 91, pp. 5, 8–9, 14–16, 39, 63, 70, 75, 132–133, 152–153, 195, 198–200, 305.

67.[103] Catalytic hydrogenation of the nitro group gave an amine that spontaneously cyclized to a lactam. Formation of this lactam required acid hydrolysis to give baclofen, *68*.[103] 3-(4-Bromophenyl)-, 3-(4-tolyl)-, and the 3-(4-hydroxyphenyl)-derivatives, as well as 3-pyridyl and 2-pyridyl-4-aminobutanoic acids were similarly prepared.[103] It is noted that homologs of baclofen, specifically 5-amino-4-(4-chlorophenyl)pentanoic acid and 5-amino-3-(4-chlorophenyl)pentanoic acid, were prepared in separate work but found to be inactive or much less potent.[104]

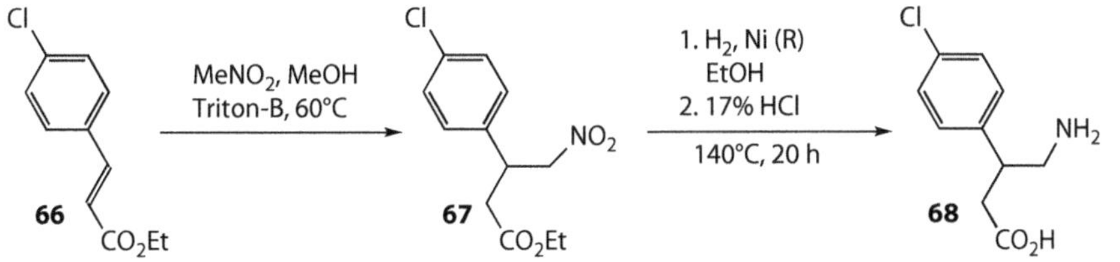

A direct asymmetric synthesis of baclofen (*68*) involved the conjugate addition of nitromethane to *69* in the presence of a chiral cinchonium catalyst.[105] The product *70* was isolated as an 85:15 *R*:*S* mixture that was purified to a 97.5:2.5 mixture after recrystallization, and then converted to *68* by reduction of the nitro group to complete the asymmetric synthesis.

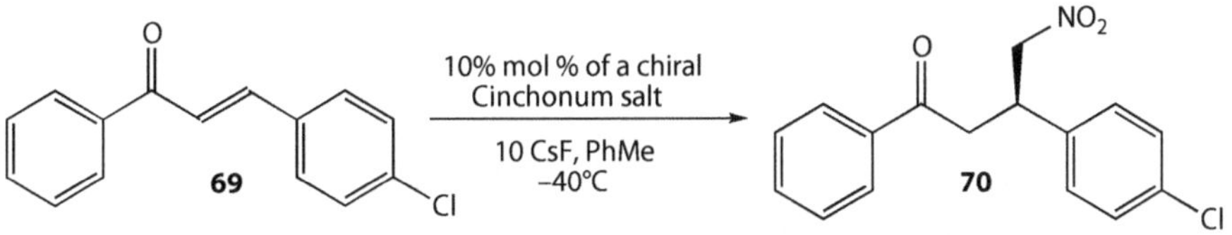

A different Michael addition[106] route used conjugated lactam *71* in a reaction with 4-chlorophenylboronic acid to give *72* with good asymmetric induction.[107] This reaction was catalyzed by a Pd complex with the chiral diene shown. Lactam *72* was converted to (*R*) baclofen (*68*). Similarly, *71* reacted via conjugate addition with 3-(cyclopentyloxo)-4-methoxyphenylboronic acid (*73*) to give a lactam that was deprotected to give (*R*)-rolipram, *74*,[107] which is a selective phosphodiesterase inhibitor and has been used to treat depression.

[103] Uchimaru, F.; Sato, M.; Kosasayama, E.; Shimizu, M.; Takashi, H. *Jpn*. 70 16,692 [*Chem. Abstr.* ***1970***, *73*: P77617w]. Also see Karla, R.; Ebert, B.; Thorkildsen, C.; Herdeis, C.; Johansen, T.N.; Nielsen, B.; Krogsgaard-Larsen, P. *J. Med. Chem.* ***1999***, *42*, 2053.

[104] Karla, R.; Ebert, B.; Thorkildsen, C.; Herdeis, C.; Johansen, T.N.; Nielsen, B.; Krogsgaard-Larsen, P. *J. Med. Chem.* ***1999***, *42*, 2053. Also see Karla, R.; Ebert, B.; Thorkildsen, C.; Herdeis, C.; Johansen, T.N.; Nielsen, B.; Krogsgaard-Larsen, P. *J. Med. Chem.* ***1999***, *42*, 2053.

[105] Corey, E.J.; Zhang, F.-Y. *Org. Lett.* ***2000***, *2*, 4257.

[106] (a) Michael, A. *J. Prakt. Chem.* ***1887***, *35*, 379; (b) Bergmann, E.D.; Gingberg, D.; Pappo, R. *Org. React.* ***1959***, *10*, 179; (c) Perlmutter, P. *Conjugative Addition Reactions in Organic Synthesis*. Pergamon Press, Oxford, ***1992***; (d) see Smith, M.B. *Organic Synthesis*, 3rd ed. Wavefunction, Inc./Elsevier, Irvine, CA/London, England, ***2010***, pp. 877–888; (e) Smith, M.B. *March's Advanced Organic Chemistry*, 7th ed. John Wiley & Sons, Hoboken, NJ, ***2013***, pp. 943–949.

[107] Shao, C.; Yu, H.-J.; Wu, N.-Y.; Tian, P.; Wang, R.; Feng, C.-G.; Lin, G.-Q. *Org. Lett.* ***2011***, *13*, 788.

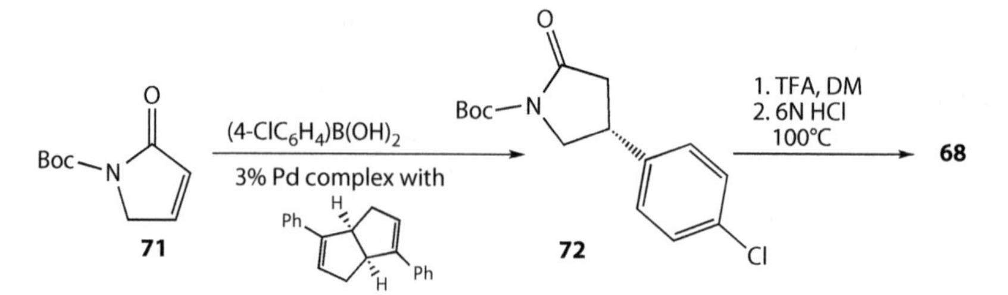

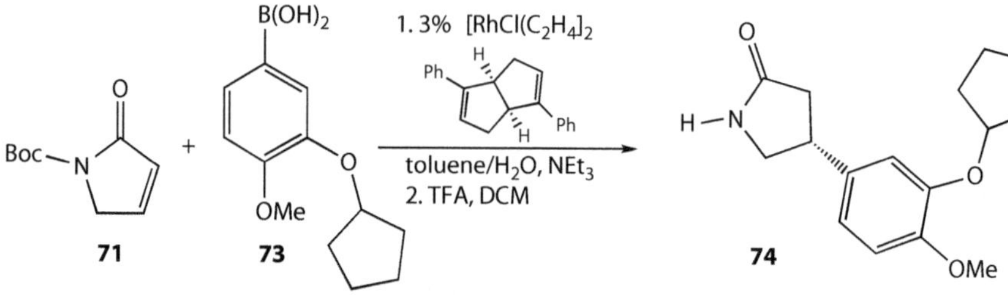

Another methodology has been used to generate GABAergic compounds. The reaction of ***75*** with phenyl isocyanate gave a nitrile *N*-oxide, and *in situ* addition to 1-hexene gave hydroisoxazole ***76***.[108] Hydrolysis of the ester and catalytic hydrogenation opened the ring to give 4-amino-6-hydroxydecanoic acid, ***77***. When compared to GABA, ***77*** was shown to be a more lipophilic GABAergic compound[91] that inhibited GABA binding to synaptic membranes. Similarly prepared were 8-methyl-4-amino-6-hydroxynonanoic acid, 4-amino-6-hydroxydecanoic acid, 4-amino-6-hydroxydodecanoic acid, 4-amino-6-hydroxytetradecanoic acid, 4-amino-6-hydroxypentadecanoic acid, and 4-amino-6-hydroxyhexadecanoic acid.[108] In general, all of these amino acids were more lipophilic than GABA and retained GABAergic activity.[108]

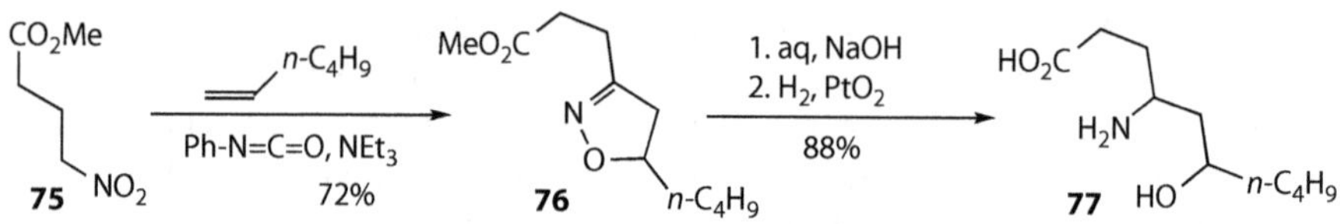

A fluorinated derivative was prepared by conversion of keto-ester ***78*** to enamino-ester ***79*** as a 1:1 mixture of *cis:trans* isomers.[109] Reduction of this enamine and removal of the protecting groups gave ***80***, which was shown to be an inhibitor[109] of the enzyme GABA transaminase (GABA-T).[110] A transaminase is an enzyme that transfers nitrogenous groups and catalyzes the reaction with 2-oxoglutarate to give succinate semialdehyde.[111] Inhibitors of GABA-T of this type are of general interest as anticonvulsant agents.[112]

---

[108] Borea, P.A.; Bonora, A.; Baraldi, P.G.; Simoni, D. *Farmaco Ed. Sci.* ***1983***, *38*, 411.

[109] Bey, P.; Gerhart, F.; Jung, M.; Schirlin, D. *Eur. Pat. Appl.* EP 46,711 [*Chem. Abstr.* ***1982***, *97*: P39394f].

[110] Roberts, E. *Biochem. Pharm.* ***1974***, *23*, 2637.

[111] (a) Scott, E.M.; Jakoby, W.B. *J. Biol. Chem.* ***1959***, *234*, 932; (b) Schousboe, A.; Wu, J.Y.; Roberts, E. *Biochemistry* ***1973***, *12*, 2868.

[112] Hammond, E.J.; Wilder, B.J. *Gen. Pharmacol.* ***1985***, *16*, 441.

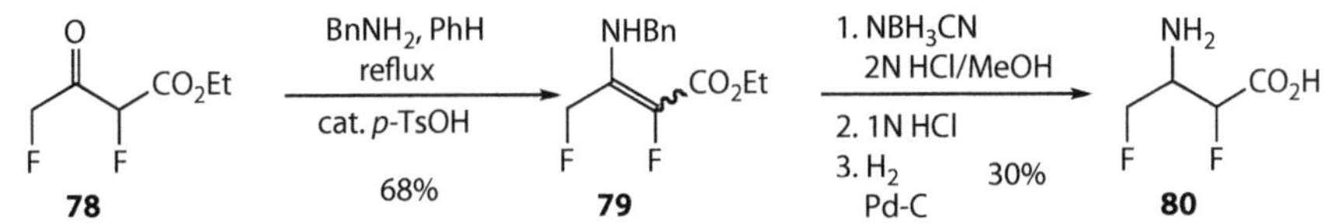

2-Pyrrolidinone derivatives are useful synthons for the preparation of 4-aminobutanoic acid derivatives of all types. Ethyl pyroglutamate (***81***), derived from *L*-glutamic acid, was converted to cyanomethyl-2-pyrrolidinone ***82***.[113] Reduction to the amine and Cope elimination[114] gave ***83***, which was hydrolyzed to 4-aminohex-5-enoic acid (***84***; known as vigabatrin). It is known that ***84*** is an irreversible inhibitor of GABA-T. Due this mode of action, vigabatrin has been shown to increase levels of GABA in the brain. Vigabatrin, sold as Sabril, has been approved for the treatment of infantile spasms in children ages 1 month to 2 years, and for the treatment of complex partial seizures in adults.[115] This mode of action is important because GABA acts on extrasynaptic receptors, and these receptors act as a kind of "master switch" that is capable of turning off a range of differently induced activities.[116]

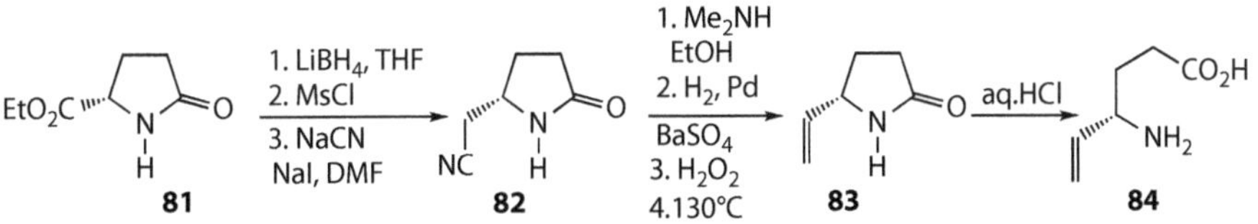

An alternative synthesis of ***84*** converted ethyl pyroglutamate (***81***) to aldehyde ***85***, using a *N*-butenyl protecting group that inhibited racemization at C6.[117] Reduction of the ester was followed by condensation with butanal to give the *N*-alkenyl lactam. Generation of aldehyde ***85*** allowed Wittig olefination[118,119] to give the alkene, and subsequent acid hydrolysis gave ***84***.[120] It is noted that ***85*** has been used as a key synthon for the asymmetric synthesis of several chiral, nonracemic pyrrolizidine alkaloids.[121]

---

113 (a) Silverman, R.B.; Levy, M.A. *J. Org. Chem.* ***1980***, *45*, 815; (b) Grieben, W.; Gerhart, F. *Br. UK Pat. Appl.* GB 2,133,002 [*Chem. Abstr.* ***1984***, *101*: P231027j].

114 (a) Cope, A.C.; Foster, T.T.; Towle, P.H. *J. Am. Chem. Soc.* ***1949***, *71*, 3929; (b) Cope, A.C.; Pike, R.A.; Spencer, C.F. *J. Am. Chem. Soc.* ***1953***, *75*, 3212; (c) DePuy, C.H.; King, R.W. *Chem. Rev.* ***1960***, *60*, 431.

115 Hopkins, C.R. *ACS Chem. Neurosci.* ***2010***, *1*, 475.

116 Nasrallah, F.A.; Balcar, V.J.; Rae, C.D. *J. Neurosci. Res.* ***2011***, *89*, 1935. Also see Leppik, I.E. *Epilepsia* ***1998***, *39*, 2.

117 Smith, M.B.; Keusenkothen, P.F.; Dembofsky, B.; Wang, C.-J.; Zezza, C.A.; Kwon, T.W.; Sheu, J.; Son, Y.C.; Menezes R.F.; Fay, J.N. *Chem. Lett.* ***1992***, 247.

118 (a) Wittig, G.; Rieber, M. *Ann.* ***1949***, *562*, 187; (b) Wittig, G.; Geissler, G. *Ann.* ***1953***, *580*, 44; (c) Wittig, G.; Schöllkopf, U. *Chem. Ber.* ***1954***, *87*, 1318; (d) Gensler, W.J. *Chem. Rev.* ***1957***, *57*, 191 (see p. 218).

119 See (a) Smith, M.B. *Organic Synthesis*, 3rd ed. Wavefunction, Inc./Elsevier, Irvine, CA/London, England, ***2010***, pp. 729–739; (b) Smith, M.B. *March's Advanced Organic Chemistry*, 7th ed. John Wiley & Sons, Hoboken, NJ, ***2013***, pp. 1165–1173.

120 Kwon, T.W.; Keusenkothen, P.; Smith, M.B.; Zezza, C.A. *J. Org. Chem.* ***1992***, *57*, 6169.

121 (a) Keusenkothen, P.F.; Smith, M.B. *Tetrahedron* ***1992***, *48*, 2977; (b) Smith, M.B. *Tetrahedron Lett.* ***1989***, *30*, 3369; (c) Smith, M.B. *J. Chem. Soc. Perkin Trans. I* ***1994***, 2485; (d) Smith, M.B. *J. Mol. Structure* ***1994***, *319*, 289.

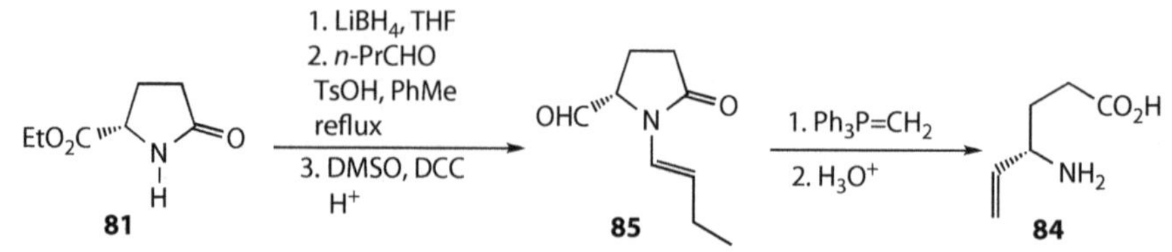

A different approach to ***84*** involved the preparation of epoxy-alcohol ***86***[122] via a Sharpless asymmetric epoxidation reaction using known methodology. Opening the chiral epoxide with benzhydrylamine in the presence of titanium tetraisopropoxide, and then hydrogenation in the presence of Boc anhydride, gave Boc-protected amino diol ***87***.[123] Protection of the diol moiety allowed oxidation of the benzene ring to a carboxylic acid, which was converted to methyl ester ***88***. This interesting transformation allows a benzene ring to function as a surrogate for a carboxyl group. Deprotection was followed by conversion to the thionocarbonate, which allowed a Corey-Hopkins deoxygenation[124] with the diazaphospholidine shown to give vigabatrin, ***84***.

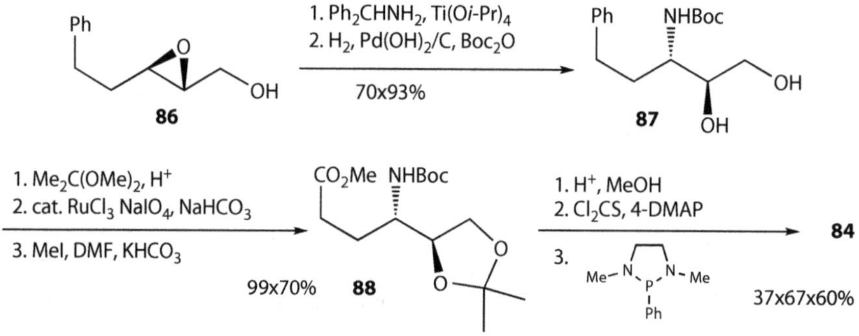

Alkynyl amino acids such as ***93*** also show GABAergic activity. The synthesis of ***93*** involved the reaction of 3-amino-1-propyne (***89***) with benzaldehyde to give ***90***.[125] Reaction with ethylmagnesium bromide gave the alkyne anion, which reacted with chlorotrimethylsilane to give the C-SiMe$_3$ derivative, ***91***. Treatment with *n*-butyllithium gave the carbanion, and subsequent conjugate addition with methyl acrylate gave ***92***. Acid hydrolysis liberated 4-aminohex-5-ynoic acid, ***93***.[125,126] This amino acid shows properties as a sedative, an antidepressant, and a GABA-T inhibitor.[125]

[122] (a) Nuñez, M.T.; Martín, V.S. *J. Org. Chem.* ***1990***, *55*, 1928; (b) Poch, M.; Alcon, M.; Moyano, A.; Pericàs, M.A.; Riera, A. *Tetrahedron Lett.* ***1993***, *34*, 7781.

[123] Alcón, M.; Poch, M.; Moyano, A.; Pericàs, M.A. *Tetrahedron Asym.* ***1997***, *8*, 2967.

[124] Corey, E.J.; Hopkins, P.B. *Tetrahedron Lett.* ***1982***, *23*, 1979.

[125] Metcalf, B.W.; Casar, P. *Tetrahedron Lett.* ***1975***, 3337. Also see Estreicher, H.; Flint, D.H.; Silverman, R.B. *U.S. Pat.* 4,582,529 [*Chem. Abstr.* ***1986***, *105*; P56355y]; (a) Lundkvist, J.R.M.; Hacksell, U. *Acta Pharm. Suec.* ***1986***, *23*, 416; (b) Lundkvist, J.R.M.; Ringdahl, B.; Hacksell, U. *J. Med. Chem.* ***1989***, *32*, 863

[126] (a) Metcalf, B.W.; Jung, M. *U.S. Pat.* 3,959,356 [*Chem. Abstr.* ***1976***, *85*: P142651s]; (b) Metcalf, B.W.; Jung, M. *U.S. Pat.* 4,041,041 [*Chem. Abstr.* ***1977***, *87*: P200807b].

Palladium-catalyzed hydrogenation[127] of an analog of ***93*** [4-(*N*-acetylamino)hex-5-ynoic acid] gave ***84***.[128]

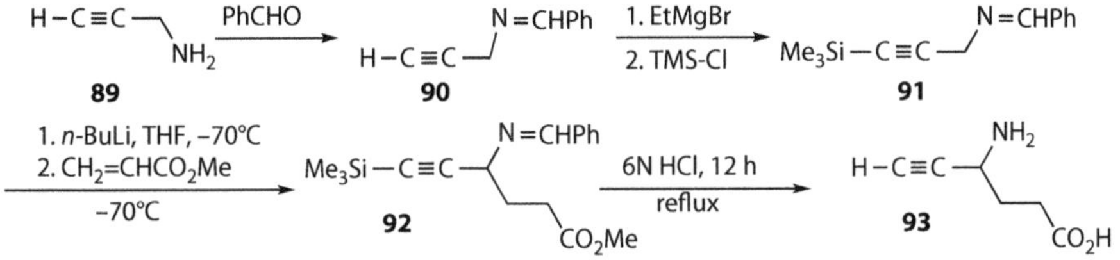

4-Aminohex-5-ynoic acid, ***93***, was also prepared in eight steps from the commercially available diethyl ester of (*S*)-glutamic acid.[129] Initial conversion to the imine using 2,4-dimethoxybenzaldehyde was followed by reduction of the benzylic protecting group. Heating led to cyclization to the 2-pyrrolidinone, and sodium borohydride effectively reduced the ester moiety to a hydroxymethyl group (see ***94***). Swern oxidation[130] of the alcohol in ***94*** allowed conversion to the alkyne moiety in ***95***, and acid hydrolysis of the lactam afforded *S*-***93***. In this synthesis, the authors indicated that the key step was conversion of the aldehyde to the alkyne moiety using diethylmethyldiazophosphonate.

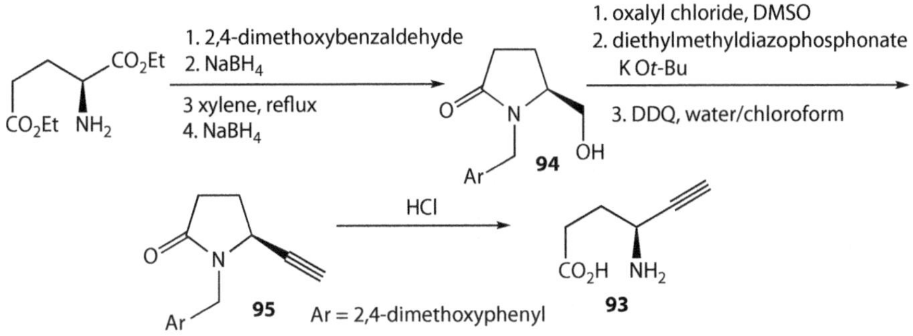

A GABA-amino transferase inactivator, (1*S*,3*S*)-3-amino-4-difluoromethylenyl-1-cyclopentanecarboxylic acid (***96***), was prepared and shown to give inhibition of cocaine-induced increases in extracellular dopamine and in synaptic dopamine in the nucleus accumbens.[131] Amino acid ***96*** may be useful to treat addiction with fewer side effects than vigabatrin (see ***84***).

---

[127] (a) Lindlar, H. *Helv. Chim. Acta* ***1952***, *35*, 446; (b) Fieser, L.F.; Fiesser, M. *Reagents for Organic Synthesis*, vol. 1. Wiley, New York, ***1967***, p. 566.

[128] (a) Nishizawa, R.; Saino, A.; Setani, K. *Jpn. Kokai Tokkyo Koho* 80 00,355 [*Chem. Abstr.* ***1980*** *92*: P163584g]; (b) Metcalf, B.W. *U.S. Pat.* 3,960,927 [*Chem. Abstr.* ***1976***, *85*: P143512j].

[129] McAlonan, H.; Stevenson, P.J. *Tetrahedron Asym.* ***1995***, *6*, 239.

[130] (a) Mancuso, A.J.; Swern, D. *Synthesis* ***1981***, 165. See (b) Smith, M.B. *Organic Synthesis*, 3rd ed. Wavefunction, Inc./Elsevier, Irvine, CA/London, England, ***2010***, pp. 241–243; (c) Smith, M.B. *March's Advanced Organic Chemistry*, 7th ed. John Wiley & Sons, Hoboken, NJ, ***2013***, pp. 1447–1448.

[131] Pan, Y.; Gerasimov, M.R.; Kvist, T.; Wellendorph, P.; Madsen, K.K.; Pera, E.; Lee, H.; Schousboe, A.; Chebib, M.; Bräuner-Osborne, H.; Craft, C.M.; Brodie, J.D.; Schiffer, W.K.; Dewey, S.L.; Miller, S.R.; Silverman, R.B. *J. Med. Chem.* **2012**, *55*, 357.

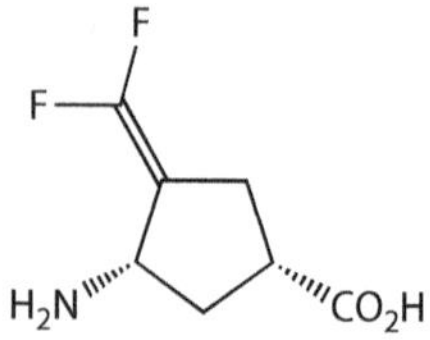

96

There are many antibiotics that contain a 4-aminobutanoic acid moiety. In some cases, the molecule is not used clinically as an antibiotic since the other biological properties (such as being an antiviral or an anticancer agent) are more important.

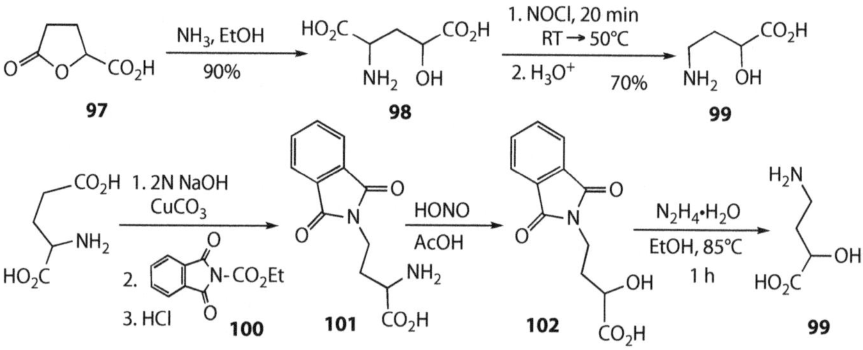

Lactone ***97*** (α-carboxy-γ-butyrolactone) is an interesting and useful synthetic intermediate[132] that is readily converted to 2-hydroxy glutamic acid (***98***) and subsequent treatment with NOCl to give 2-hydroxy-4-aminobutanoic acid (***99***).[133] This simple amino acid is biologically important since it is a structural component of the antibiotic butirosin B[134] and also of amikacin BBK-8.[135] It is noted that the carboxyl moiety in ***99*** can be reduced to give the diol derivative.[135] That diol was converted to 5-amino-3-hydroxy-2-methylpentanoic acid.[136] An alternative synthesis prepared amino acid ***99*** from glutamic acid. Initial reaction of glutamic acid with basic cupric carbonate and then ***100*** gave ***101***.[137] The amino group was converted to a hydroxyl group by reaction with HONO (giving ***102***), and removal of the phthalimidoyl group by reaction with hydrazine gave ***99***.

4-Amino-4-deoxy-*D*-erythronic acid (4-amino-2,3-dihydroxybutanoic acid, ***105***) was used in a synthesis of eritadenine, which has hypocholesterolemic activity.[138] The carbohydrate derivative, 2,3-*O*-isopropylidene-*D*-erythronolactone (***103***),

---

[132] Coppola, G.M.; Shuster, H.F. *Asymmetric Synthesis*. Wiley, New York, ***1987***, pp. 237–242.

[133] (a) Naito, T.; Nakagawa, S. *Brit.* 1,466,001 [*Chem. Abstr.* ***1977***, *87*: P102649g]; (b) Naito, T.; Nakagawa, S. *Jpn. Kokai* 74 24,914 [*Chem. Abstr.* ***1974***, *81*: P4264j]; (c) Naito, T.; Nakagawa, S., *U.S.* 3,823,187 [*Chem. Abstr.* ***1974***, *81*: P91929w].

[134] Woo, P.W.K.; Dion, H.W.; Bartz, O.R. *Tetrahedron Lett.* ***1971***, 2617.

[135] Kawaguchi, H.; Naito, T.; Nakagawa, S.; Fujisawa, K. *J. Antibiot.* ***1972***, *25*, 695.

[136] Herdeis, C.; Lütsch, K. *Tetrahedron Asym.* ***1993***, *4*, 121.

[137] Horiuchi, Y.; Akita, E.; Ito, T. *Agric. Biol. Chem.* ***1976***, *40*, 1649.

[138] Sugiyama, K.; Akachi, T.; Yamakawa, A. *J Nutr.* ***1995***, *125*, 2134.

was prepared from isoascorbic acid[139] and then converted to phthaloyl derivative ***104***.[140] Deprotection provided ***105***.

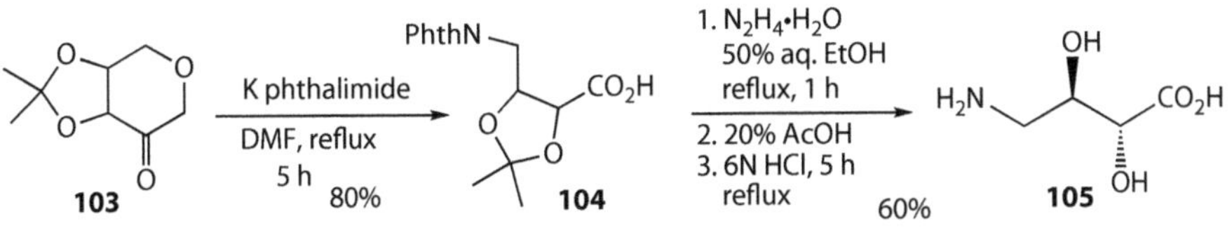

Amino acid ***108*** is an acid hydrolysis product of bleomycin A_2[141,142] and a structural component that must be incorporated in a total synthesis. Although bleomycin can be categorized as an antibiotic, it is probably best known for its anticancer activity.[141,142] Oxazolinone ***106***, which functions as a chiral auxiliary for the propanoic acid moiety, was converted to its enolate anion and condensed with *N*-Boc alanine anhydride,[143] to give ***107***. Reduction of the ketone moiety and acid hydrolysis led to 2*S*,3*S*,4*R*-4-amino-3-hydroxy-2-methylpentanoic acid (***108***).[142] It was observed that modifying the C3-hydroxyl and C2-methyl substituents on the 4-aminobutanoic acid subunit had little effect on DNA cleavage efficiency or selectivity of the corresponding bleomycin analog.[144]

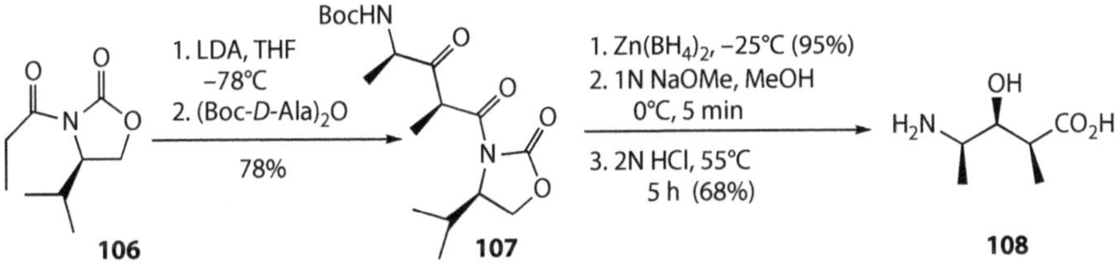

The protected amino acid derivative ***111*** was also a structural component of bleomycin.[142] Carbohydrates have been used as chiral templates in previous sections, and Hecht and coworkers converted 5-deoxy-*L*-arabino-γ-lactone (***109***) to ***110***.[142e] Reaction

---

[139] Mitchell, D.L. *Can. J. Chem.* ***1963***, *41*, 214.

[140] Kamiya, T.; Saito, Y.; Hashimoto, M.; Seki, H. *Tetrahedron*, ***1972***, *28*, 899.

[141] (a) Ikekawa, T.; Iwami, F.; Hiranaka, H.; Umezawa, H. *J. Antibiot.* (Tokyo) ***1964***, *17A*, 194; (b) Umezawa, H.; Maeda, K.; Takeuchi, T.; Okami, Y. *J. Antibiot.* (Tokyo) ***1966***, *19A*, 200; (c) Umezawa, H. Suhara, Y.; Takita, T.; Maeda, K. *J. Antibiot.* (Tokyo) ***1966***, *19A*, 210.

[142] (a) McGowan, D.A.; Jordis, U.; Minster, D.K.; Hecht, S.M. *J. Am. Chem. Soc.* ***1977***, *99*, 8078; (b) Arai, H.; Hagman, W.K.; Suguna, H.; Hecht, S.M. *J. Am. Chem. Soc.* ***1980***, *102*, 6631; (c) Levin, M.D.; Subrahamanian, K.; Katz, H.; Smith, M.B.; Burlett, D.J.; Hecht, S.M. *J. Am. Chem. Soc.* ***1980***, *102*, 1452; (d) Hecht, S.M.; Rupprecht, K.M.; Jacobs, P.M. *J. Am. Chem. Soc.* ***1979***, *101*, 3982; (e) Ohgi, T.; Hecht, S.M. *J. Org. Chem.* ***1981***, *46*, 1232; (f) Pozsgay, V.; Ohgi, T.; Hecht, S.M. *J. Org. Chem.* ***1981***, *46*, 3761; (g) Aoyagi, Y.; Suguna, H.; Murugesan, N.; Ehrenfeld, G.M.; Chang, L.-H.; Ohgi, T.; Shekhani, M.S.; Kirkup, M.P.; Hecht, S.M. *J. Am. Chem. Soc.* ***1982***, *104*, 5237; (h) Aoyagi, Y.; Katano, K.; Suguna, H.; Primeau, J.; Chang, L.-H.; Hecht, S.M. *J. Am. Chem. Soc.* ***1982***, *104*, 5537. Also see Yoshioka, T.; Hara, T.; Takita, T.; Umezawa, H. *J. Antibiot.* ***1974***, *27*, 356.

[143] DiPardo, R.M.; Bock, M.G. *Tetrahedron Lett.* ***1983***, *24*, 4805.

[144] Boger, D.L.; Colletti, S.L.; Teramoto, S.; Ramsey, T.M.; Zhou, J. *Bioorg. Med. Chem.* ***1995***, *3*, 1281.

with azide, hydrogenation of the azide, and an unusual *N*-benzylation reaction gave the acetonide of *N*-benzyl 4-(*N*-benzylamino)-2,3-dihydroxypentanamide, ***111***.

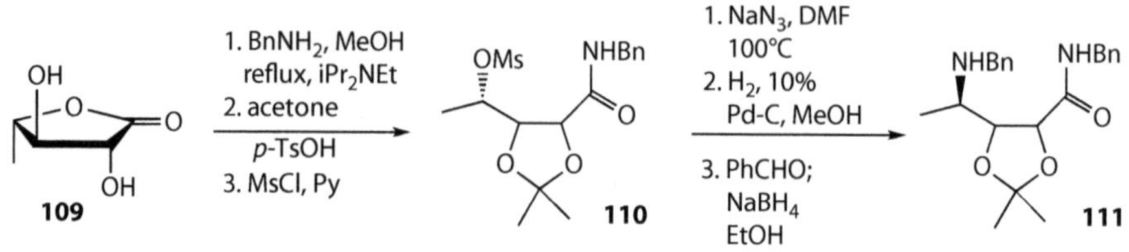

GABA derivatives are strucutral components of anticancer drugs. Several phosphopeptide mimics, such as ***119***, targeted to the SH2 domain of Stat3 were prepared, as putative anticancer drugs.[145]

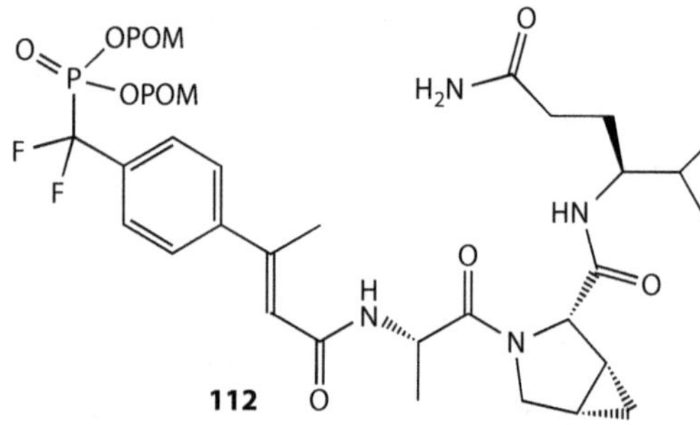

When *D*-pantolactone (***113***)[146] was treated with potassium phthalimide, racemic ***114*** was obtained.[147] Resolution with *p*-nitrophenyl-2-amino-1,3-propanediol, and separation and isolation of both resulting diastereomers of the ***114*** racemate were followed by hydrolysis and removal of the phthalimido group with hydrazine to give 4-amino-2-hydroxy-3,3,-dimethylbutanoic acid (***115***), the 4-amino-4-deoxy analog of pantothenic acid.[147] Pantothenic acid is vitamin B_5, necessary for the biosynthesis of coenzyme-A (CoA), and it is believed to play a role in several important biosynthetic pathways.

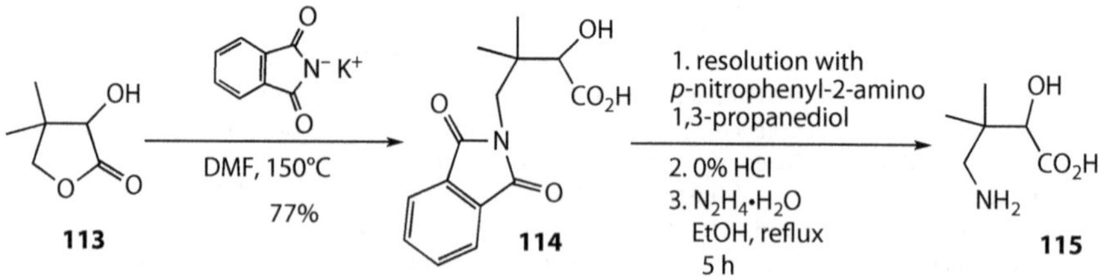

[145] Mandal, P.K.; Gao, F.; Lu, Z.; Ren, Z.; Ramesh, R.; J. Birtwistle, S.; Kaluarachchi, K.K.; Chen, X.; Bast, R.C., Jr.; Liao, W.S.; McMurray, J.S. *J. Med. Chem.* **2011**, *54*, 3549.

[146] Saito, Y.; Hashimoto, M.; Seki, H.; Kamiya, T. *Tetrahedron Lett.* ***1970***.

[147] (a) Kopelevich, V.M.; Bulanova, L.N.; Gunar, V.I. *Tetrahedron Lett.* ***1979***, 3893; (b) Bulanova, L.N. *Khim. Biokhim. Funkts. Primen. Pantotenovoi Kisloty Mater. Grodn. Simp. 4th* ***1977***, 24 [*Chem. Abstr.* ***1979***, *91*: 21033g].

Cyanuric chloride (***116***) was used as a starting material to produce hapten ***118***. Initial amination with 4-aminobutanoic acid led to ***117***, and subsequent hydroxylation of the aromatic chloride moiety gave ***118***.[148] Hapten ***118*** was developed as an indirect competitive enzyme-linked immunosorbent assay (ELISA) to produce polyclonal antibodies against dealkylated hydroxytriazines.[149] These compounds are degradation products derived from dealkylated hydroxytriazines, and may have a significant impact in the environment. It is noted that other haptens have been prepared using aminopropanoic acid linkers.[150]

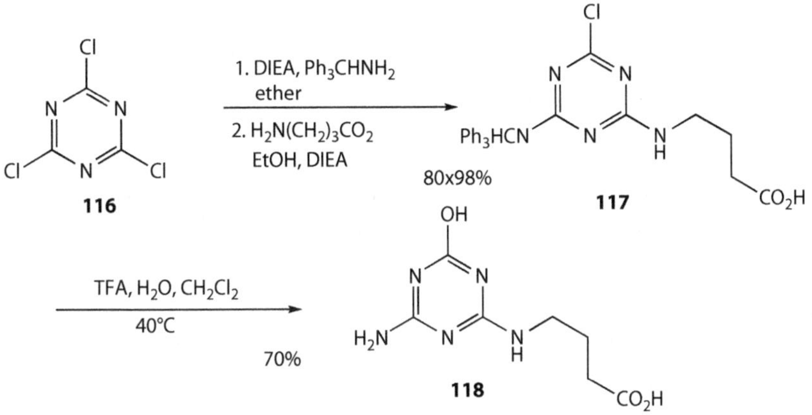

116 117 118

Other hapten derivatives have been prepared that involve non-α-amino acids. Amino acids ***119*** (n = 2–6) were prepared as derivatives of the pesticide carbofuran, ***120***.[151] These new compounds were used to develop antibody-based immunoassays.

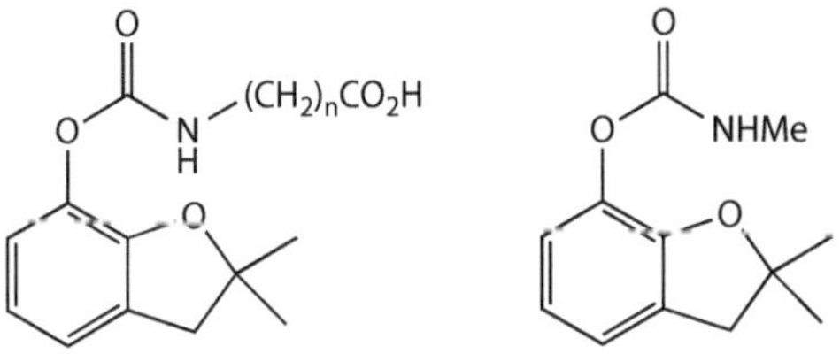

119 120

6.4 GABOB AND CARNITINE

3-Hydroxy GABA analogs have been shown to be especially important for their biological applications, so they have been separated into a separate section, with a focus on 4-amino-3-hydroxybutanoic acid (GABOB, ***121***) and its trimethylammonium

148 Sanvicens, N.; Pichon, V.; Hennion, M.-C.; Marco, M.-P. *J. Agric. Food Chem.* **2003**, *51*, 156.
149 (a) Goolsby, D.A.; Thurman, E.M.; Clark, M.L.; Pomes, M.L. *ACS Symp. Ser.* **1990**, *451*, 86; (b) Scribner, E.A.; Thurman, E.M.; Zimmerman, L.R. *Sci. Total Environ.* **2000**, *248*, 157.
150 Lee, N.; McAdam, D.P.; Skerritt, J.H. *J. Agric. Food Chem.* **1998**, *46*, 520.
151 Abad, A.; Moreno, M.J.; Montoya, A. *J. Agric. Food Chem.* **1999**, *47*, 2475.

derivative (carnitine, ***122***). GABOB is generally prepared by the methods described in previous chapters, but its biological importance has led to a number of other synthetic approaches. GABOB has been the object of a remarkable number of syntheses. For that reason, several different syntheses are shown in this section to illustrate the variety of approaches to this molecule. Discussions of these syntheses also serve to illustrate many of the synthetic approaches presented in Chapters 1–5.

R-(–)-GABOB is an inhibitory neurotransmitter[152] that has been used in the treatment of human epilepsy.[153] It also has hypertensive properties.[154] Interestingly, GABOB was first prepared by Tomita in 1923.[155]

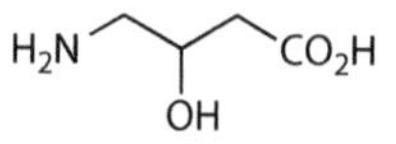

121

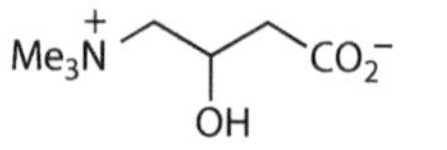

122

Carnitine is simply the *N,N,N*-trimethylammonium derivative of GABOB, and it is an amino acid found in red meat and dairy products, used by the body to convert fat into energy, and it maintains the ratio of acetyl CoA/CoA constant within the cells.[156] Carnitine is used to treat myocardial ischemia[157] and carnitine deficiency.[28a] *L*-Carnitine is important for the transport of long-chain fatty acids through the mitochondrial membrane[158] and is used for the treatment of myopathic deficiency.[159] It is sold as a dietary supplement, available at health food stores. This section will primarily focus on the synthesis of GABOB since it is clear that the trimethylammonium derivative of GABOB (carnitine, ***122***) can be prepared from ***121*** by exhaustive methylation, or by several alternative synthetic routes.

This point is illustrated in the following examples. There are several synthetic approaches that rely on enzymatic transformations for the preparation of chiral, nonracemic GABOB or carnitine. In one, *threo* β-hydroxyl glutamic acid (***123***)

---

[152] (a) Roberts, E.; Krause, D.N.; Wong, E.; Mori, A. *J. Neurosci.* ***1981***, *1*, 132; (b) Otsuka, M.; Obata, K.; Miyata, Y.; Tanaka, Y. *J. Neurosci.* ***1971***, *18*, 287; (c) Otsuka, M.; Miyata, Y. *Advances in Biochemical Psychopharmacology*, vol. 6, Raven Press, New York, ***1972***, p. 61; (d) Yamamoto, I.; Absalom, N.; Carland, J.E.; Doddareddy, M.R.; Gavande, N.; Johnston, G.A.R.; Hanrahan, J.R.; Chebib, M. *ACS Chem. Neurosci.* ***2012***, *3*, 665.

[153] (a) *Unlisted Drugs* ***1964***, *16*, 6k; (b) *Unlisted Drugs* ***1973***, *25*, 141d; (c) Hayashi, T.; Nagai, K., in *Nervous Inhibition: Proceedings of the 2nd Faraday Harbor Symposium*, Forey, ed. Pergamon Press, London, ***1961***, pp. 389–394; (d) Hayashi, T., in *Nervous Inhibition: Proceedings of the 2nd Faraday Harbor Symposium*, Florey, E., ed. Pergamon Press, London, ***1961***, pp. 378–388.

[154] (a) DeMaio, D.; Madeddu, A.; Faggioli, L. *Acta Neurol.* ***1961***, *16*, 366; (b) Buscaino, G.A.; Ferrari, E. *Acta Neurol.* ***1961***, *16*, 748.

[155] Tomita, M. *Z. Physiol Chem.* ***1923***, *124*, 253.

[156] See (a) Harmeyer, J. *The Physiological Role of L-Caritine Lohman Information* ***2002***, *27*, pp. 15–21; (b) Steiber, A.; Kerner, J.; Hoppel, C. *Mol. Aspects Med.* ***2004***, *25*, 455; (c) Bremer, J. *Physiol. Rev.* ***1983***, *63*, 1420.

[157] (a) Bahl, J.J.; Bressler, R. *Ann. Rev. Pharmacol. Toxicol.* ***1987***, *27*, 257; (b) DiPalma, J.R.; Ritchie, D.M.; McMichael, R.F. *Arch. Int. Pharmacodyn. Ther.* ***1975***, *217*, 246; (c) Rebuzzi, A.G.; Schiavoni, G.; Amico, C.M.; Montenero, A.S.; Meo, F.; Manzoli, U. *Drugs Exp. Clin. Res.* ***1984***, *10*, 219; (d) Imai, S.; Matsui, K.; Nakazawa, M.; Takatsuka, N.; Takeda, K.; Tamatsu, H. *Br. J. Pharmacol.* ***1984***, *82*, 533.

[158] Bremer, J. *Physiol. Rev.* ***1983***, *63*, 1420.

[159] Borum, P.R. *Nutr. Rev.* ***1981***, *39*, 385.

was synthesized as the starting material since it was not commercially available. Once in hand, ***123*** was selectively decarboxylated by *E. coli* to give GABOB (***121***).[160] Exhaustive methylation gave carnitine (***122***). This synthesis clearly illustrates that ***122*** is readily available after a synthesis of ***121***.

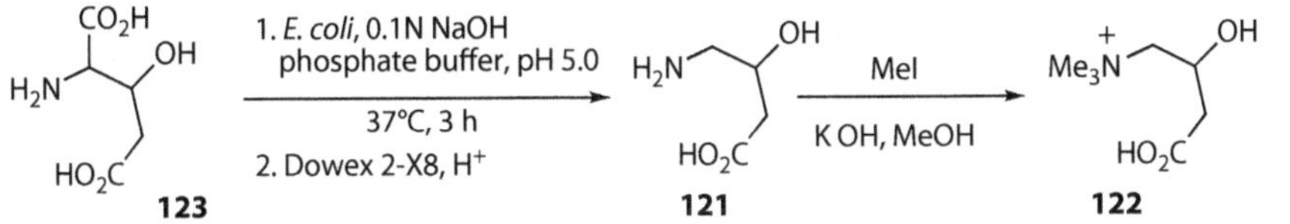

123 **121** **122**

Selective reduction of ketones to give chiral, nonracemic alcohols is possible with enzymatic reactions. Reduction of ***124*** with baker's yeast provided ***125*** with high selectivity.[161] Reaction with trimethylamine followed by acid hydrolysis gave carnitine (***122***). In this case trimethylamine is used as a nucleophile, but ammonia could be used to give GABOB rather than carnitine.

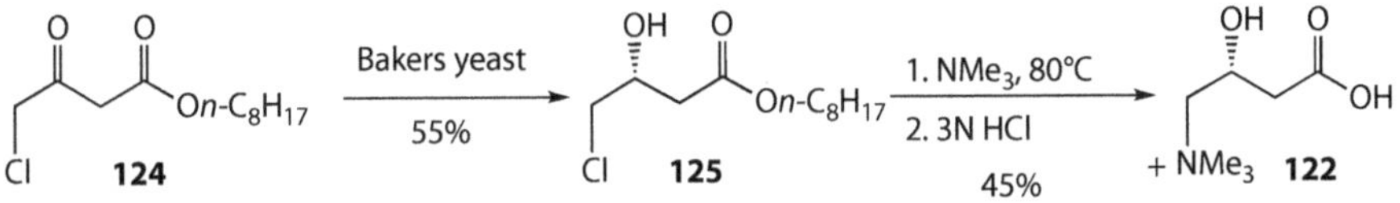

124 **125** **122**

This statement is illustrated by reduction of the ketone moiety in keto-amide ***126*** with sodium borohydride, and subsequent reduction with ammonia gave ***127***.[162] Aqueous acid hydrolysis led to GABOB, ***121***. A related ketone precursor was an α-bromo-ketone.[163] These latter two examples illustrate the fact that slight modification of a fundamental synthetic route can produce either GABOB or carnitine.

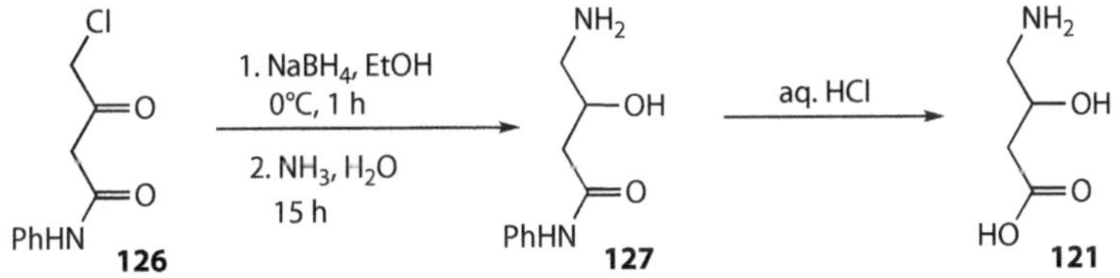

126 **127** **121**

A common synthetic approach to GABOB uses epichlorohydrin (***128***) as a starting material. In a simple example, ***128*** was treated with potassium phthalimide and sodium cyanide to give ***121***, after hydrolysis and removal of the phthalimidoyl

---

160 Kaneko, T.; Toshida, R. *Bull Chem. Soc. Jpn.* ***1962***, *35*, 1153.

161 Zhou, B.; Gopalan, A.S.; Van Middlesworth, F.; Shieh, W.-R.; Sih, C.J. *J. Am. Chem. Soc.* ***1983***, *105*, 5925.

162 Kurono, M.; Shigeoka, S.; Miyamoto, S.; Imaki, K. *Jpn. Kokai* 76 39,634 [*Chem. Abstr.* ***1976***, *85*: P62690p].

163 D'Alo, F.; Masserini, A. *Farmaco Ed. Sci.* ***1964***, *19*, 30.

group.[164] In one variation of this reaction, the intermediate phthalimidoyl product was isolated prior to treatment with cyanide.[165]

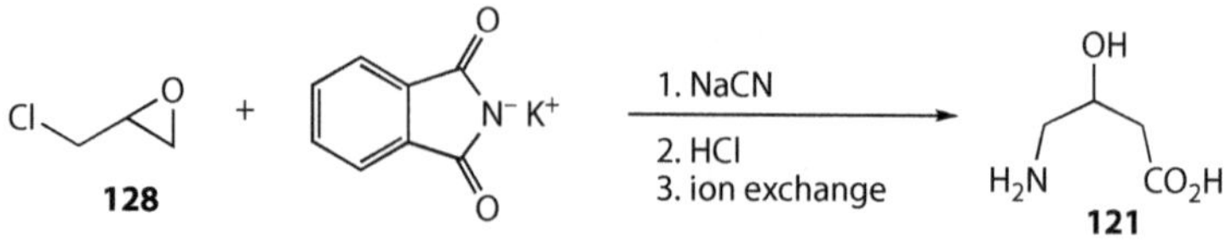

A different strategy reacted ***128*** with methanolic carbon monoxide to give ***129*** via displacement at the epoxy carbon. When the chlorine moiety in ***129*** was displaced by ammonia, spontaneous cyclization occurred to give lactam ***130***.[166] Subsequent reaction with aqueous sodium hydroxide gave GABOB (***121***). The cyano analog of ***129*** (replace CO_2Me with CN) has also been prepared, from allyl bromide, and converted to carnitine.[167]

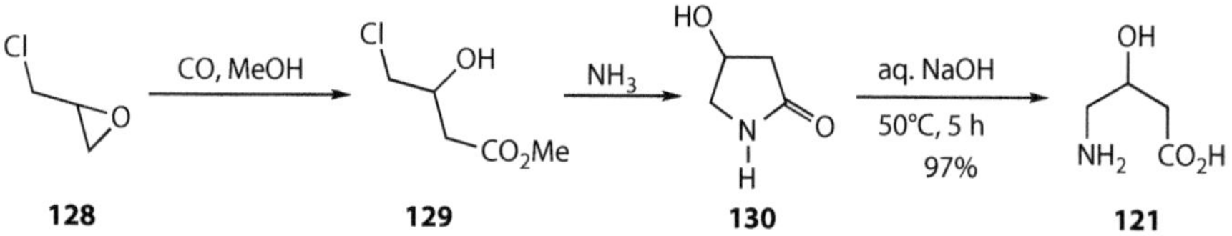

Halo-acid derivatives are useful precursors to GABOB. Interestingly, chlorination of 4-aminobutanoic acid (GABA) gave 3-chloro-4-aminobutanoic acid (***131***) in good yield and with high regioselectivity.[168] Subsequent treatment with aqueous hydroxide and passage through an ion exchange resin led to refunctionalization of the chlorine to a hydroxyl group, giving GABOB (***121***). A related study used azide as an amine surrogate to displace the chloride moiety.[169]

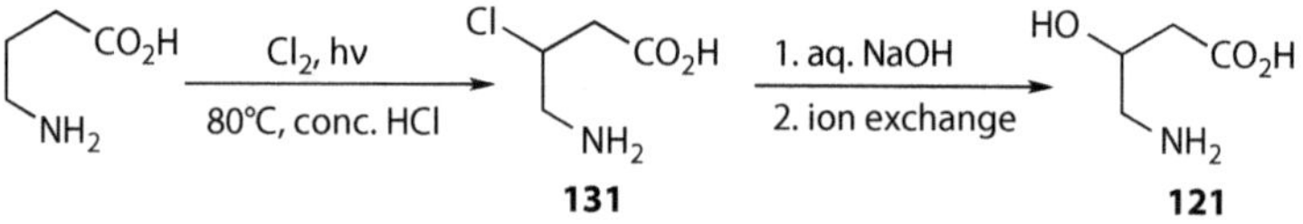

Conjugated acids such as crotonic acid can be used in this type of reaction with *N*-bromosuccinimide to give an allylic bromide, and subsequent reaction with

[164] Hebron, S.A. *Sp. Pat.* 391,718 [*Chem. Abstr.* ***1974***, *80*: P47461t]. Also see (a) Kaken Chem. Co. Ltd. *Fr. Pat.* 1,360,648 [*Chem. Abstr.* ***1964***, *61*: P11895f]; (b) Sakai, S.; Miyaji, Y.; Furutani, H.; Kobata, M.; Hachisuka, T.; Nakayama, A.; Takad, S.; Hayashi, T. *Jpn.* 12,64 ('62) [*Chem. Abstr.* ***1963***, *59*: P9805e]; Kurono, M.; Shigeoka, S.; Sakai, Y.; Itoh, H. *Ger. Offen.* 2,323,043 [*Chem. Abstr.* ***1974***, *80*: P37450z].

[165] Antonio Gallardo, S.A. *Sp. Pat.* 278,780 [*Chem. Abstr.* ***1964***, *60*: P2779a].

[166] Kenki, K.; Kogyo, K.K. *Jpn. Kokai Tokkyo Koho* JP 57,183,749 [*Chem. Abstr.* ***1983***, *98*: P106811d].

[167] Kolb, H.C.; Bennani, Y.L.; Sharpless, K.B. *Tetrahedron Asym.* ***1993***, *4*, 133.

[168] Fujimoto, Y.; Koshimoto, S. *Jpn.* 71 08,682 [*Chem. Abstr.* ***1971***, *75*: 36680j]. Also see Denki Kagaku Kogyo, K.K. *Jpn. Kokai Tokkyo Koho* 81 68,649 [*Chem. Abstr.* ***1981***, *95*: P169788k].

[169] Casati, P.; Carmeno, M.; Benedetti, A.; Fuganti, C.; Grasselli, P. *Eur. Pat. Appl.* EP 169,614 [*Chem. Abstr.* ***1986***, *105*: P173041d].

ammonium hydroxide followed by reaction with an acidic resin in water to convert the alkenyl moiety to an alcohol gave GABOB.[170]

A "heterocycle" strategy can also be used to prepare GABOB. A [3+2]-cycloaddition[171] reaction between ***132*** (prepared from glyoxylic acid)[172] and *N*-acetyl allylamine generated dihydroisoxazole ***133***.[173] Treatment with methanolic potassium carbonate gave lactam ***134***, and catalytic hydrogenation over Raney nickel opened the ring to give the *N*-acetyl methyl ester of GABOB (***135***). Acid hydrolysis liberated GABOB.[173]

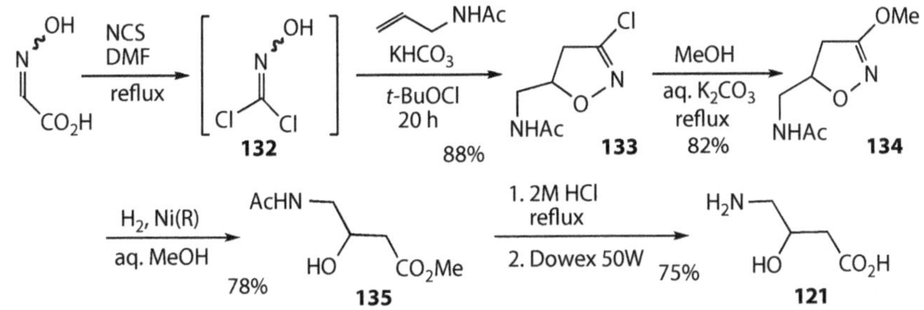

The use of enzymes to prepare chiral, nonracemic GABOB and carnitine was introduced early in this section, to show the synthetic relationship between the two targets. Other enzymatic strategies are available. The kinetic resolution of key intermediates is a common enzymatic transformation applied to the synthesis of GABOB. The diethyl ester of citric acid (***136***) was similarly converted to half-ester ***137*** with *Arthobacter* sp. (ATCC 19140).[174] Acetylation of the alcohol and conversion to the acid chloride were followed by reaction with sodium azide to give diazoketone ***138***. Heating induced a Curtius rearrangement,[175,176] and acid hydrolysis of the resulting product led to GABOB (36% overall yield from ***121***). Carnitine (***122***) was prepared by exhaustive methylation. Similar results were obtained with citric acid when ***137*** was formed in 70% yield using *Cornebacterium equi* (IFO-3730).[174]

---

170 Pinza, M.; Pifferi, G. *J. Pharm. Sci.* ***1978***, *67*, 120.

171 See (a) Smith, M.B. *Organic Synthesis*, 3rd ed. Wavefunction, Inc./Elsevier, Irvine, CA/London, England, ***2010***, pp. 1101–1115; (b) Smith, M.B. *March's Advanced Organic Chemistry*, 7th ed. John Wiley & Sons, Hoboken, NJ, ***2013***, pp. 1014–1020.

172 Wieland, H. *Ber. Dtsch. Chem. Ges.* ***1910***, *43*, 3362.

173 Halling, K.; Thomsen, I.; Torssell, K.B.G. *Liebigs Ann. Chem.* ***1989***, 985.

174 Gopalan, A.S.; Sih, C.J. *Tetrahedron Lett.* ***1984***, *25*, 5235. Also see Lu, Y.; Miet, C.; Kunesch, N.; Poisson, J. *Tetrahedron Asym.* ***1990***, *1*, 707.

175 (a) Curtius, T. *J. Praakt. Chem.* ***1894***, *50*, 275; (b) Smith, P.A.S. *Org. React.* ***1946***, *3*, 337; (c) Saunders, J.H.; Slocombe, R.J. *Chem. Rev.* ***1948***, *43*, 203; (d) Mundy, B.P.; Ellerd, M.G. *Name Reactions and Reagents in Organic Synthesis*. Wiley, New York, ***1988***, pp. 60–61.

176 See (a) Smith, M.B. *Organic Synthesis*, 3rd ed. Wavefunction, Inc./Elsevier, Irvine, CA/London, England, ***2010***, pp. 192, 1051; (b) Smith, M.B. *March's Advanced Organic Chemistry*, 7th ed. John Wiley & Sons, Hoboken, NJ, ***2013***, pp. 1361–1362.

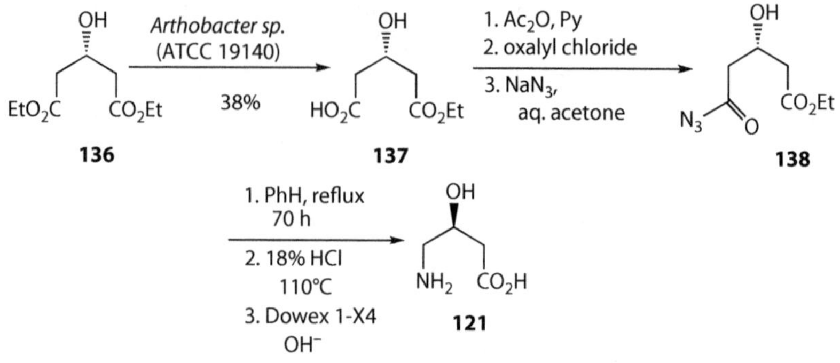

Asymmetric synthesis is possible using chiral auxiliaries. Enantiomer ***139*** contains a chiral phenethylamine auxiliary, and reaction with iodine gave a 1:1 mixture of ***140*** and ***141***, which were chromatographically separated.[177] The iodide moiety in ***140*** was reduced by treatment with tri-*n*-butyltin hydride. Removal of the auxiliary-protecting group from nitrogen, hydrolysis to open the ring, and conversion of the ester to an acid gave GABOB, ***121***.

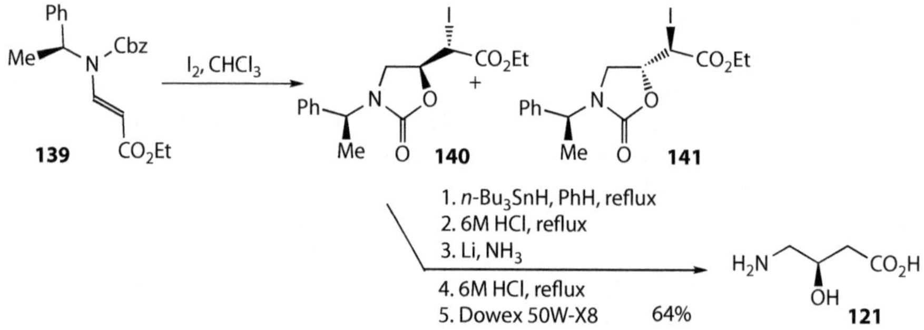

Enolate condensation reactions constitute another route to GABOB. The reaction of Cbz-protected glycinal with the ester enolate of ***142*** (bearing a chiral auxiliary) gave ***143***.[178] Hydrolysis of the ester removed the auxiliary, and removal of the Cbz group by hydrogenation over palladium gave GABOB in 82%ee.

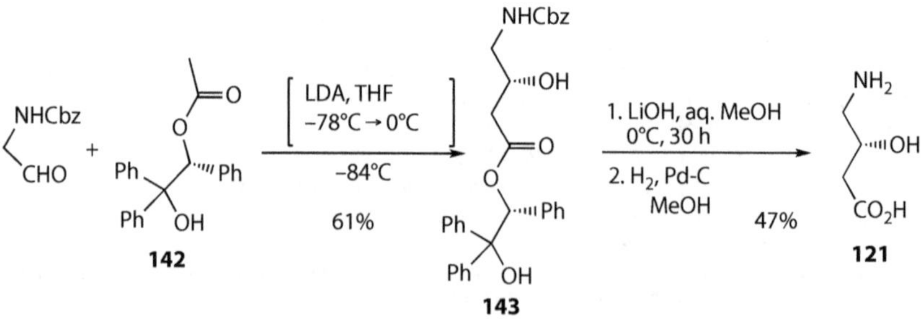

177 Bongini, A.; Cardillo, G.; Orena, M.; Porzi, G.; Sandri, S. *Tetrahedron* ***1987***, *43*, 4377.
178 Braun, M.; Waldmüller, D. *Synthesis* ***1989***, 856.

The potassium salt of *D*-erythronate (***144***) is an example of a carbohydrate template that was converted to ***145***.[179] Subsequent reaction with sodium azide, catalytic hydrogenation to reduce it to the amine, and acid hydrolysis led to *S*-GABOB. Using potassium *L*-erythronate as a precursor likewise led to *R*-GABOB in 55% overall yield.[179]

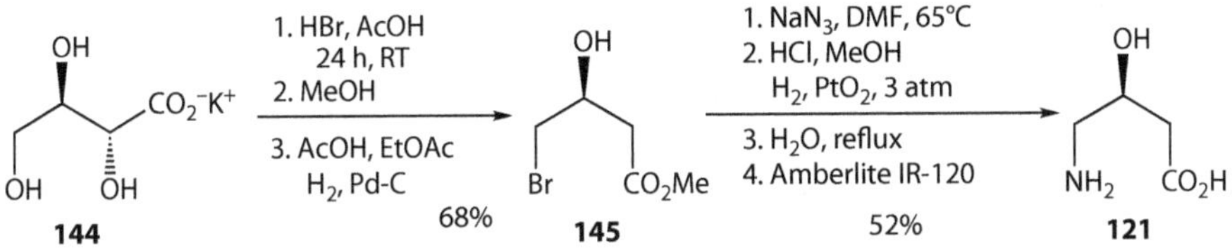

Ascorbic acid (***146***) is a carbohydrate template that was converted to ***147*** in eight steps. Conversion to the mesylate and reaction with potassium azide gave ***148***. Hydrogenation of the azide moiety and hydrolysis gave GABOB.[180]

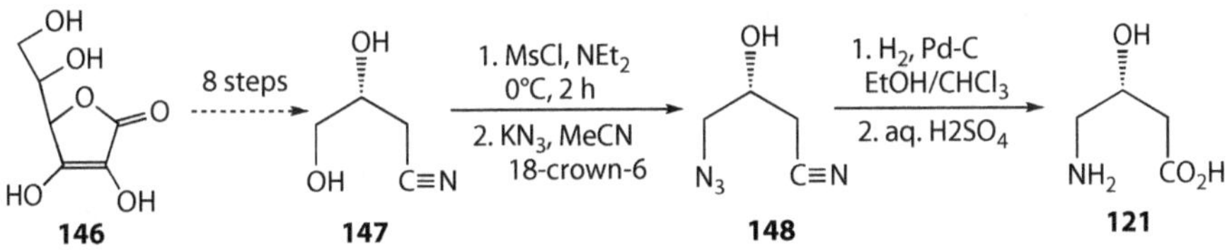

The methyl ester of 4-hydroxyproline (***149***) was prepared[181] and used as a chiral template. Deprotonation and treatment with the hypochlorite gave an *N*-chloramine, which was dehydrohalogenated to imine ***150*** by treatment with triethylamine.[182] The ester was hydrolyzed and the resulting imino acid was oxidized with hydrogen peroxide to "add" an oxygen and effect decarboxylation to generate lactam ***R-130***. Subsequent ring opening with aqueous acid gave GABOB (***121***), and the overall yield was 29%.

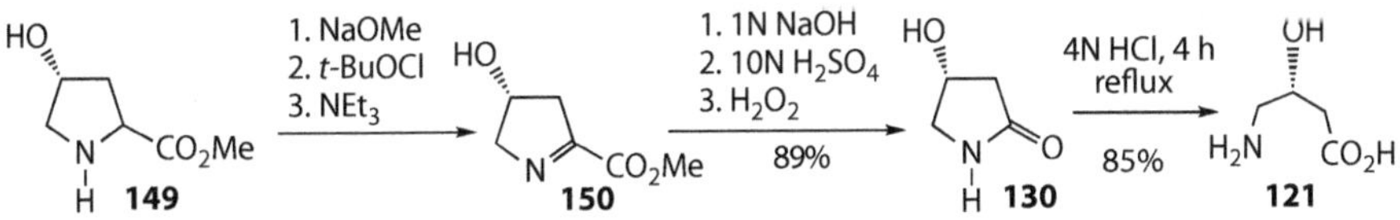

A chiral, nonracemic oxazolidine derivative (***151***) was obtained by resolution of the racemic material prepared from serine.[183] Electrolytic conversion of the acid moiety to an *O*-acetate (***152***) followed by thermal elimination of acetic acid gave ***153***. Reaction with HCl gave ***154***, which was condensed with the enolate anion of dibenzyl malonate to give ***155*** after deprotection and decarboxylation (25% yield for both

179 Bock, K.; Lundt, I.; Pedersen, C. *Acta Chem. Scand. Ser. B*, ***1983***, *B37*, 341.
180 Jung, M.E.; Shaw, T.J. *J. Am. Chem. Soc.* ***1980***, *102*, 6304.
181 Häusler, J.; Schmidt, U. *Liebigs Ann. Chem.* ***1979***, 1881.
182 Mitteilung, K.; Häusler, J. *Monatsh. Chem.* ***1987***,*118*, 865.
183 Stucky, G.; Seebach, D. *Chem. Ber.* ***1989***, *122*, 2365.

steps, with the displacement step proceeding in only 27%). Opening the ring with aqueous acid gave GABOB.

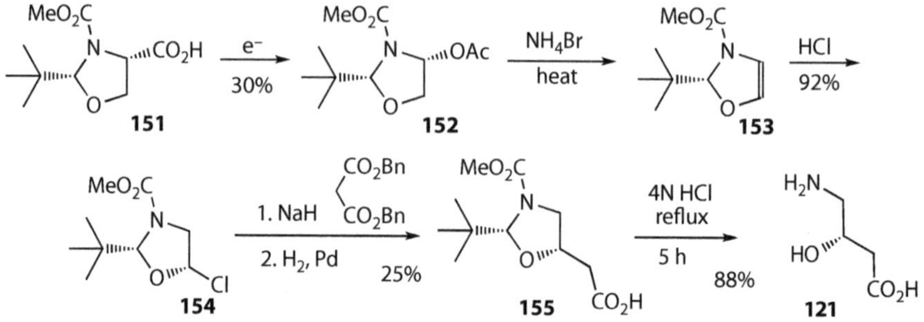

An asymmetric synthesis of *R*-(–)-GABOB was reported in which β-pinene was converted to ***156*** in eight steps.[184] Oxidative cleavage of the cyclohexene ring led to ***157***, which was hydrolyzed to ***158*** in good enantiomeric purity. Ester ***158*** was converted into either GABOB via base hydrolysis or carnitine via reaction with dimethylamine followed by subsequent *N*-methylation.

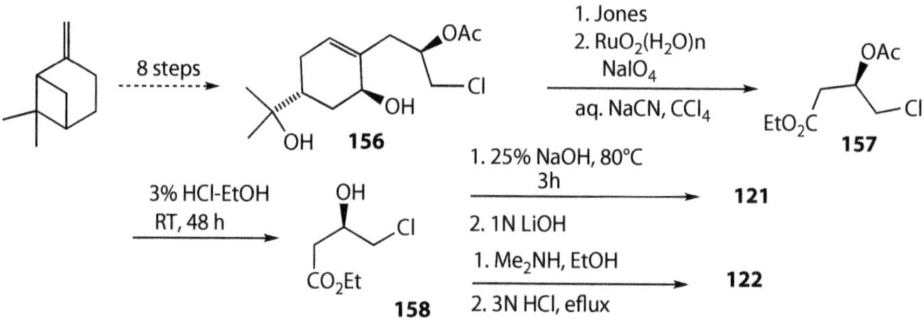

Allylic alcohol ***159*** was prepared in about seven steps from tartaric acid, and once in hand, treatment with sodium hydride and benzyl bromide gave ***160***.[185] Hydroboration of the alkene moiety and Jones oxidation[186] of the resultant alcohol gave a carboxylic acid (***161***). Deprotection and ring opening gave GABOB.[185]

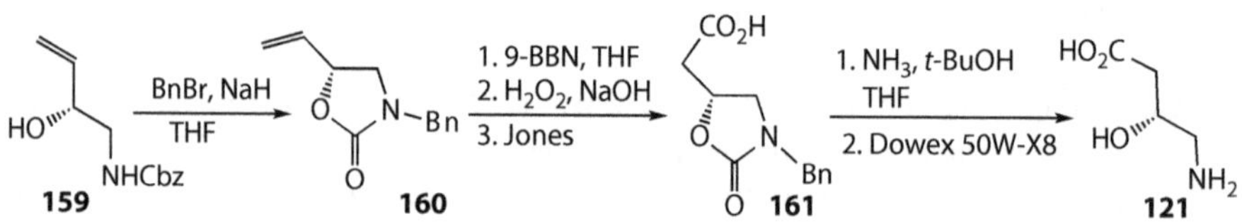

[184] Pelagata, R.; Dosi, I.; Villa, M.; Lesma, G.; Palmisano, G. *Tetrahedron* ***1985***, *41*, 5607.
[185] Bose, S.; Gurjar, M.K. *Synth. Commun.* ***1989***, *19*, 3313.
[186] See (a) Smith, M.B. *Organic Synthesis*, 3rd ed. Wavefunction, Inc./Elsevier, Irvine, CA/London, England, ***2010***, pp. 233–234; (b) Smith, M.B. *March's Advanced Organic Chemistry*, 7th ed. John Wiley & Sons, Hoboken, NJ, ***2013***, pp. 1142–1145.

To complete this section it is important to note that GABOB derivatives other than carnitine are known. For example, two cyclic peptides, solomonamides A and B, were isolated from the marine sponge *Theonella swinhoei*. Solomonamide A (***162***) showed in vivo anti-inflammatory activity, and it contained the 2-methyl-3-hydroxy-4-aminobutanoic acid residue, as marked in ***162***. This amino acid residue is essentially 2-methyl GABOB.[187]

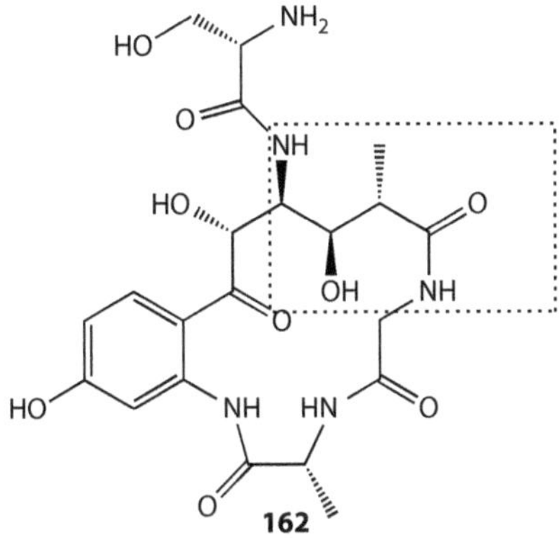

162

6.5 DAVA (AVA) AND AHX

This section will cover two important amino acids, 5-aminopentanoic acid and 6-aminohexanoic acid, as well as derivatives of those compounds. These two amino acids appear quite often in biologically important molecules, so this section will illustrate both the biological inactivity and different synthetic approaches.

5-Aminopentanoic acid (δ-aminovaleric acid, DAVA) is a neurotransmitter that is used for the treatment of neuromuscular disease. In at least one study, 3-alkyl-4-aminobutanoic acid derivatives were shown to be in vitro activators of *L*-glutamic acid decarboxylase, and they showed anticonvulsant activity.[188] Other studies have shown DAVA derivatives to be GABA-B antagonists.[189]

DAVA and DAVA derivatives appear in many biologically important systems. Several analogs of human melanin-concentrating hormone were prepared and shown to be nonselective agonists for receptors 1 and 2 (hMCH-1R and hMCH-2R). These peptide analogs incorporated 5-aminopentanoic acid in segments of the disulfide ring.[190] Signal transducers and activators of transcription 3 (Stat3) are targets for anticancer

---

187 (a) Festa, C.; De Marino, S.; Sepe, V.; D'Auria, M.V.; Bifulco, G.; Débitus, C.; Bucci, M.; Vellecco, V.; Zampella, A. *Org. Lett.* ***2011***, *13*, 1532; Also see (b) Kashinath, K.; Vasudevan, N.; Reddy, D.S. *Org. Lett.* ***2012***, *14*, 6222.

188 Taylor, C.P.; Vartanian, M.G.; Andruszkiewicz, R.; Silverman, R.B. *Epilepsy Res.* ***1992***, *11*, 103 [*Chem. Abstr.* ***1992***, *117*: 783s].

189 Frydenvang, K.; Kristiansen, U.; Frølund, B.; Herdeis, C.; Krogsgaard-Larsen, P. *Chirality* ***1995***, *7*, 526.

190 Bednarek, M.A.; Hreniuk, D.L.; Tan, C.; Palyha, O.C.; Douglas J. MacNeil, D.J.; Van der Ploeg, L.H.Y.; Howard, A.D.; Feighner, S.D. *Biochemistry* **2002**, *41*, 6383.

drug design, and phosphopeptide mimics have been prepared to target the SH2 domain of Stat3. Lactam ***163*** and analogs were examined, and pertinent to this discussion, they contained ω-amino carboxylic acid units (aminobutanoic, aminopentanoic, and aminohexanoic acids).[191] Note that POM is pivaloyloxymethyl.

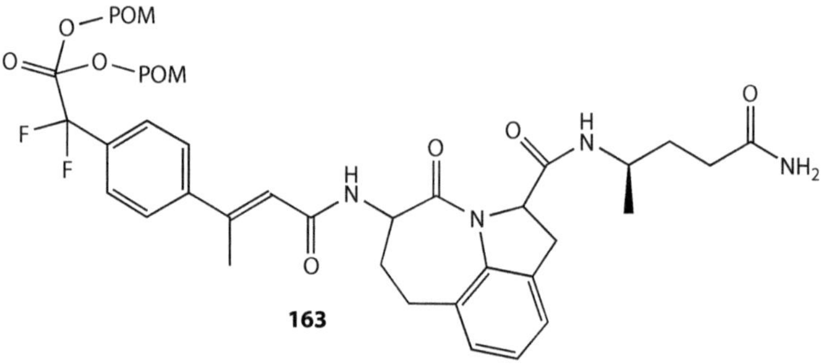

163

Alpha-7 nicotinic acetylcholine receptors (α7 nAChR) are important for the modulation of many cognitive functions. This observation led to development of α7 nAChR agonists as therapeutic candidates for the treatment of Alzheimer's disease and schizophrenia. 5-Amino acid derivatives such as ***164*** led to a study that identified ***165*** as an improved lead. Further work prepared 5-(4-acetyl[1,4]diazepan-1-yl) pentanoic acid [5-(4-methoxyphenyl)-1H-pyrazol-3-yl] amide (***166***) as a novel, full agonist of the α7 nAChR.[192]

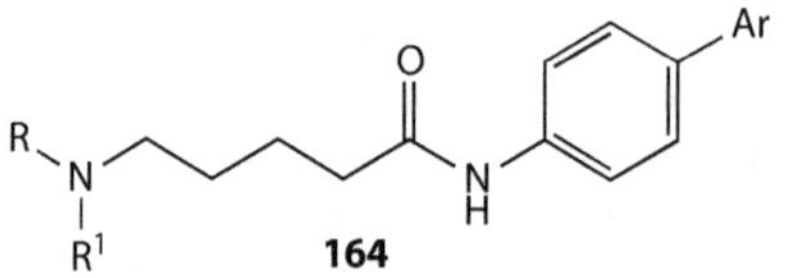

164

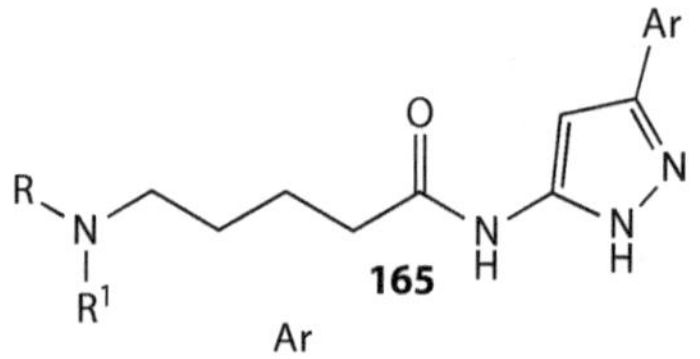

165

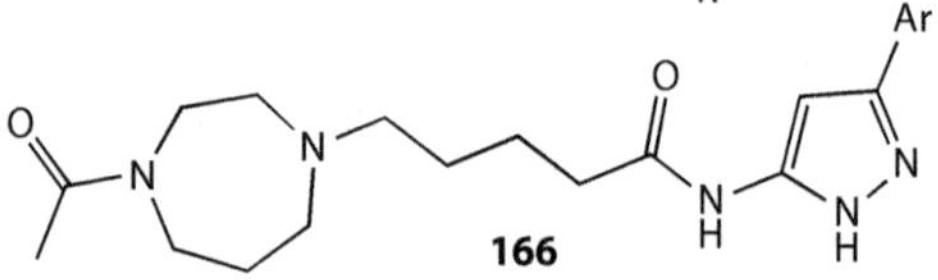

166

Work has been reported to develop selective *a7* nACh agonists with the goal of improving cognition in schizophrenia and Alzheimer's disease. One example of an agonist of this type is ***167***, with a pyrazole moiety coupled to a 5-aminopentanamide.[193] Interestingly, the naturally occurring compound obyanamide, isolated from the

[191] Mandal, P.K.; Gao, F.; Lu, Z.; Ren, Z.; Ramesh, R.; Birtwistle, J.S.; Kaluarachchi, K.K.; Chen, X.; Bast, R.C., Jr.; Liao, W.S.; McMurray, J.S. *J. Med. Chem.* ***2011***, *54*, 3549.

[192] Zanaletti, R.; Bettinetti, L.; Castaldo, C.; Cocconcelli, G.; Comery, T.; Dunlop, J.; Gaviraghi, G.; Ghiron, C.; Haydar, S.N.; Jow, F.; Maccari, L.; Micco, I.; Nencini, A.; Scali, C.; Turlizzi, E.; Valacchi, M. *J. Med. Chem.* ***2012***, *55*, 4806.

[193] Zanaletti, R.; Bettinetti, L.; Castaldo, C.; Cocconcelli, G.; Comery, T.; Dunlop, J.; Gaviraghi, G.; Ghiron, C.; Haydar, S.N.; Jow, F.; Maccari, L.; Micco, I.; Nencini, A.; Scali, C.; Turlizzi, E.; Valacchi, M. *J. Med. Chem.* ***2012***, *55*, 4806.

marine cyanobacterium *Lyngbya conervoides* and shown to exhibit cytotoxic activity, contains 3-aminopentanoic acid as one amino acid residue.[194] Obyanamide also contains a thiazole amino acid residue (see heterocyclic amino acids in Chapter 7).

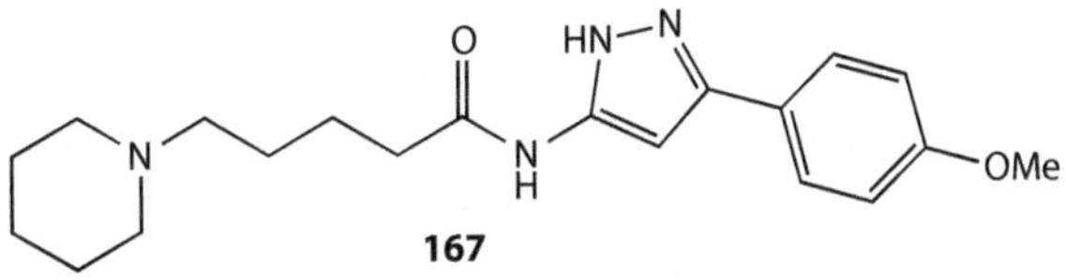

167

BACE-1 (β-site amyloid precursor protein cleaving enzyme) protease inhibitors such as ***172*** that contain non-α-amino acid residues have been prepared from chiral carbohydrates.[195] These inhibitors show high BACE-1 potency and good selectivity against cathepsin D.[196] Initial conversion of ***168*** to furanone ***169*** was followed by ring opening of the lactone moiety to give ***170***. Reduction of the acid and coupling with carboxylic acid ***171*** gave ***172***.[196]

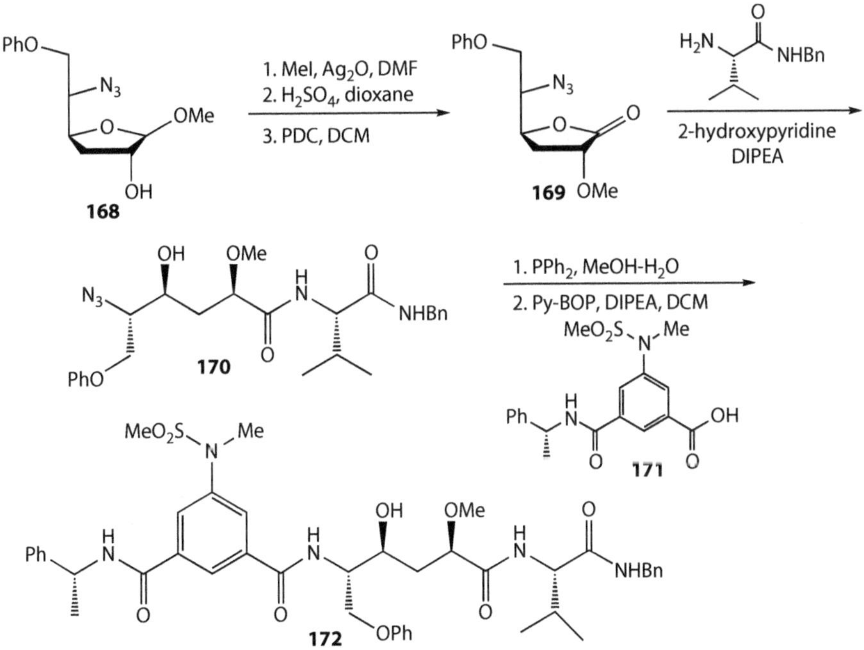

[194] Williams, P.G.; Yoshida, W.Y.; Moore, R.E.; Paul, V.J. *J. Nat. Prod.* ***2002***, *65*, 29.

[195] Björklund, C.; Oscarson, S.; Benkestock, K.; Borkakoti, N.; Jansson, K.; Lindberg, J.; Vrang, L.; Hallberg, A.; Rosenquist, Å.; Samuelsson, B. *J. Med. Chem.* ***2010***, *53*, 1458.

[196] Liaudet-Coopman, E.; Beaujouin, M.; Derocq, D.; Garcia, M.; Glondu-Lassis, M.; Laurent-Matha, V.; Prébois, D.; Rochefort, H.; Vignon, F. *Cancer Lett.* ***2006***, *237*, 167.

6.6 STATINE AND RELATED COMPOUNDS

Statine (4-amino-3-hydroxy-6-methylheptanoic acid, ***173***) is an important hydroxy amino acid with biological activities that have attracted considerable attention. Indeed, there are many syntheses of this important amino acid. Statine is a key constituent of pepstatin and related inhibitors and also of the blood pressure regulatory enzyme renin.[197] It is a potential antihypertensive medicinal agent since it is a component of renin inhibitors.[197] It is also a constituent of cathepsin D, an aspartic protease.[198] Statine is a component of didemnin cyclodipeptide.[199] The didemnins are known to inhibit the growth of RNA and DNA viruses and are cytotoxic to L1210 leukemia cells.[197,200]

The most important method for preparing statine probably involves the conversion of leucine to leucinal derivatives (***174***). This aldehyde could then be condensed with a variety of nucleophilic reagents or with an ylid. In both approaches, the carbon chain is extended and an acid moiety is incorporated.

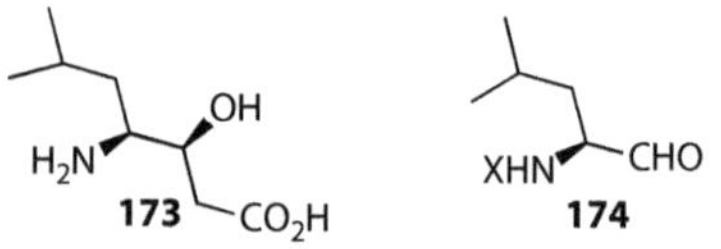

173 **174**

Natural products contain the statine moiety. A statine unit-containing linear decadepsipeptide, grassystatin A (***175***), which selectively inhibited cathepsins D and E, was isolated from the cyanobacterium *Lyngbya* cf. *confervoides*.[201]

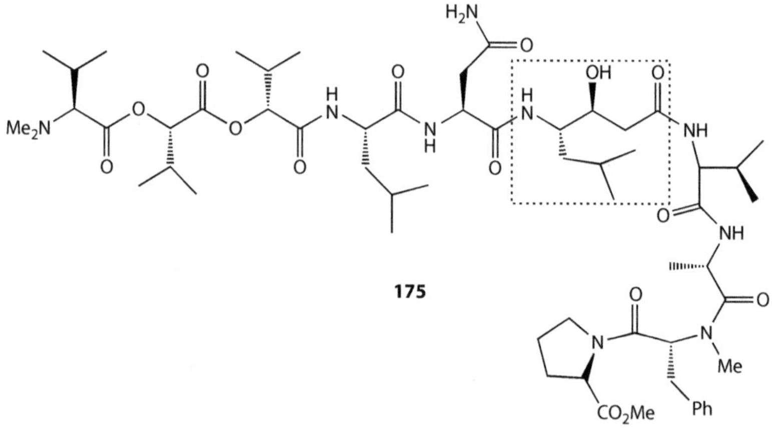

175

[197] (a) Rinehart, K.L., Jr.; Gloer, J.B.; Hughes, R.G., Jr.; Renis, H.E.; McGovern, J.P.; Swynenberg, E.B.; Stringfellow, D.A.; Kuentzel, S.L.; Li, L.H. *Science* ***1981***, *212*, 933; (b) Rinehart, K.L., Jr.; Gloer, J.B.; Cook, J.C. Jr. *J. Am. Chem. Soc.* ***1981***, *103*, 1857; (c) Umezawa, H.; Aoyagi, T.; Morishima, H.; Masuzaki, M.; Hamada, M.; Takenchi, T. *J. Antibiot.* ***1970***, *23*, 259.

[198] (a) Boger, J.; Lohr, N.S.; Ulm, E.H.; Poe, M.; Blaine, E.H.; Fanelli, G.M.; Lin, T.-Y.; Payne, L.S.; Schorn, T.W.; LaMont, B.I.; Vassil, T.C.; Stabilito, I.I.; Veber, D.F. *Nature* ***1983***, *303*, 81; (b) Rich, D.H. *J. Med. Chem.* ***1985***, *28*, 263.

[199] Rinehart, K.L., Jr.; Gloer, J.B.; Cook, J.C., Jr. *J. Am. Chem. Soc.* ***1981***, *103*, 1857.

[200] Rinehart, K.L., Jr.; Gloer, J.B.; Hughes, R.G.; Renis, H.E.; McGovern, J.P. Swynenberg, E.B.; Stringfellow, A.; Kuentzel, S.L.; Li, L.H. *Science* ***1981***, *212*, 933.

[201] Kwan, J.C.; Eksioglu, E.A.; Liu, C.; Paul, V.J.; Luesch, H. *J. Med. Chem.* ***2009***, *52*, 5732.

An early and typical synthesis shows the conversion of *N*-Boc-*L*-leucine methyl ester to aldehyde ***176***.[202] This aldehyde was condensed with the lithium enolate of ethyl acetate to give ***177***. Alcohol ***177*** was produced as a mixture of diastereomers, and one of the important challenges of this strategy was to control the diastereoselectivity of the condensation. In this example, the 3*S*,4*S* diastereomer of ***177*** was separated and isolated (in 40% yield based on ***176***) and acid hydrolysis gave *N*-Boc statine (***178***).[202]

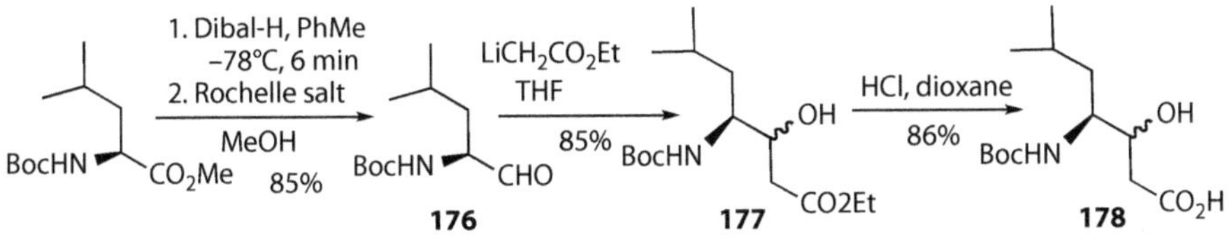

Leucinal derivatives can react with carbonyl-bearing ylids to form an alkenyl amino acid. The alkene moiety can be converted to an alcohol, allowing a synthesis of statine. Conjugated ester ***179*** was prepared from *N*-Cbz-leucinal with good selectivity (>30:1 *cis:trans*).[203] The ester moiety was reduced to an aldehyde and the double bond epoxidized, allowing reduction with Red-Al to give ***180***. The primary alcohol moiety was oxidized with oxygen and platinum oxide to give *N*-Cbz statine (***181***). A number of other derivatives were prepared by this method, including 4-amino-3-hydroxypentanoic acid, 4-amino-3-hydroxy-7-methyloctanoic acid, and 4-amino-3-hydroxy-5-phenylpentanoic acid.[203]

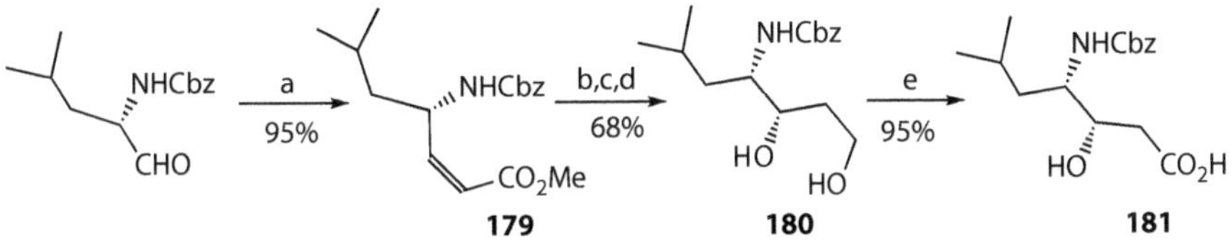

(a) $[CF_3CH_2O]_2P(O)CH_2CO_2Me$, TMS_2NK, THF, −78°C (b) Dibal-H, CH_2Cl_2, −78°C (c) mCPBA (d) Red-Al, THF (e) PtO_2, O_2, $NaHCO_3$

The Mukaiyama reaction[204] has been used as a key step for the preparation of statine. *L*-Leucine was converted to the *N*-isopropylcarbamate leucinal (***182***).[205] Mukaiyama reaction with the trimethylsilyl enolate of methyl acetate, catalyzed by titanium tetrachloride, gave ***183*** with good diastereoselectivity (94:6 *anti:syn*). The *anti* diastereomer was converted to the ethyl ester of statine (***173***) in 88% yield.

---

202 Rich, D.H.; Sun, E.T.; Boparai, A.S. *J. Org. Chem.* ***1978***, *43*,3624.

203 Kogen, H.; Nishi, T. *J. Chem. Soc. Chem. Commun.* ***1987***, 311. Also see Misiti, D.; Zappia, G. *Tetrahedron Lett.* ***1990***, *31*, 7359.

204 (a) Mukaiyama, T. *Org. React.* ***1982***, *28*, 203; (b) Mukaiyama, T. *Angew. Chem. Int. Ed. Engl.* ***1977***, *16*, 817; (c) see Smith, M.B. *Organic Synthesis*, 3rd ed. Wavefunction, Inc./Elsevier, Irvine, CA/London, England, ***2010***, pp. 837–841; (d) Smith, M.B. *March's Advanced Organic Chemistry*, 7th ed. John Wiley & Sons, Hoboken, NJ, ***2013***, pp. 1151–1153.

205 (a) Ohta, T.; Shiokawa, S.; Sakamoto, R.; Nozie, S. *Tetrahedron Lett.* ***1990***, *31*, 7329; (b) Takemoto, Y.; Matsumoto, T.; Ito, Y.; Terashima, S. *Chem. Pharm. Bull.* ***1991***, *39*, 2425.

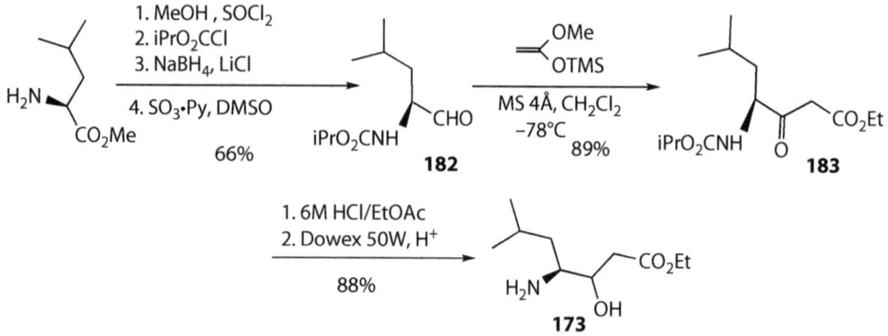

Another approach used a leucinal derivative in a Reformatsky reaction.[206] When the *N*-phthaloyl derivative ***184*** reacted with zinc and bromo-*tert*-butyl acetate, ***185*** was obtained as a mixture of diastereomers.[207] Separation of the requisite diastereomer was followed by conversion to statine (***173***) by deprotection of the amine and the carboxyl groups. A modification of this sequence converted *N*-phthaloyl leucine to the acid chloride (with thionyl chloride), and this protection step was followed by reaction with the magnesium salt of mono-ethyl malonate.[208] The resultant ketone was reduced and decarboxylated to give statine, but in only a 6.2% overall yield.[207] This strategy had been reported earlier in the patent literature.[209]

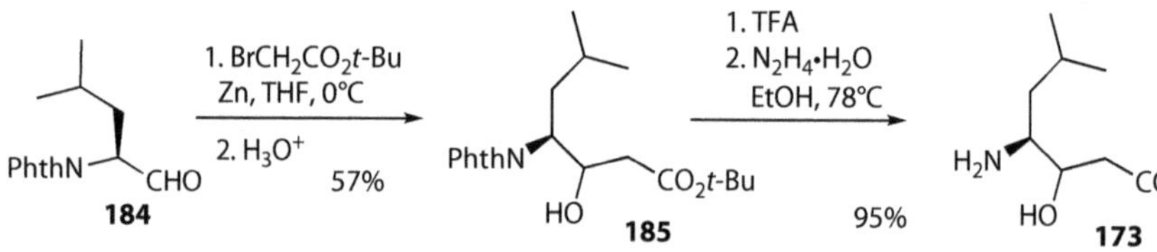

There are several syntheses that involve the use of transition metal-catalyzed coupling strategies. Reaction of leucinal (***186***) with 1,3-butanediol gave a mixture of ***187*** (in 9% yield) and ***188*** (in 71% yield).[210] After the chromatographic separation of ***188***, the titanium tetrachloride-catalyzed reaction with 3-trimethylsilyl-1-propene gave ***189***. Ozonolysis and oxidation gave lactam ***190***, which was converted to ***190*** in five steps (overall yield was 32% from ***190***). Note that in this case, the coupling reaction with sodium cyanide required the use of ultrasound [))))))].

---

[206] (a) Reformatsky, S. *Ber.* ***1887***, *20*, 1210; (b) Rathke, M.W. *Org. React.* ***1975***, *22*, 423; (c) Diaper, D.G.M.; Kuksis, A. *Chem. Rev.* ***1959***, *59*, 89; (d) see Smith, M.B. *Organic Synthesis*, 3rd ed. Wavefunction, Inc./Elsevier, Irvine, CA/London, England, ***2010***, pp. 885–887; (e) Smith, M.B. *March's Advanced Organic Chemistry*, 7th ed. John Wiley & Sons, Hoboken, NJ, ***2013***, pp. 1129–1131.

[207] Liu, W.-S.; Smith, S.C.; Glover, G.I. *J. Med. Chem.* ***1979***, *22*, 577.

[208] Liu, W.-S.; Glover, G.I. *J. Org. Chem.* ***1978***, *43*, 754.

[209] Atsumi, T.; Yamamoto, H. *Jpn. Kokai* 78 05,114 [*Chem. Abstr.* ***1978***, *88*: P190098a].

[210] Andrew, R.G.; Conrow, R.E.; Elliott, J.D.; Johnson, W.S. *Tetrahedron Lett.* ***1987***, *28*, 6535.

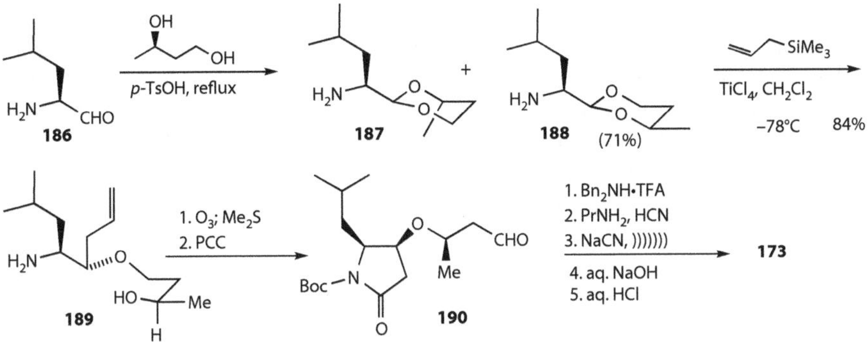

Other amino acid precursors have been used for synthesis. The α-amino acid valine was converted to statine via initial condensation of the ethyl ester of valine with ***191*** to give ***192***.[211] Heating this adduct in methanol led to lactam ***193***, which was reduced to ***194***. Hydrolysis gave statine, in 63% overall yield from valine.

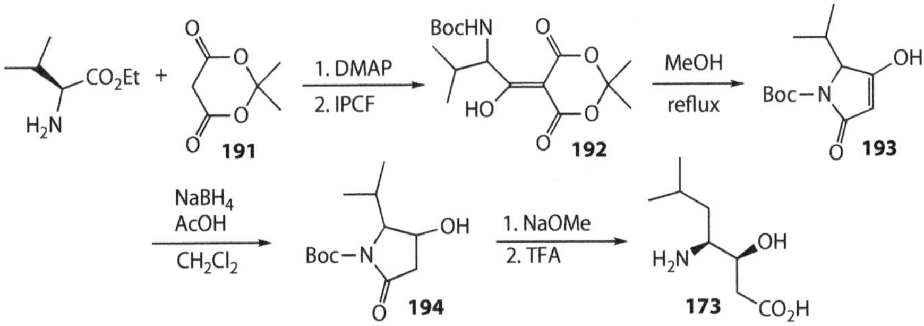

An oxidative route involved allylpalladium-catalyzed oxazolidinone formation of carbamate ***195***. This reaction gave a 15:1 mixture of ***196:197***.[212] Hydroboration of the vinyl group was followed by hydrolysis to give ***198***. The primary alcohol moiety was oxidized to the acid moiety using platinum oxide and oxygen to give *N*-Boc statine, ***199***.

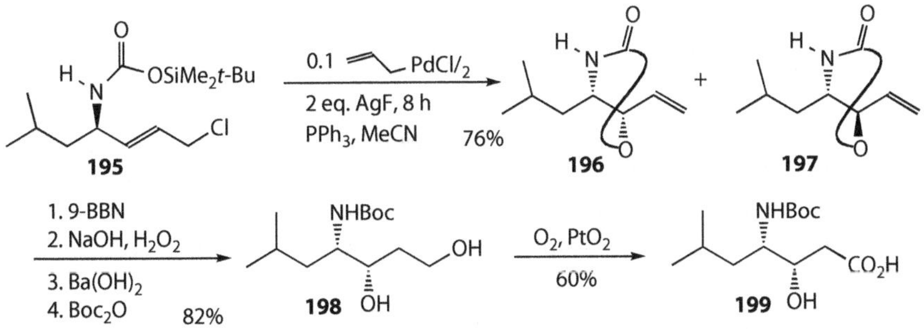

[211] Jouin, P.; Castro, B.; Poncet, J.; Nisato, D. *Rept. Proc. Eur. Pept. Symp. 19th*, ***1986***, (Pub. 1987), 659 [*Chem. Abstr.* ***1988***, *108*: 38381m].

[212] Sakaitni, M.; Ohfune, Y. *J. Am. Chem. Soc.* ***1990***, *112*, 1150.

Carbohydrates are important chiral, nonracemic precursors (chiral templates) to statine and its derivatives. Aldehyde ***200***[213] was converted to alcohol ***201***, for example.[214] After conversion to the benzoate (***202***), reaction with sodium azide gave ***203***. The tetrahydrofuran ring was opened and oxidation gave ***204***.[214] Oxidation of the aldehyde to the carboxylic acid and then reduction of the azide moiety led to (+)-statine (***173***). Reaction of ***201*** with mesyl chloride and then with sodium benzoate gave the benzoate (***202***) of opposite absolute configuration, which was converted to (–)-statine by the same sequence.

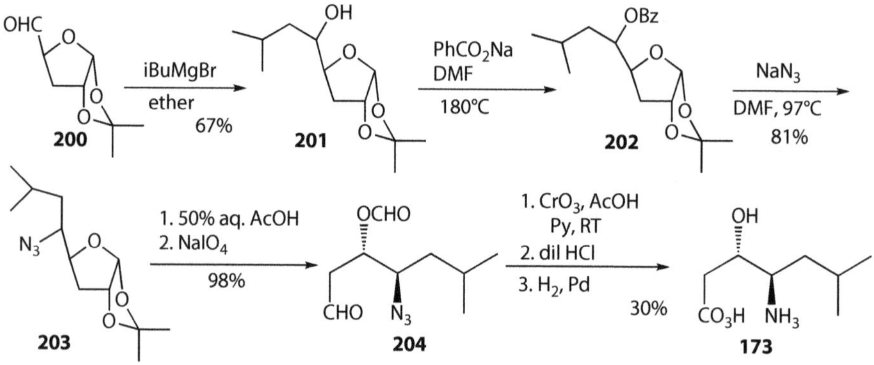

Enolate anion condensation reactions have been used. Wuts showed that ***205*** (which bears a chiral auxiliary) condensed with ***184*** to give an 8:1 mixture of ***206*** and ***207***.[215] Saponification gave a 9.2:1 mixture of diastereomeric *N*-Boc statine derivatives, ***199*** and ***208***.

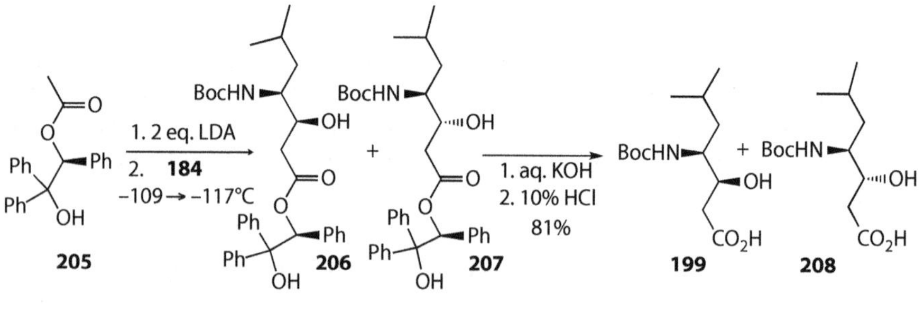

A lactam-driven synthesis reacted imide ***209*** with isobutenylmagnesium chloride to give ***210***.[216] Catalytic hydrogenation gave ***211*** along with 18% of ***212***. Separation of these products allowed the conversion of ***211*** into statine. This fundamental approach was later used by Speckamp except that a vinylsilane was used in the condensation

[213] Murrary, D.H.; Prokop, J. *J. Pharm. Sci.* ***1965***, *54*, 1468.

[214] (a) Kinoshita, M.; Hagiwara, A.; Aburaki, S. *Bull. Chem. Soc. Jpn.* ***1975***, *48*, 570; (b) Kinoshita, M.; Aburaki, S.; Hagiwra, A.; Imai, J. *J. Antibiot.* ***1973***, *26*, 249. Also see Yanagisawa, H.; Kanazaki, T.; Nishi, T. *Chem. Lett.* ***1989***, 687.

[215] Wuts, P.G.M.; Putt, S.R. *Synthesis* ***1989***, 951.

[216] Ohta, T.; Shiokawa, S.; Sakamoto, R.; Nozie, S. *Tetrahedron Lett.* ***1990***, *31*, 7329.

reaction with the imide, to form an isobutylene moiety.[217] Speckamp introduced a modification that allowed stereocontrol of the amine and alcohol moieties.

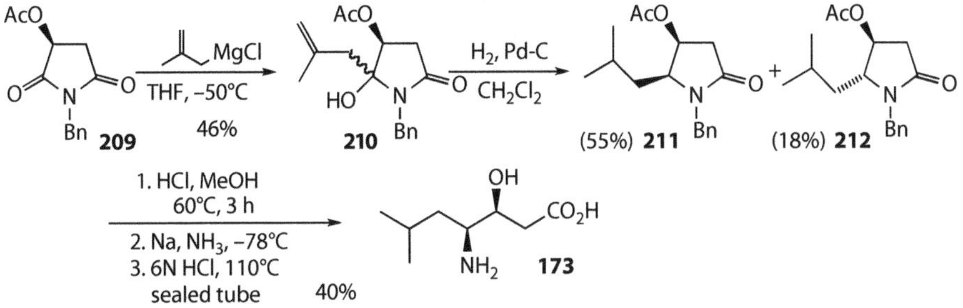

An oxazolidone-driven synthesis reacted ***213*** with benzyl isocyanate to give ***214***.[218] Reduction to the aldehyde to an alcohol and adjustment of the pH led to formation of ***215***, but with poor diastereoselectivity (isolated as a 1:1 *cis:trans* mixture). Incorporation of the isobutenyl group and ring opening gave ***216***. Hydrogenation of the alkenyl moiety was followed by formation of ***217***. A $RuCl_2/NaIO_4$ oxidation of the phenyl group gave ***218***,[218] and hydrolysis gave ***173***.[219]

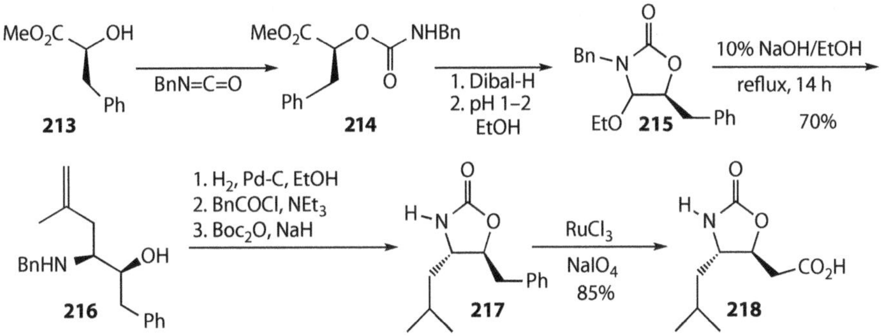

Another oxazolidone-driven synthesis used the hetereocycle as a protected carboxyl (an amide), and it also served as a chiral auxiliary. Treatment of ***219*** with a boron triflate, for example, gave boron enolate reagent ***220*** in situ, and it reacted with *N*-Boc leucinal to form ***223***.[220] Boron enolates have been used in many syntheses as an alternative to aldol reactions.[221] Hydrogenolysis of the methylthio moiety gave a methylene, and treatment of the amide auxiliary with base gave ***173*** (24% overall yield).

[217] Koot, W.-J.; Van Ginkel, R.; Kranenburg, M.; Hiemstra, H.; Louwrier, S.; Moolenaar, M.J.; Speckamp, W.N. *Tetrahedron Lett.* ***1991***, *32*, 401.

[218] Kano, S.; Yuasa, Y.; Yokomatsu, T.; Shibuya, S. *J. Org. Chem.* ***1988***, *53*, 3865.

[219] Kano, S.; Yokomatsu, T.; Iwasawa, H.; Shibuya, S. *Tetrahedron Lett.* ***1987***, *28*, 6331.

[220] Woo, P.W.K. *Tetrahedron Lett.* ***1985***, *26*, 2973.

[221] See (a) Van Horn, D.E.; Masamune, S. *Tetrahedron Lett.* ***1979***, 2229; (b) Evans, D.A.; Nelson, J.V.; Vogel, E.; Taber, T.R. *J. Am. Chem. Soc.* ***1981***, *103*, 3099.

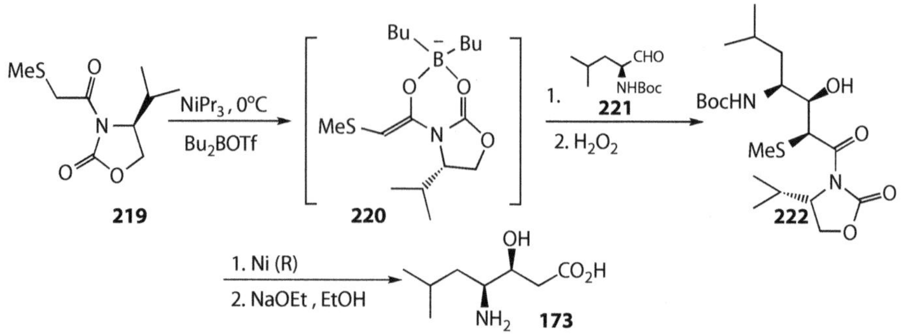

A different heterocyclic strategy used an isoxazoline intermediate. When 5-methyl-2*E*-hexen-1-ol was treated with trichloroacetonitrile and heated, amide ***225*** was formed.[222] Reaction with dichlorooxime led to a 6.4:1 mixture of ***226:227***. Conversion to the methoxy derivative allowed ring opening by catalytic hydrogenation to give ***228***. Acid hydrolysis gave the *threo* derivative ***173*** (84% yield from ***226***). Likewise, after separation from ***226***, ***227*** was converted to the *erythro* diastereomer, ***229***.

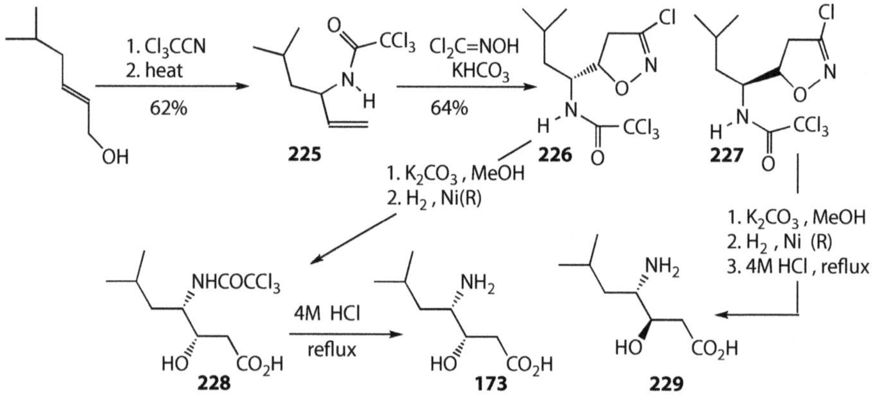

Statine derivatives are available by more or less classical reaction sequences. The reaction of epichlorohydrin with cyanide gave 3-hydroxyglutaronitrile (***230***). A nitrilase-catalyzed desymmetrization gave ***231*** with 100% conversion and 99%ee.[223] Esterification gave ethyl (*R*)-4-cyano-3-hydroxybutyrate, ***232***, a potential intermediate in the synthesis of atorvastatin (Lipitor). This sequence was done on a kilogram scale.[223] An alternative desymmetrization sequence that generated ***231*** used halohydrin dehalogenase.[224] It is noted that a scalable synthesis of statin derivatives has been reported.[225]

[222] Halling, K.; Torssell, K.B.G.; Hazell, R.G. *Acta Chem. Scand.* ***1991***, *45*, 736.

[223] Bergeron, S.; Chaplin, D.A.; Edwards, J.H.; Ellis, B.S.W.; Hill, C.L.; Holt-Tiffin, K.; Knight, J.R.; Mahoney, T.; Osborne, A.P.; Ruecroft, G. *Org. Process Res. Dev.* ***2006***, *10*, 661.

[224] Majerić Elenkov, M.; Tang, L.; Hauer, B.; Janssen, D.B. *Org. Lett.* ***2006***, *8*, 4227.

[225] Hobson, L.A.; Akiti, O.; Deshmukh, S.S.; Harper, S.; Katipally, K.; Lai, C.J.; Livingston, R.C.; Lo, E.; Miller, M.M.; Ramakrishnan, S.; Shen, L.; Spink, J.; Tummala, S.; Wei, C.; Yamamoto, K.; Young, J.; Parsons, R.L., Jr. *Org. Process Res. Dev.* ***2010***, *14*, 441.

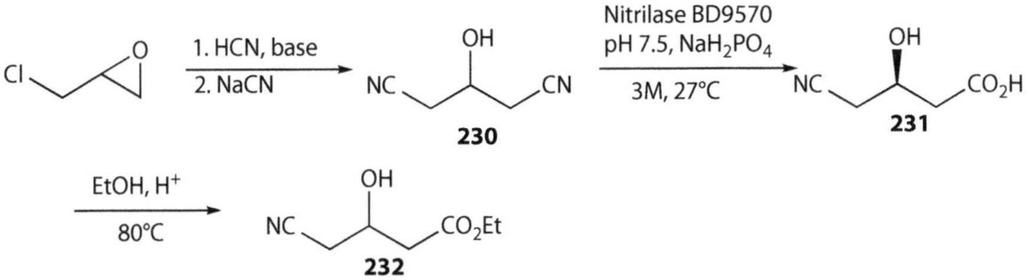

The importance of statine as a synthetic target has led to development of analogs. One such analog is a statine derivative known as isostatine (***237***), which has an EtCHMe moiety rather than the Me_2CHCH_2 moiety found in statine. The amino moiety of isoleucine was converted to a hydroxyl group (see ***233***).[226] Conversion to the azide, reduction to the amine, and condensation of the ester with lithio methyl (2-trimethylsilyloxycarboxy) acetate gave ***234***. Sodium borohydride reduction gave ***235***, along with 9% of ***236***. Hydrolysis and reprotection of ***235*** gave *N*-Boc isostatine, ***236***.

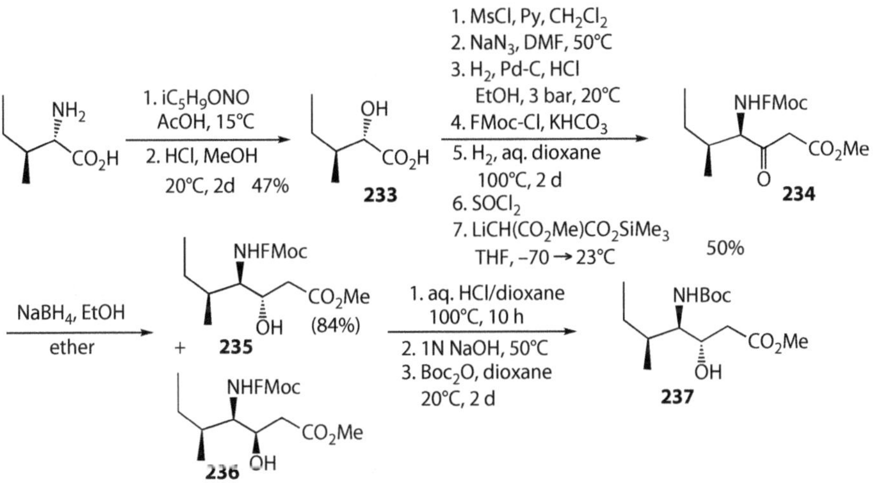

Other isostatine derivatives have been prepared from isoleucine. In one synthesis, *L*-isoleucine was protected as the *N*-Boc derivative, reduced to the alcohol, and oxidized to the isoleucinal. Subsequent condensation with the lithium enolate of ethyl acetate gave ***238***.[227] Similarly, *L*-*allo*-isoleucine (***239***) was converted to ***240***. Both *D*-leucine and *D*-*allo*-isolecuine were converted to the corresponding isosteres.[227a]

---

[226] Schmidt, U.; Kroner, M.; Griesser, H. *Synthesis* ***1989***, 832.

[227] (a) Rinehart, K.L.; Sakai, R.; Kishore, V.; Sullins, D.W. *J. Org. Chem.* ***1992***, *57*, 3007; (b) Rinehart, K.L.; Kishore, V.; Bible, K.C.; Sakai, R.; Sullins, D.W.; Li, K.-M. *J. Nat. Prod.* ***1988***, *51*, 1. Also see Jouin, P.; Poncet, J. Dufour, M.-N.; Maugras, I.; Pantaloni, A.; Castro, B. *Tetrahedron Lett.* ***1988***, *29*, 2661.

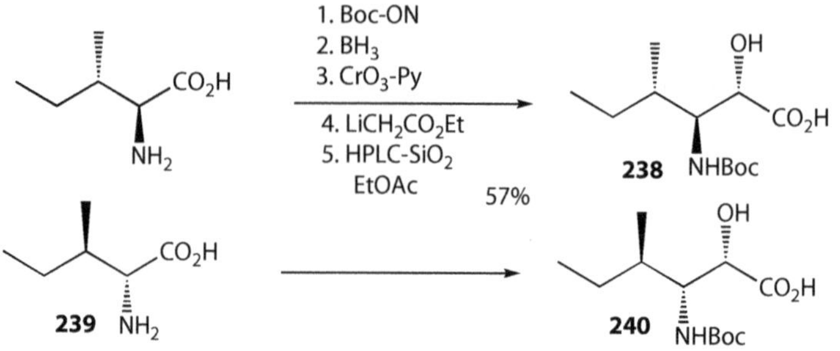

238 240 239

Bestatin (***244***) is another important statine analog, and it has been shown to be a competitive inhibitor of aminopeptidase B and of leucine aminopeptidase.[228] Bestatin is a dipeptide mimetic and a potent inhibitor of the malarial PfA-M1 metallo-aminopeptidase. The *Plasmodium falciparum* malarial M1 alanyl-aminopeptidase (PfA-M1) is an enzyme involved in the terminal stages of hemoglobin digestion and the generation of an amino acid pool within the parasite. It was prepared from the *N*-Cbz derivative of phenylalanine, which was converted to ***241*** and then reduced to give alcohol ***242***.[229] Reaction with sodium thiosulfate and then potassium cyanide gave the cyanohydrin, which was hydrolyzed to 4-amino-2-hydroxy-5-phenylpentanoic acid (***243***).[229] This was converted to bestatin, ***244***.

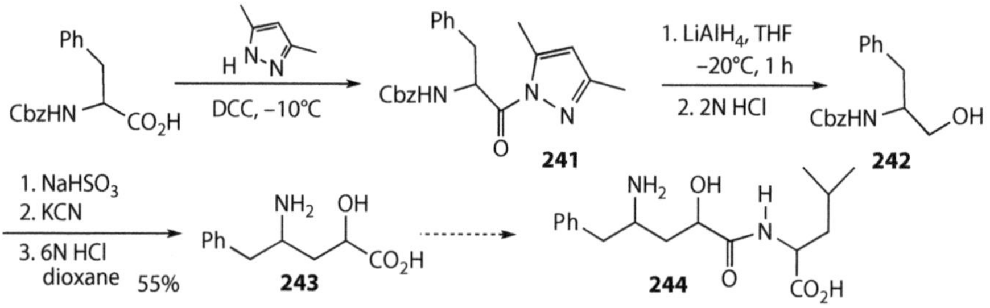

241 242 243 244

The malarial PfA-M1 metallo-aminopeptidase is a potential drug target. Bestatin derivatives (see ***244***) were synthesized with variants at the side chain of either the α-hydroxy-β-amino acid or the adjacent natural α-amino acid. Using a Wang resin A[230], a diverse library of derivatives have been prepared.[231]

An intramolecular acylnitrene-mediated aziridination leads to a key bicyclic aziridine, which was opened with various reagents to give substituted oxazolidinones. Such oxazolidinones are subsequently converted to bestatin as well as bestatin analogs. Chiral allylic alcohol ***245*** was converted to ***246***, for example, and then cyclized

[228] Umezawa, H.; Aoyagi, T.; Suda, H.; Hamada, M.; Takesuchi, T. *J. Antibiot.* ***1976***, *29*, 97.
[229] Nishizawa, R.; Saino, T. *J. Med. Chem.* ***1977***, *20*, 570.
[230] (a) Hermkens, P.H.H.; Ottenheijm, H.C.J.; Rees, D.C. *Tetrahedron* ***1997***, *53*, 5643; (b) Hermkens, P.H.H.; Ottenheijm, H.C.J.; Rees, D.C. *Tetrahedron* ***1996***, *52*, 4527.
[231] Velmourougane, G.; Harbut, M.B.; Dalal, S.; McGowan, S.; Oellig, C.A.; Meinhardt, N.; Whisstock, J.C.; Klemba, M.; Greenbaum, D.C. *J. Med. Chem.* ***2011***, *54*, 1655.

to give aziridine ***247***.[232] Reaction with several organolithium reagents led to several compounds ***248***, and bestatin (***244***) was prepared via reaction of ***247*** with phenyllithium followed by several subsequent steps. This route prepared the amino moiety of the functionalized amino acid.

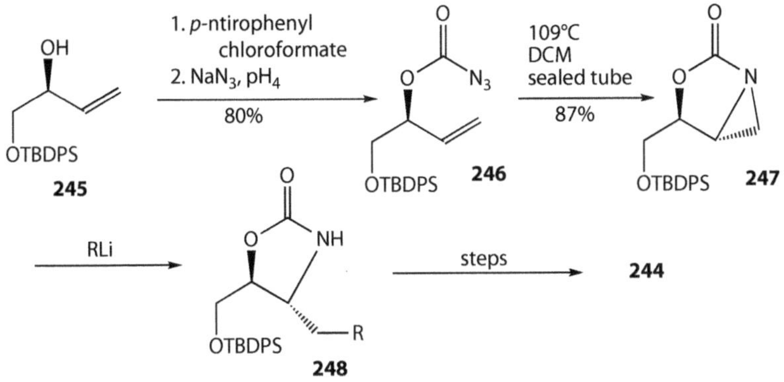

A chemical isostere is defined as two or more molecules with similarities in physiochemical properties of atoms, groups, and radicals with similar electronic structures.[233] Other properties that contribute to being an isostere include basicity, bond angles, acidity, electronegativity, polarizability, size, shape of molecular orbitals, electron density, and partition coefficients. The term essentially refers to replacing one group in a molecule with another that will give similar chemical or physiological properties.[233a] Cyclohexyl derivative (***249***) is an isostere of statine (***173***), for example, and it has been prepared from phenylalanine.[234,235] Indeed, highly substituted hydroxy amino acids may be peptidase inhibitors, and several analogs have been prepared as potential inhibitors.

A phenylalanal prepared ***250*** via an olefination reaction with a substituted phenylalanal derivative followed by reduction of the phenyl ring. The alcohol moiety

---

[232] Bergmeier, S.C.; Stanchina, D.M. *J. Org. Chem.* ***1999***, *64*, 2852.

[233] (a) Foye, W.O. *Principles of Medicinal Chemistry*, 3rd ed. Lea and Febiger, Philadelphia, ***1989***, pp. 66–75; (b) Langmuir, I. *J. Am. Chem. Soc.* ***1919***, *41*, 868; (c) Langmuir, I. *J. Am. Chem. Soc.* ***1919***, *41*, 1543; (d) Burger, A., in *Medicinal Chemistry*, Burger, A., ed., 3rd ed. New York, Wiley, ***1970***, p. 74.

[234] Boger, J.; Payne, L.S.; Perlow, D.S.; Lohr, N.S.; Poe, M.; Blaine, E.H.; Ulm, E.H.; Schorn, T.W.; LaMont, B.I.; Lin, T.-Y.; Kawai, M.; Rich, D.H.; Veber, D.F. *J. Med. Chem.* ***1985***, *28*, 1779.

[235] Nishi, T.; Saito, F.; Nagahori, H.; Kataoka, M.; Morisawa, Y.; Yabe, Y.; Sakurai, M.; Higashida, S.; Shoji, M.; Matsushita, Y.; Iijima, Y.; Ohizumi, K.; Koike, H. *Chem. Pharm. Bull.* ***1990***, *38*, 103.

was converted to the *N*-phthaloyl derivative (***251***)[236] by a Mitsunobu inversion.[237,238] Conjugate addition with a higher-order silyl cuprate gave ***252***.[236] The phenylsilyl moiety was converted to an alcohol (***253***) by treatment with tetrafluoroboric acid and then KF/*m*-chloroperoxybenzoic acid. Removal of the phthalimidoyl group with hydrazine led to lactam ***254***.[236] Hydrolysis gave 4-amino-5-cyclohexyl-3-hydroxy-pentanoic acid (***255***) in 60% yield.

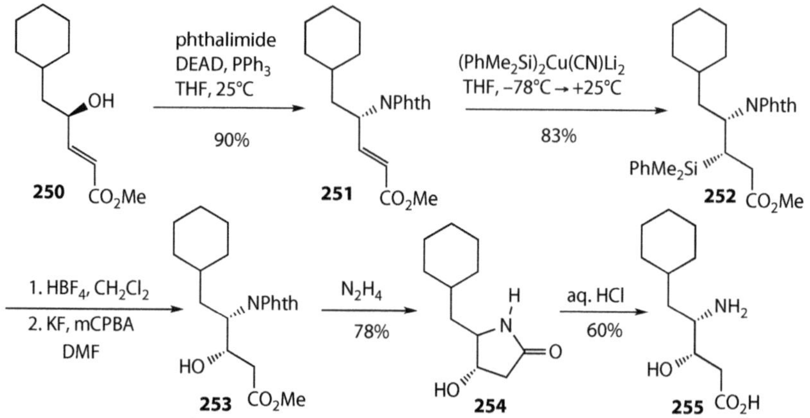

A different isostere was prepared by initial conversion of keto-alkyne ***256*** to carboxylic acid ***257*** in four steps.[239] Subsequent formation of lactone ***258*** allowed displacement of the bromide to give azide ***259***. Treatment with butylamine and hydrogenation gave *N*-butyl 5-amino-6-cyclohexyl-2-isopropylhexanamide, ***260***. This amino acid product is considered a hydroxy-ethylene dipeptide isostere.[239,240]

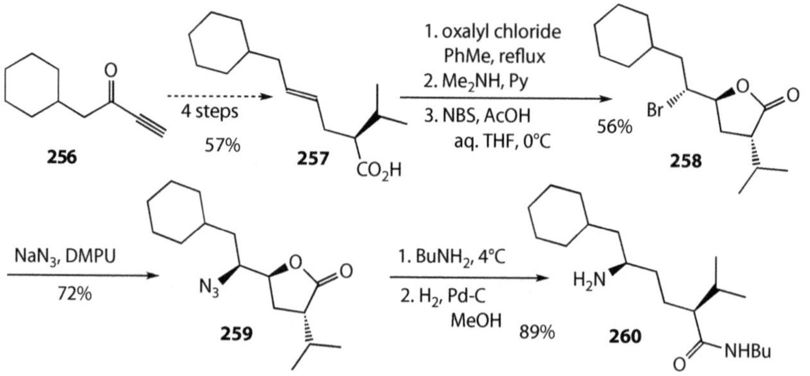

---

[236] Burgess, K.; Cassidy, J.; Henderson, I. *J. Org. Chem.* ***1991***, *56*, 2050.

[237] Mitsunobu, O. *Synthesis* ***1981***, 1.

[238] See (a) Smith, M.B. *Organic Synthesis*, 3rd ed. Wavefunction, Inc./Elsevier, Irvine, CA/London, England, ***2010***, pp. 1213–126; (b) Smith, M.B. *March's Advanced Organic Chemistry*, 7th ed. John Wiley & Sons, Hoboken, NJ, ***2013***, pp. 469–470.

[239] Herold, P.; Duthaler, R.; Rihs, G.; Angst, C. *J. Org. Chem.* ***1989***, *54*, 1178.

[240] Wolfenden, R. *Nature* ***1969***, *223*, 704.

6.7 OTHER BIOLOGICALLY IMPORTANT AMINO ACID DERIVATIVES

There are other biologically important molecules that have a non-α-amino acid as an important structural component, or use the amino acid as an important synthetic precursor. This section collects several different types of these targets. As the name of this section implies, the applications are difficult to categorize, but this section will illustrate a variety of applications that are exhibited by non-α-amino acids. They are loosely categorized by their biological activity. In addition to biological applications, there are synthetic applications. Indeed, some non-α-amino acids have been used as precursors for the synthesis of heterocyclic compounds, particularly heterocyclic three-membered rings.[241]

Several azasterols were prepared in order to evaluate activity against *Trypanosoma brucei rhodesiense*, *T. cruzi*, *Leishmania donovani*, and *Plasmodium falciparum*, the causative agents of human African trypanosomiasis, Chagas disease, leishmaniasis, and malaria, respectively.[242] The commercially available sterol derivative ***261*** was converted to amino acid derivative ***262*** by the reaction of amino-esters $H_2N(CH_2)_nCO_2Me$, where n = 1–7, with the benzotriazolyl derivative of ***261*** to give ***262*** (21–34%). Modification of this sequence also produced other derivatives. Several of these compounds showed antiparasitic activity.

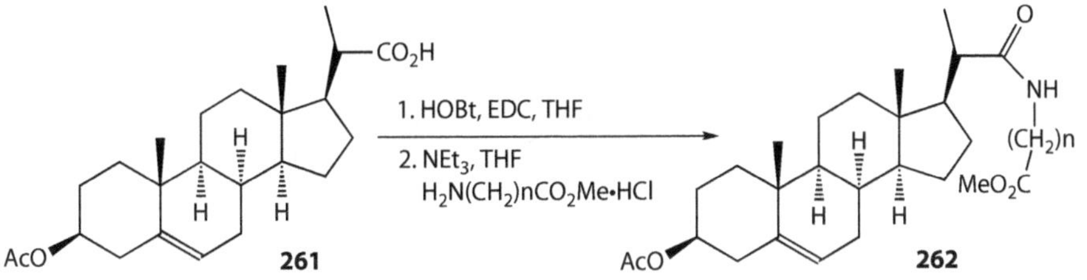

Alkenyl amino acid ***267*** was shown to be hepatotoxic and a component of microcyctins.[243] An oxazolidinone derivative was used in the synthesis, with an initial conversion of ***263*** to ***264*** followed by ring opening to give ***265***.[244] Oxidation to the aldehyde and Wittig olefination[245] gave ***266***. This conjugated ester was "chain extended" in six steps (18% yield) to give methyl 3-(*N*-Boc-amino)-2,6,8-trimethyl-9-methoxy-10-phenylnonan-4,6-dienoate, ***267***. Amino acid ***267*** and other related amino acids produced in this study were also shown to be components of nodularin.[246]

---

241 Tišcler, M.; Kolar, P. *Adv. Heterocyclic Chem.* ***1995***, *64*, 1.

242 Yardley, L.R.; Wharton, H.; Little, S.; Croft, S.L.; Ruiz-Perez, L.M.; Gonzalez-Pacanowska, D.; Gilbert, I.H. *J. Med. Chem.* ***2006***, *49*, 6094.

243 Rinehart, K.L.; Harada, K.; Namikoshi, M.; Chen, C.; Harvis, C.A.; Munro, M.H.G.; Blunt, J.W.; Mulligan, P.E.; Beasley, V.R.; Dahlem, A.M.; Carmichael, W.W. *J. Am. Chem. Soc.* ***1988***, *110*, 8557.

244 Beatty, M.F.; Jenings-White, C.; Avery, M.A. *J. Chem. Soc. Chem. Commun.* ***1991***, 351.

245 See (a) Smith, M.B. *Organic Synthesis*, 3rd ed. Wavefunction, Inc./Elsevier, Irvine, CA/London, England, ***2010***, pp. 729–739; (b) Smith, M.B. *March's Advanced Organic Chemistry*, 7th ed. John Wiley & Sons, Hoboken, NJ, ***2013***, pp. 1165–1173.

246 Namikoshi, M.; Rinehart, K.L.; Dahlem, A.M.; Beasley, V.R.; Carmichael, W.W. *Tetrahedron Lett.* ***1989***, *30*, 4349.

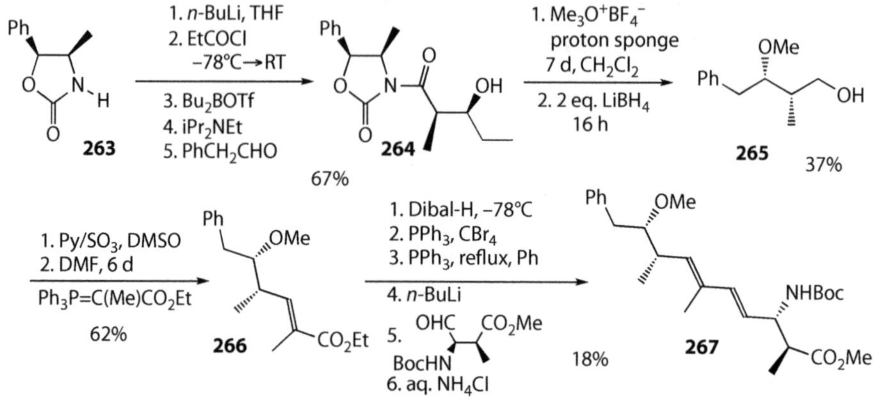

Analogs based on betulinic acid have been shown to inhibit human immunodeficiency virus (HIV)-1 entry. Derivatives of ***268***, for example, were prepared with 3-aminopropanoic acid, 8-aminooctanic acid, and 11-undecanoic acid linkers, attached to a leucine residue.[247]

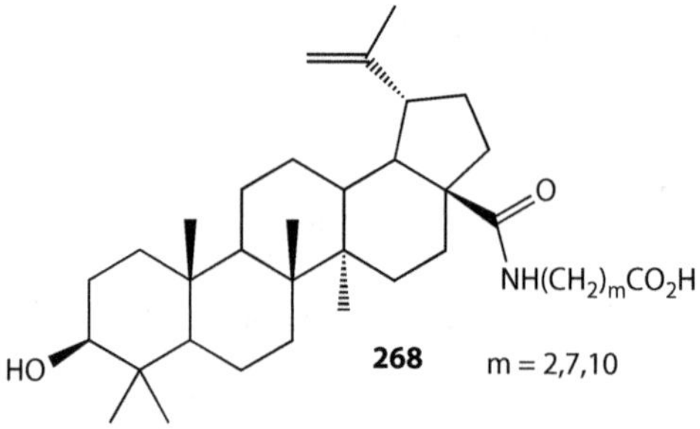

A series of ω-aminoalkanoic acid derivatives of betulinic acid were synthesized and evaluated for their activity against HIV. Acid chlorides ***269*** were treated with ω-amino-esters to give ***270***, which was further converted to the compounds of interest.[248]

[247] Sun, I.-C.; Chen, C.-H.; Kashiwada, Y.; Wu, J.-H.; Wang, H.-K.; Lee, K.-H. *J. Med. Chem.* ***2002***, *45*, 4271. Also see Soler, F.; Poujade, C.; Evers, M.; Carry, J.-C.; Hénin, Y.; Bousseau, A.; Huet, T.; Pauwels, R.; De Clercq, E.; Mayaux, J.-F.; Le Pecq, J.-B.; Dereu, N. *J. Med. Chem.* ***1996***, *39*, 1069.

[248] Soler, F.; Poujade, C.; Evers, M.; Carry, J.-C.; Hénin, Y.; Bousseau, A.; Huet, T.; Pauwels, R.; De Clercq, E.; Mayaux, J.-F.; Le Pecq, J.-B.; Dereu, N. *J. Med. Chem.* ***1996*** *39*, 1069.

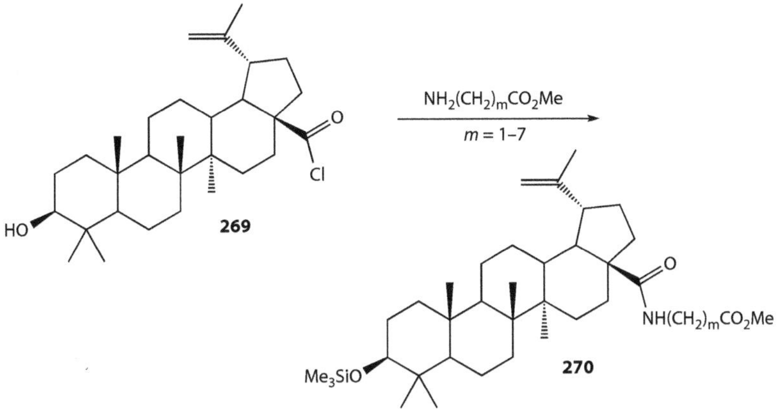

(–)-Galantinic acid (6-amino-3,5,7-trihydroxyheptanoic acid, ***277***) is a structural component of galantin I, a peptide antibiotic.[249] Amino acids can be converted to aldehydes, as amply demonstrated in previous sections. These precursors can also be used for the preparation of polyhydroxy amino acids. Horner-Wadsworth-Emmons olefination[250,251] of ***271*** gave ***272***.[252] Epoxidation, oxidation, and Wittig olefination[253] gave ***273***. Isomerization of the double bond out of conjugation, to give ***274***, was followed by treatment with base and gave lactone ***275***. Introduction of the β-hydroxyl group (see ***276***) led to the final product, ***277***.

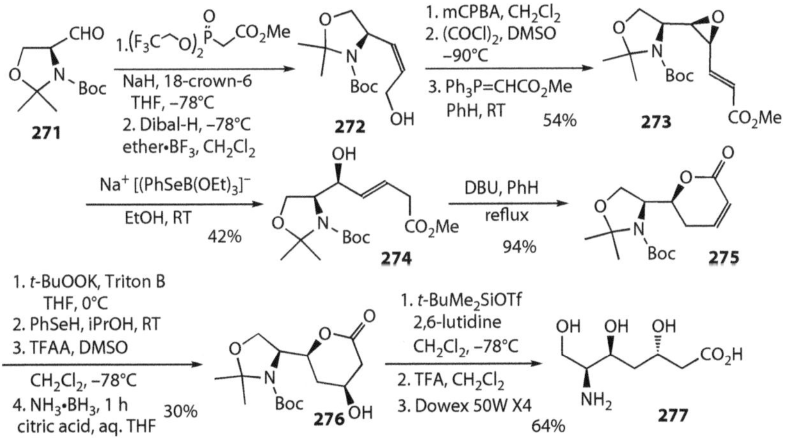

---

[249] Shohi, J.; Sakazaki, R.; Wakisaka, Y.; Koizumi, K.; Mayama, M.; Matsuura, S. *J. Antiobiot.* ***1975***, *28*, 122.

[250] (a) Horner, L.; Hoffmann, H.; Wippel, J.H.; Klahre, G. *Ber.* ***1959***, *92*, 2499; (b) Wadsworth, W.S., Jr.; Emmons, W.D. *J. Am. Chem. Soc.* ***1961***, *83*, 1733; (c) Boutagy, J.; Thomas, R. *Chem. Rev.* ***1974***, *74*, 87.

[251] See (a) Smith, M.B. *Organic Synthesis*, 3rd ed. Wavefunction, Inc./Elsevier, Irvine, CA/London, England, ***2010***, pp. 739–744; (b) Smith, M.B. *March's Advanced Organic Chemistry*, 7th ed. John Wiley & Sons, Hoboken, NJ, ***2013***, pp. 1169–1170.

[252] Sakai, N.; Ohfune,Y. *Tetrahedron Lett.* ***1990***, *31*, 4151.

[253] See (a) Smith, M.B. *Organic Synthesis*, 3rd ed. Wavefunction, Inc./Elsevier, Irvine, CA/London, England, ***2010***, pp. 729–739; (b) Smith, M.B. *March's Advanced Organic Chemistry*, 7th ed. John Wiley & Sons, Hoboken, NJ, ***2013***, pp. 1165–1173.

α-Kainic acid is used as a tool in neuropharmacology[254] and exhibits anthelmintic[255] and neurotransmitting[256] activities. The simple amino alcohol (3-aminopropanol) was used as a starting material in a reaction with ethyl bromoacetate to give ***278***.[257] Swern oxidation[258] of the alcohol gave an aldehyde, and Wittig olefination[259] gave ethyl 5-(*N*-benzoyl-*N*-carboethoxymethyl)pent-2*E*-enolate (***279***), which was converted to α-kainic acid.

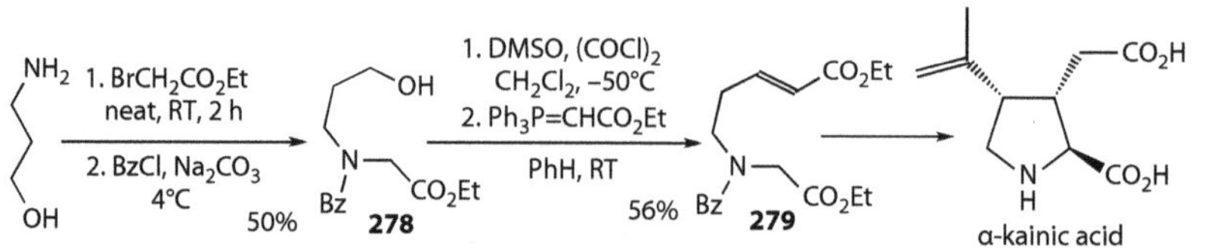

A β-glucuronidase-responsive albumin-binding prodrug for the selective delivery of doxorubicin to cancerous tumors was developed as a prodrug. Prodrug ***280*** contains an aminohexanoic acid side chain, with the amino moiety incorporated as a maleimide unit.[260] It was shown to inhibit tumor growth in a manner similar to that of doxorubicin, but side effects induced by the free drug were avoided.

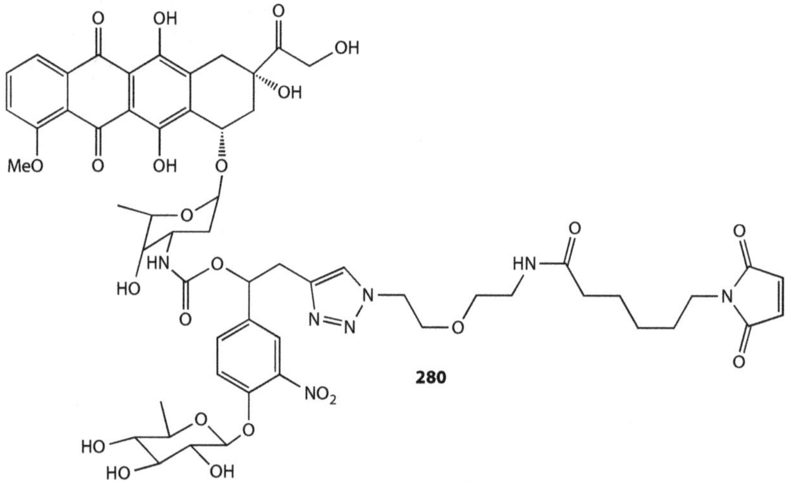

[254] MacGeer, E.G.; Olney, J.W.; MacGeer, P.L. *Kainic Acid as a Tool in Neurobiology*, Raven, New York, ***1978***.

[255] Watase, H.; Tomiie, Y.; Nitta, I. *Nature* (London) ***1958***, *181*, 761.

[256] See Cantrell, B.E.; Zimmerman, D.M.; Monn, J.A.; Kamboj, R.L.; Hoo, K.H.; Tizzno, J.P.; Pullar, I.A.; Farell, L.N.; Bleakman, D. *J. Med. Chem.* ***1996***, *39*, 3617.

[257] Yoo, S.; Lee, S.-H.; Kim, N.-J. *Tetrahedron Lett.* ***1988***, *29*, 2195.

[258] (a) Mancuso, A.J.; Swern, D. *Synthesis* ***1981***, 165. See (b) Smith, M.B. *Organic Synthesis*, 3rd ed. Wavefunction, Inc./Elsevier, Irvine, CA/London, England, ***2010***, pp. 241–243; (c) Smith, M.B. *March's Advanced Organic Chemistry*, 7th ed. John Wiley & Sons, Hoboken, NJ, ***2013***, pp. 1447–1448.

[259] See (a) Smith, M.B. *Organic Synthesis*, 3rd ed. Wavefunction, Inc./Elsevier, Irvine, CA/London, England, ***2010***, pp. 729–739; (b) Smith, M.B. *March's Advanced Organic Chemistry*, 7th ed. John Wiley & Sons, Hoboken, NJ, ***2013***, pp. 1165–1173.

[260] Legigan, T.; Clarhaut, J.; Renoux, B.; Tranoy-Opalinski, I.; Monvoisin, A.; Berjeaud, J.-M.; Guilhot, F.; Sébastien Papot, S. *J. Med. Chem.* **2012**, *55*, 4516.

A series of β-turn mimics, common features of peptide and protein secondary structures, were prepared from caprolactam derivatives such as ***281***.[261] Reaction with Boc-glycine-alanine led to β-turn mimics such as ***282***, containing the 6-aminohexanoic acid residue.

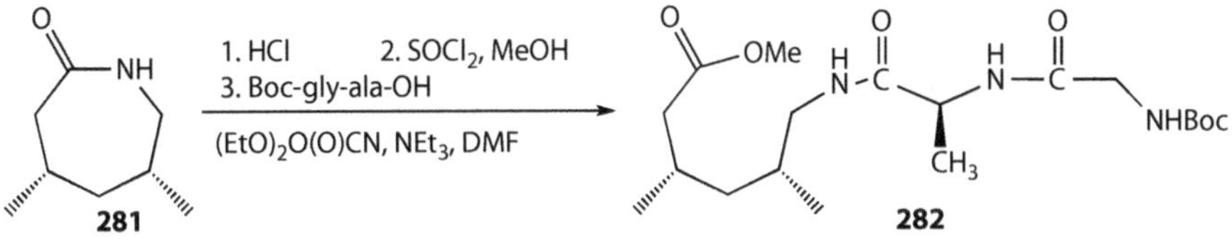

281 **282**

Inhibitors of histone deacetylase modulate transcription and induce apoptosis or differentiation in cancer cells. Analogs of trichostatin A were prepared and shown to inhibit maize HD-2 and induce terminal cell differentiation in Friend leukemic cells. As shown in ***283***, the analogs incorporated amino acids ranging from 3-aminopropanoic acid to 7-aminoheptanoic acid.[262]

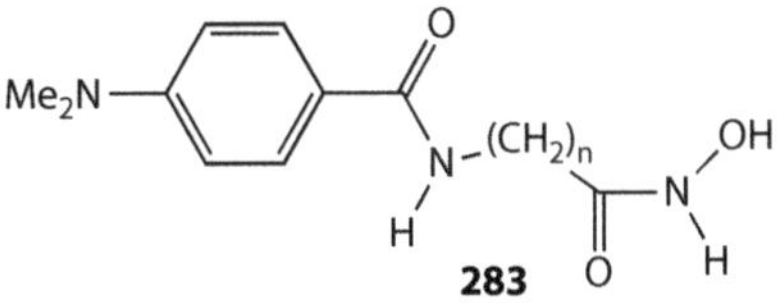

283 m = 3–7

Ambrucitin S (***284***) is an important antifungal agent isolated from the myxobacterium *Sorangium cellulosum* in 1970, and it has been synthesized.[263] Subsequently, several new antifungal metabolites (the ambrucitins VS-1 to VS-5) were isolated from *Sorangium cellulosum* and identified as (5*S*,6*R*)-5-amino-6-hydroxypolyangioic acids.[264] The amino acid units are attached to one of the carbohydrate moieties. One specific example is ambrucitin VS-3, ***285***. Other ambrucitin VS analogs have been prepared via semisynthesis.[265]

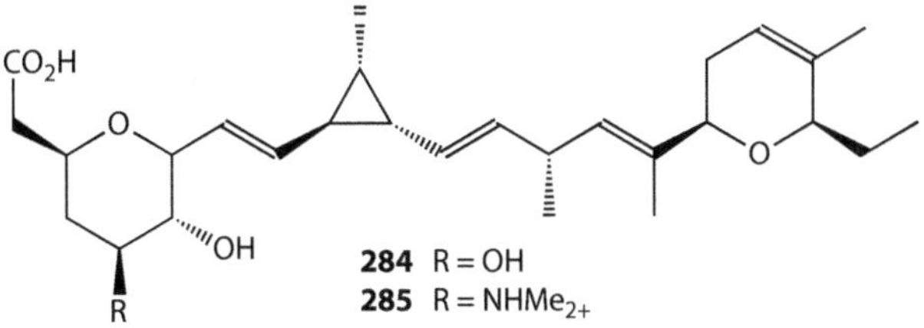

284 R = OH
285 R = $NHMe_{2+}$

---

[261] Kitagawa, O.; Velde, D.V.; Dutta, D.; Morton, M.; Takusagawa, F.; Aube, J. *J. Am. Chem. Soc.* ***1995***, *117*, 5169.

[262] Jung, M.; Brosch, G.; Kölle, D.; Scherf, H.; Gerhäuser, C.; Loidl, P. *J. Med. Chem.* ***1999***, *42*, 4669.

[263] See Höfle, G.; Steinmetz, H.; Gerth, K.; Reichenbach, H. *Liebigs Ann. Chem.* ***1991***, 941. For a synthesis, see Kende, A.S.; Fujii, Y.; Mendoza, J.S. *J. Am. Chem. Soc.* ***1990***, *112*, 9645.

[264] Knauth, P.; Reichenbach, H. *J. Antibiot.* ***2000***, *53*, 1182.

[265] Xu, Y.; Wang, Z.; Tian, Z.-Q.; Li, Y.; Shaw, S.J. *ChemMedChem* ***2006***, *1*, 1063.

Amino acid moieties have been used as structural components of polymeric contrast agents that are used for magnetic resonance angiography. Contrast agent ***286*** is used for scintigraphic pharmacokintetic analysis in vivo, and contains aminohexanoic fragments as well as 3-aminopropanoic acid residues.[266]

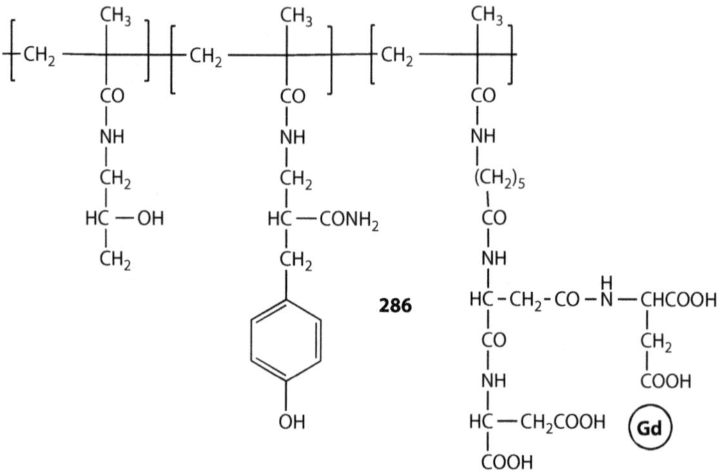

286

6.8 NON-α-AMINO ACIDS IN CATALYSIS

There does not appear to be much direct literature in which non-α-amino acids are used as catalyst ligands. Reduction of amino acids to amino alcohols, or conversion to diamines, generates compounds that may be used as chiral ligands in various catalytic reactions. One or two examples will be presented to illustrate the concept.

Chiral 1,3-diaminosulfonamides have been prepared from β-amino acids and used as chiral ligands for the enantioselective reaction of diethylzinc with aldehydes to produced chiral alcohols.[267] Conversion of (−)-*cis*-2-benzamidocyclohexanecarboxylic acid (***287***) to the amide was followed by reduction, which converted the *N*-benzoyl group to the *N*-benzylamine moiety, and converted the carboxyl group to the aminomethyl moiety shown in ***288***. Tosylation of the primary amine gave the requisite sulfonamide, and subsequent debenzylation by catalytic hydrogenation allowed conversion to the carbamate and reduction to the *N*-methylamine in ***289***. Several analogs of ***289*** were prepared by this general route.

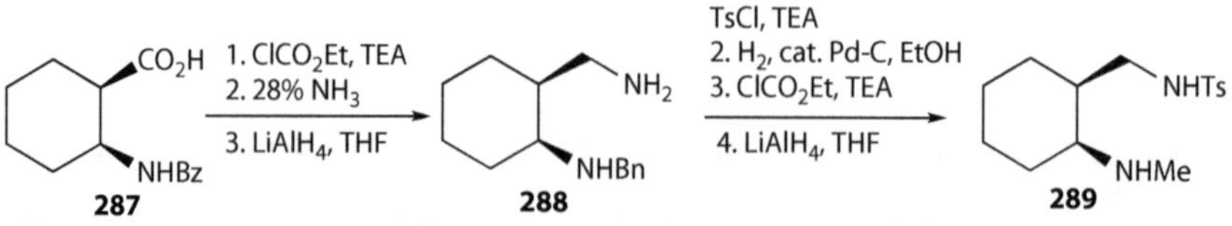

[266] Kiessling, F.; Heilmann, M.; wan Lammers, T.; Ulbrich, K.; Subr, V.; Peschke, P.; Waengler, B.; Mier, W.; Schrenk, H.-H.; Bock, M.; Schad, L.; Semmler, W. *Bioconjugate Chem.* **2006**, *17*, 42.

[267] Hirose, T.; Sugawara, K.; Kodama, K. *J. Org. Chem.* **2011**, *76*, 5413.

An amino-ester derived from pinene was used to prepare *N*-Boc-protected amino-ester ***290***. This amino-ester was used as a starting material to prepare amino alcohols ***291***, ***292***, and ***293***, which were used as ligand for the reaction of benzaldehyde, and other aromatic aldehydes, with diethylzinc.[268] As shown, the enantioselectivity achieved with these ligands was rather modest in reactions with benzaldehyde, although it was somewhat improved when substituents were attached to the benzene ring.

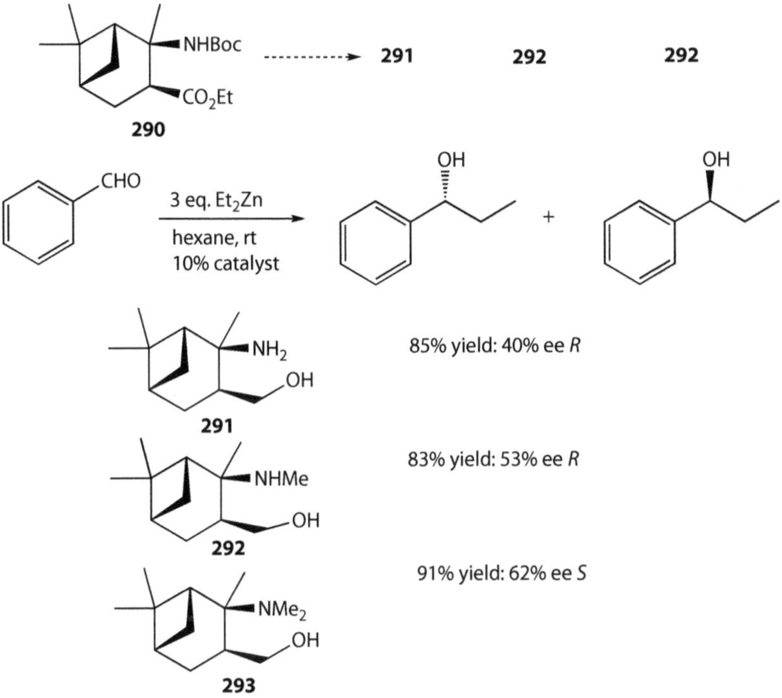

[268] Szakonyi, Z.; Balázs, Á.; Martinek, T.A.; Fülöp, F. *Tetrahedron Asym.* **2006**, *17*, 199.

7 Aminocyclic and Heterocyclic Amino Acids

This chapter will discuss cycloalkanes that have an amino group and a carboxyl group on the ring or attached to a ring via pendant alkyl units. The place to begin a discussion of such molecules is perhaps with the three-, four-, five-, and six-membered ring derivatives. Of the two possible cyclopropaneamino acids, the 1-amino derivative is an α-amino acid, but ***1*** is a β-amino acid. Apart from the α-amino acid, there are two non-α-amino acid structures for aminocyclobutanoic acids, ***2*** (2-aminocyclobutane-1-carboxylic acid) and ***3*** (3-aminocyclobutane-1-carboxylic acid). The non-α-amino acid cyclopentanecarboxylic acid

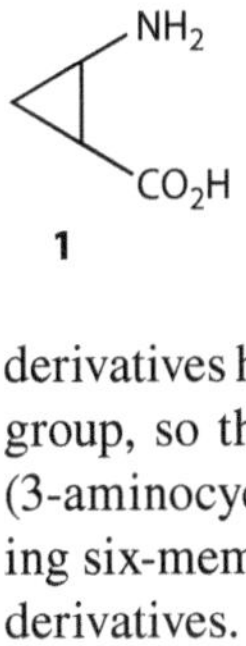
1

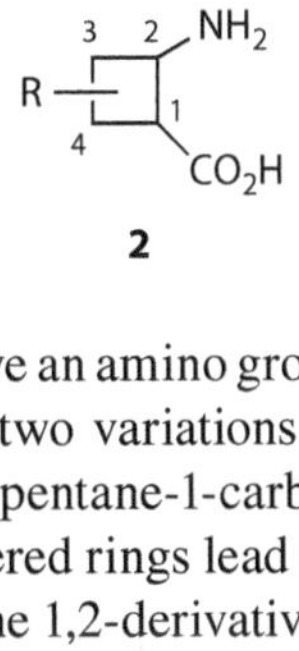
2

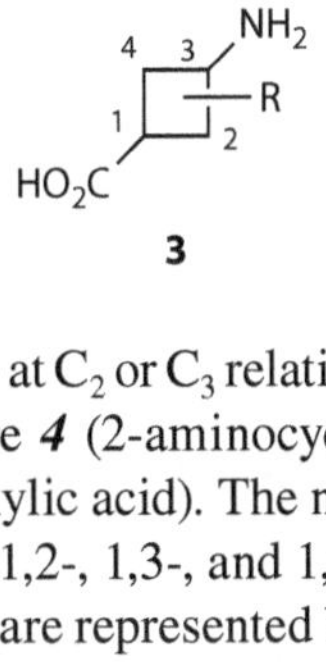
3

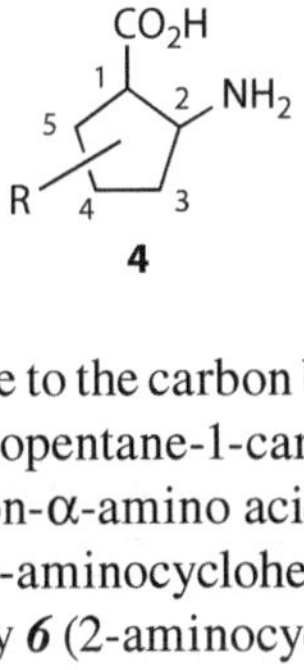
4

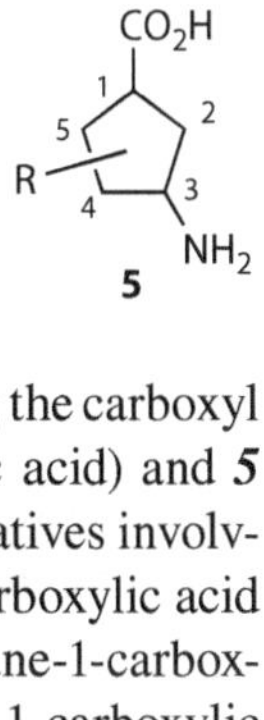
5

derivatives have an amino group at C_2 or C_3 relative to the carbon bearing the carboxyl group, so the two variations are ***4*** (2-aminocyclopentane-1-carboxylic acid) and ***5*** (3-aminocyclopentane-1-carboxylic acid). The non-α-amino acid derivatives involving six-membered rings lead to 1,2-, 1,3-, and 1,4-aminocyclohexanecarboxylic acid derivatives. The 1,2-derivatives are represented by ***6*** (2-aminocyclohexane-1-carboxylic acid), the 1,3-derivatives are represented by ***7*** (3-aminocyclohexane-1-carboxylic acid), and the 1,4-derivatives are represented by ***8*** (4-aminocyclohexane-1-carboxylic acid).[1] There are other cyclic amino acids of this type based on other rings, of course, and several will be presented in succeeding examples.

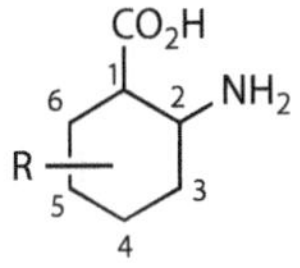
6

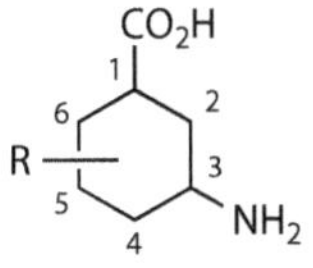
7

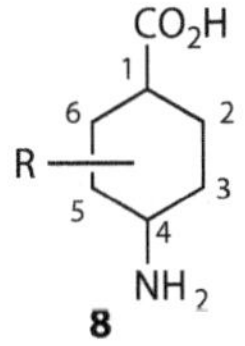
8

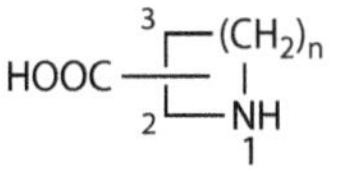
9

[1] For binding of *7* with the kringle 2 (K2) module of human plasminogen (Pgn), see Marti, D.N.; Schaller, J.; Llinás, M. *Biochemistry* ***1999***, *38*, 15741.

A different class of cyclic amino acids is derived from heterocycles, specifically reduced heterocyclic rings with a nitrogen in the ring and a pendant carboxyl group. The generic structure ***9*** shows that rings of different sizes are possible (azetidine, pyrrolidine, piperidine, etc.) with the carboxyl moiety at the α-, β-, or γ-position relative to the ring nitrogen. Formally, this class of amino acids will include aziridinecarboxylic acids, which are actually α-amino acids. The non-α-amino acids will include azetidinecarboxylic acids, pyrrolidinecarboxylic acids (including the α-amino acid proline), piperidinecarboxylic acids, and so on.

There is a class of amino acids based on acyclic compounds in which the amino and carboxyl moieties are not directly attached to the ring but rather attached to the ring via carbon "spacers." One interesting example is known as gabapentin, ***10***. It is a GABA analog developed for the treatment of epilepsy, but is also used to relieve neuropathic pain, which is often due to damage to the nerve fibers, or those fibers may be dysfunctional or injured.[2] Gabapentin (Gpn) can be used in the design of well-folded hybrid peptides, and several internally hydrogen-bonded structures have been demonstrated in the solid state for Gpn-containing peptides.[3]

Another example of this last structural class of amino acids is trans-4-(benzyloxycarbonylaminomethyl)cyclohexanecarboxylic acid (***11***),[4] which was converted to ***12***. Carbamate salt ***12*** functioned as a transdermal permeation enhancer.[4]

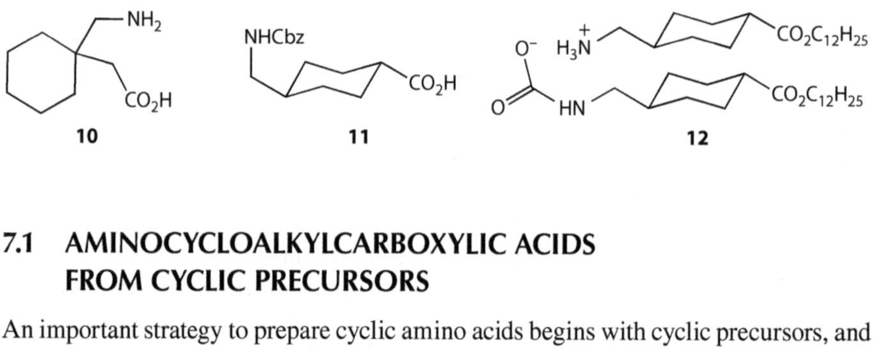

10 11 12

7.1 AMINOCYCLOALKYLCARBOXYLIC ACIDS FROM CYCLIC PRECURSORS

An important strategy to prepare cyclic amino acids begins with cyclic precursors, and subsequent refunctionalization of existing groups gives a new amino acid. If a cyclic amino acid is commercially available, this approach is quite attractive for synthetic modification, but a variety of functionalized cyclic precursors have been employed.

7.1.1 Lactam-Based Syntheses

Lactams are cyclic molecules that are obvious synthetic precursors to cyclic amino acids that have the nitrogen in the ring, usually by refunctionalization or reduction reactions. Further, β-lactams are important precursors to β-alanine derivatives, as well as cyclic amino acid derivatives.

Cyclic amino acids in which the nitrogen is not part of the ring can also be prepared, typically via cycloaddition reactions with cyclic alkenes. Reaction of

[2] See Sills, G.J. *Curr. Opin. Pharmacol.* ***2006***, *6*, 108.

[3] Vasudev, P.G.; Chatterjee, S.; Shamala, N.; Balaram, P. *Acc. Chem. Res.* ***2009***, *42*, 1628.

[4] Vávrová, K.; Hrabálek, A.; Doležal, P.; Holas, T.; Klimentová, J. *J. Controlled Release* ***2005***, *104*, 41.

cyclopentene with chlorosulfonyl isocyanate (CSI), for example, gave the β-lactam, and subsequent treatment with basic potassium iodide cleaved the sulfonyl group to give ***12***.[5,6] Acid hydrolysis gave *cis*-2-aminocyclopentane-1-carboxylic acid, ***13***. Similarly, racemic bicyclic lactam ***14*** was converted to a chiral amino acid.[7] Incubation with *Rhodococcus equi* (NCIB 40213), also known as ENZA-1,[8] led to kinetic resolution, allowing the isolation of (+)-***15***.[7] Catalytic hydrogenation of the double bond in this chiral, nonracemic bicyclic lactam was followed by a hydrolytic ring opening to give (–)-***16*** [*cis*-2-aminocyclopentane-1-carboxylic acid; also known as (–)-cispentacin], an antifungal antibiotic.[9]

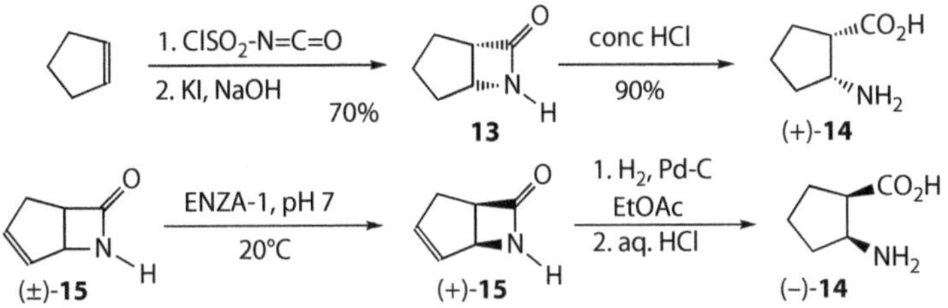

Note that formation of the β-lactam allows the synthesis or a larger ring size amino based on reaction with a preexisting ring, as in formation of ***13*** or ***15***. Similarly, condensation of cyclohexene with *N*-benzyl isocyanate gave ***16*** via [2+2]-cycloaddition.[10,11] Acid hydrolysis of the lactam ring led to 2-aminocyclohexane-1-carboxylic acid, ***17***. Other alkenes can be used in a similar transformation.[12] It is noted, however, that this method will only generate β-amino acids.

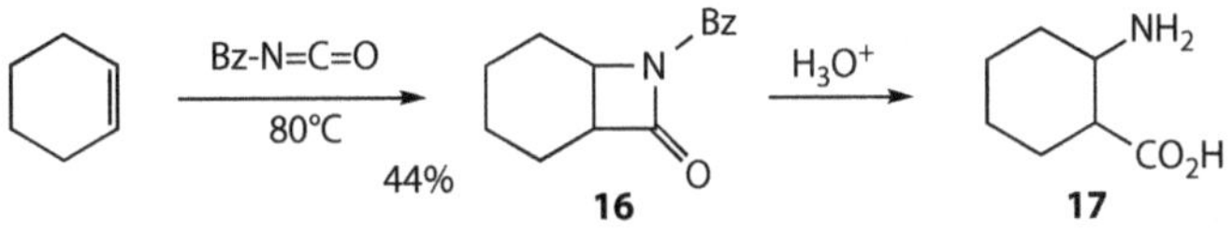

Other lactams are precursors to cyclic amino acids. The reaction of the alkene moiety in bicyclic lactam ***18*** (2-azabicyclo[7.7.1]hept-5-en-3-one)[13] led to diol ***19***.[14] Subsequent

---

5 Nativ, E.; Rona, P. *Isr. J. Chem.* ***1972***, *10*, 55.

6 Konishi, M.; Nishio, M.; Saitoh, K.; Miyaki, T.; Oki, T.; Kawaguchi, H. *J. Antibiot.* ***1989***, *42*, 1749.

7 (a) Berge, J.M.; Roberts, S.M.; Suschitzky, H.; Kemp, J.E.G. *J. Chem. Soc. S* ***1978***, 255; (b) Evans, D.A.; Biller, S.A. *Tetrahedron Lett.* ***1985***, *26*, 1907.

8 Evans, C.; McCague, R.; Roberts, S.M.; Sutherland, A.G.; Wisdom, R. *J. Chem. Soc. Perkin Trans. I* ***1991***, 2276.

9 (a) Iwamoto, T.; Tsujii, E.; Ezaki, M.; Fujie, A.; Hashimoto, S.; Okuhara, M.; Kohsaka, M.; Imanaka, H.; Kawabata, K.; Inamoto, Y.; Sakane, K. *J. Antibiot.* ***1990***, *43*, 1; (b) Kawabata, K.; Inamoto, Y.; Sakane, K.; Iwamoto, T.; Hashimoto, S. *J. Antibiot.* ***1990***, *43*, 513.

10 Arbuzov, B.A.; Zobova, N.N. *Dokl. Akad. Nauk. SSSR* ***1967***, *172*, 845 (Engl. 103).

11 See (a) Smith, M.B. *Organic Synthesis*, 3rd ed. Wavefunction, Inc./Elsevier, Irvine, CA/London, England, ***2010***, pp. 1076–1097; (b) Smith, M.B. *March's Advanced Organic Chemistry*, 7th ed. John Wiley & Sons, Hoboken, NJ, ***2013***, pp. 1040–1051.

12 Moriconi, E.J.; Kelly, J.F. *J. Org. Chem.* ***1968***, *33*, 3036.

13 Daluge, S.; Vince, R. *J. Org. Chem.* ***1978***, *43*, 2311.

14 Kam, B.L.; Oppenheimer, N.J. *J. Org. Chem.* ***1981***, *46*, 3268.

acid hydrolysis of the lactam gave 3-amino-4,5-dihydroxy-cyclopentane-1-carboxylic acid, ***20*** (see Chapter 2, Section 2.1).

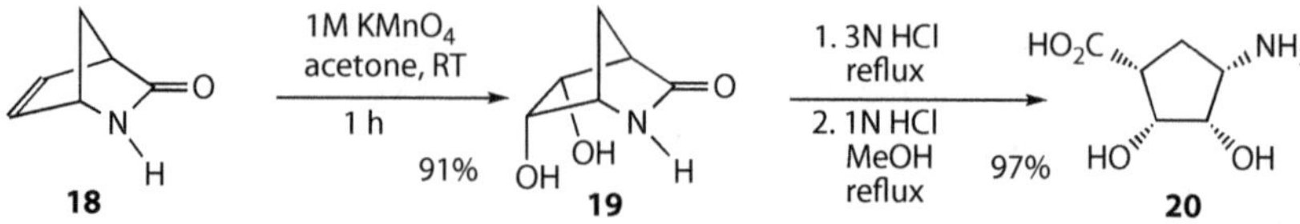

Cyclic imides are useful precursors to lactams, and therefore to amino acids. Lactam ***21*** was prepared from the corresponding succinimide derivative and then converted to ***23*** via the resaction of the lactam enolate anion with bromo-alkyne ***22***. Treatment with a Ni reagent and $NaBH_4$ led to cyclization and formation of ***24*** as a 95:5 *endo:exo* mixture.[15] Dissolving metal reduction cleaved the *N*-benzyl group, and acid hydrolysis opened the lactam ring to give ethyl 3-amino-4-ethenylcyclohexane-1-carboxylate, ***25***.[15]

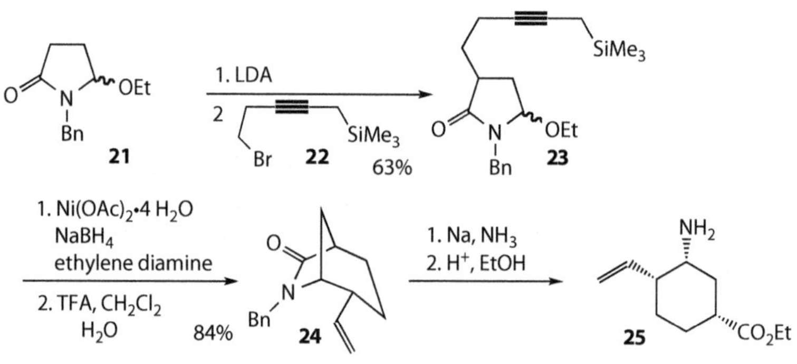

7.1.2 Dicarboxylic Acids and Related Precursors

An important diastereoselective preparation of cyclic dicarboxylic acids is via oxidative cleavage of bicyclic alkenes. *cis*-Cyclopentane dicarboxylic acid, for example, was prepared by ozonolysis of norbornene (bicyclo[2.2.1]hept-2-ene) followed by conversion to the dimethyl ester (***26***).[16] Kinetic resolution of this racemic diester with cholesterol esterase led to separation and isolation of ***27***. Subsequent conversion of the ester to an amide was followed by rearrangement to 1*S*,3*R*-3-aminocyclopentanecarboxylic acid (***28***).[16]

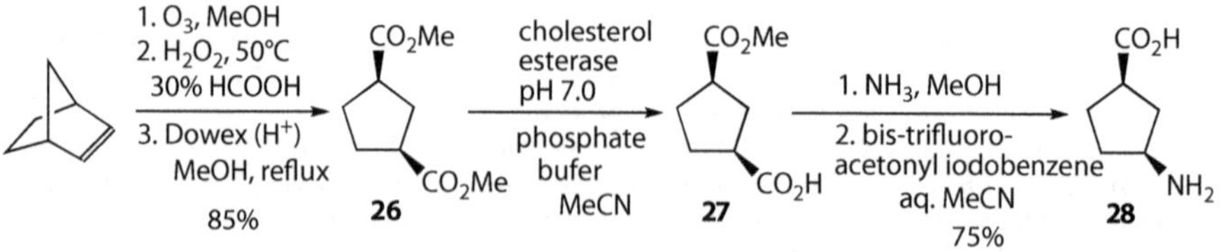

[15] Klarer, W.J.; Hiemstra, H.; Speckamp, W.N. *Tetrahedron* ***1988***, *44*, 6729.
[16] Chênevert, R.; Martin, R. *Tetrahedron Asym.* ***1992***, *3*, 199.

If a cyclic dicarboxylic acid derivative is available, refunctionalization is often straightforward. An example used *trans*-cyclobutane dicarboxylate (***29***), and selective hydrolysis of one ester liberated an acid that was reduced with borane to give ***30***.[17] Conversion of that alcohol moiety to a tosylate was followed by reaction with potassium cyanide to give nitrile ***31***.

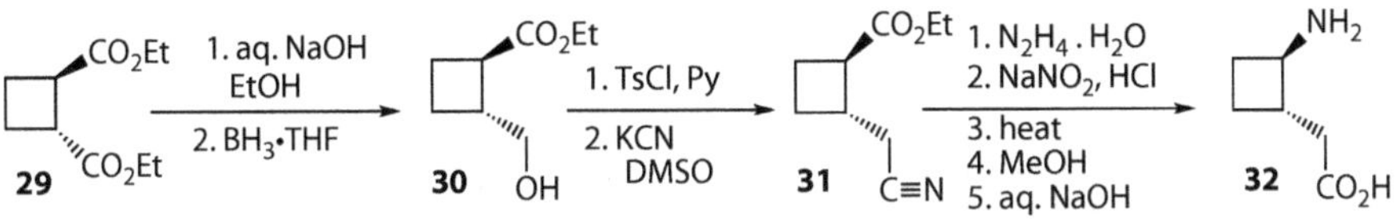

The ester was converted to an *N*-amino amide and then to an amine, allowing hydrolysis of the nitrile to give ***32***.[17] A closely related refunctionalization used the conversion of mono-ester ***33*** to isocyanate ***34***.[18] Subsequent acid hydrolysis gave methyl 2-amino-2-methylcyclohexanecarboxylate, ***35***.[19]

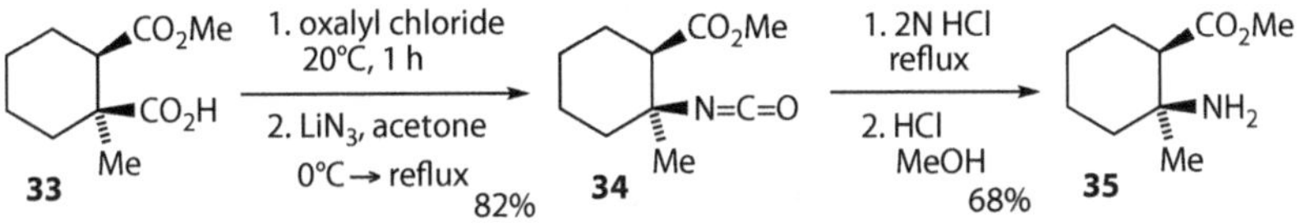

Refunctionalization of the carboxylic acid moiety via the isocyanate can be done with sodium azide as well. The reaction of ***36*** with sodium azide gave carbamate ***37*** via the isocyanate. Subsequent reduction of the alkene moiety with hydrogen in HCl led to methyl 2-amino-5-methylcyclohexane-1-carboxylate (***38***), which was shown to be a microbicide.[20]

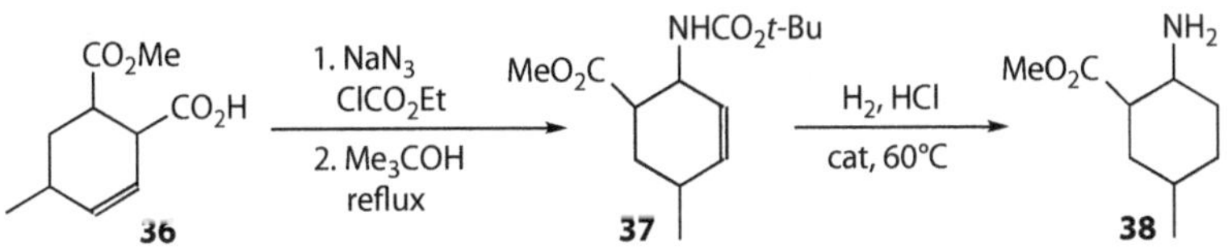

Dicarboxylic acids rather than diesters or ester-acids may be synthetic precursors using this strategy. Mono-functionalization of 1,3-dicarboxylic acid ***39*** gave γ-amino

[17] Kennewell, P.D.; Matharu, S.S.; Taylor, J.B.; Westwood, R.; Sammes, P.G. *J. Chem. Soc. Perkin Trans. I* ***1982***, *2553*, 2563.

[18] Nazarov, I.N.; Kucherov, V.F. *Izvest. Akad. Nauk. SSSR Ser. Khim.* ***1954***, 63.

[19] Armarego, W.L.F. *J. Chem. Soc. C* ***1971***, 1812.

[20] (a) Kunisch, F.; Babczinski, P.; Arlt, D.; Brandes, W.; Dehne, H.W.; Dutzmann, S.; Plempel, M. *Ger. Offen.* DE 4,033,415 [*Chem. Abstr.* ***1992***, *117*: P69489t]; (b) Kunisch, F.; Babczinski, P.; Arlt, D.; Plempel, M. *Ger. Offen.* DE 4,028,046 [*Chem. Abstr.* ***1992***, *117*: P20486a].

acid ***40*** (3-amino-5-*tert*-butylcyclohexanecarboxylic acid)[21] via a modified Schmidt rearrangement.[22]

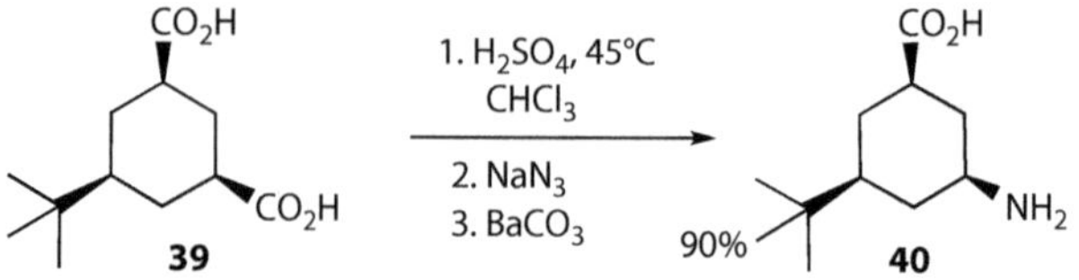

39 40

Anhydrides are another class of dicarboxylic acid derivatives as they are typically prepared from 1,2-dicarboxylic acids. Anhydrides can be synthetically modified as shown by the reaction of anhydride ***41***[23] with ammonia to give amide acid ***42***. Subsequent Hofmann rearrangement[24] of the amide led to amino acid ***43***.[25]

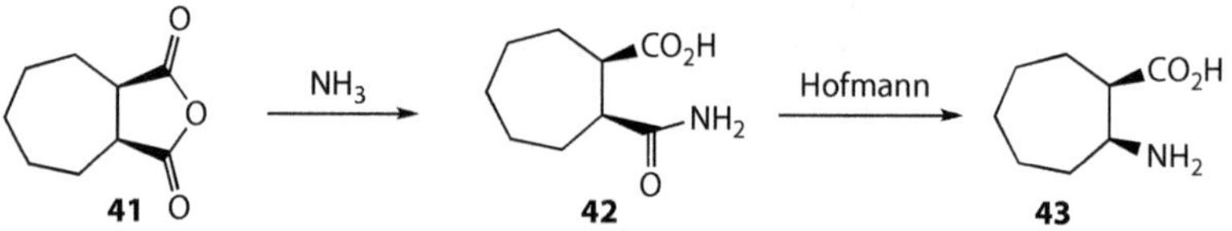

41 42 43

Lactones are cyclic esters and obvious acid derivatives, and they are often readily available or easily prepared. Lactones also serve as useful precursors to amino acids since they can be opened via nucleophilic acyl substitution reactions. Indeed, ring opening with amine surrogates provides a facile route to functionalized carboxylic acids, including non-α-amino acids. Reaction of ***44*** with the amine surrogate potassium phthalimide opened the lactone ring to give a phthalimido-acid. Conversion of the phthalimide group to an amine with aqueous methylamine gave 2-methylamino-1-phenyl-1-cyclopropanecarboxylic acid, ***25***, in 35% overall yield from ***45***.[26] Amino acid ***45*** was tested as a potential antidepressant, and several aryl analogs were prepared, including the 4-chlorophenyl, 4-methylphenyl, and 4-methoxyphenyl derivatives.

---

[21] Armitage, B.J.; Kenner, G.W.; Robinson, M.J.T. *Tetrahedron* ***1964***, *20*, 723, 747.

[22] (a) Schmidt, R.F. *Ber.* ***1924***, *57*, 704; (b) Wolff, H. *Org. React.* ***1946***, *3*, 307; (c) Koldobskii, G.I.; Terreshchenko, G.F.; Gerasimova, E.S.; Bagal, L.I. *Russ. Chem. Rev.* ***1971***, 835; (d) Beckwith, A.L.J., in *Chemistry of Amides*, Zabicky, J., ed. Interscience, London, ***1970***, pp. 137–145; (e) see Smith, M.B. *Organic Synthesis*, 3rd ed. Wavefunction, Inc./Elsevier, Irvine, CA/London, England, ***2010***, pp. 190–192; (f) Smith, M.B. *March's Advanced Organic Chemistry*, 7th ed. John Wiley & Sons, Hoboken, NJ, ***2013***, pp. 1363–1365.

[23] Sicher, J.; Sipos, F.; Jonás, J. *Coll. Czech. Chem. Commun.* ***1961***, *26*, 262.

[24] (a) Hofmann, A.W. *Ber.* ***1881***, *14*, 2725; (b) Wallis, E.S.; Lane, J.F. *Org. React.* ***1949***, *3*, 267; (c) see Smith, M.B. *Organic Synthesis*, 3rd ed. Wavefunction, Inc./Elsevier, Irvine, CA/London, England, ***2010***, pp. 190–192; (d) Smith, M.B. *March's Advanced Organic Chemistry*, 7th ed. John Wiley & Sons, Hoboken, NJ, ***2013***, pp. 1360–1361.

[25] Bernáth, G.; Göndös, Gy.; Kovács, K.; Sohár, P. *Tetrahedron*, ***1973***, *29*, 981.

[26] Bonnaud, B.; Cousse, H.; Mouzin, G.; Briley, M.; Stenger, A.; Fauran, F.; Couzinier, J.-P. *J. Med. Chem.* ***1987***, *30*, 318.

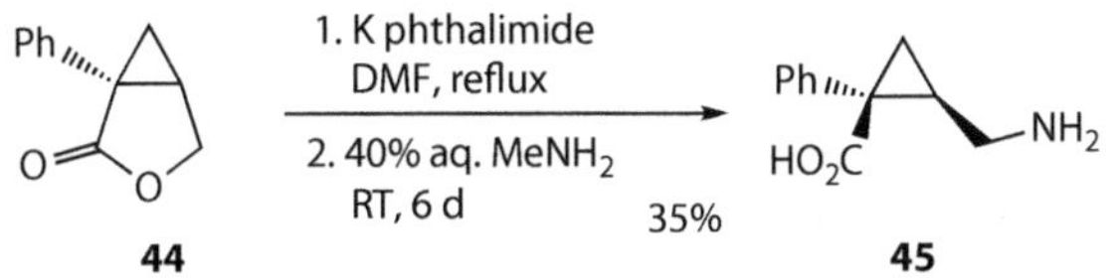

44 45

There are many cyclic precursors other than diesters, diacids, and anhydrides that may be modified to produce amino acids. One example used a keto-acid precursor and converted the ketone moiety in *cis*-pinononic acid [***46***, derived by permanganate oxidation of (–)-2-hydroxy pinocamphone] to an oxime (***47***).[27] Raney nickel reduction and Beckmann rearrangement[28] of the oxime gave ***48***. Subsequent treatment with hydrazine gave 3-amino-2,2-dimethylcyclobutanecarboxylic acid (***49***).

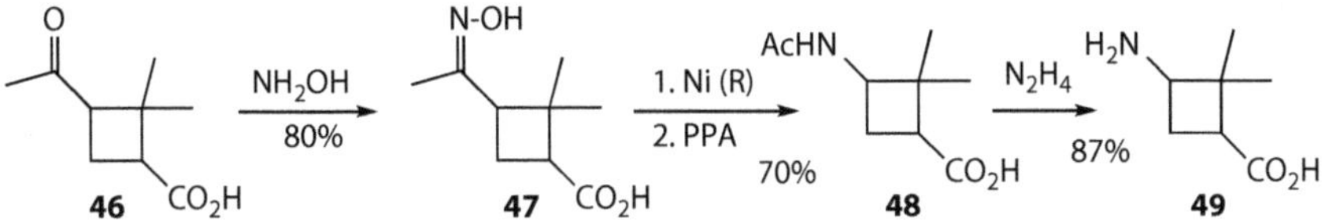

46 47 48 49

Alkene acids are another useful class of precursors to amino acids. In these compounds, the C=C moiety is refunctionalized to incorporate an amine group. An example is the Michael addition[29] of ammonia to 1-cyclopentene-1-carboxylic acid (***50***) to give 2-aminocyclopentane-1-carboxylic acid (***14***), as a mixture of *cis*- and *trans*-isomers.[30] The yield in many cases was rather low, however.

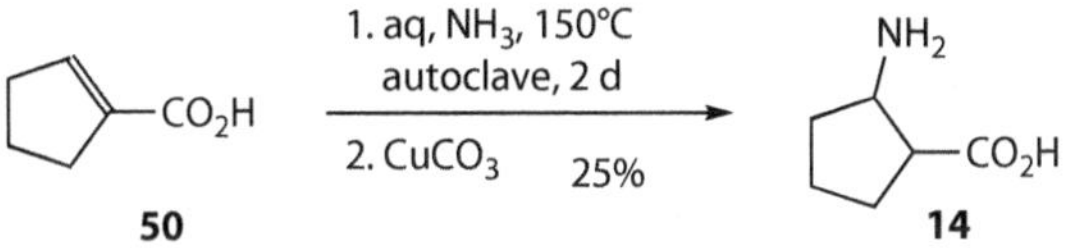

50 14

Refunctionalization of an alkene acid derivative is not restricted to conjugate acid derivatives, or converting the C=C unit to an amine. The allylic positions of alkenes are subject to reactions such as allylic halogenation, giving an alternative route to incorporate an amine group. The nonconjugated alkene ester ***51***, ethylcyclohex-

---

[27] Avotins, F.; Gilis, A.; Gudriniece, E.; Spince, B. *Latv. PSR Zinat. Akad. Vestis, Kim. Ser.* ***1984***, 339 [*Chem. Abstr.* ***1984***, *101*: 191182x].

[28] (a) Beckmann, E. *Ber.* ***1886***, *19*, 988; (b) Donaruma, L.G.; Heldt, W.Z. *Org. React.* ***1960***, *11*, 1; (c) see Smith, M.B. *Organic Synthesis*, 3rd ed. Wavefunction, Inc./Elsevier, Irvine, CA/London, England, ***2010***, pp. 188–190; (d) Smith, M.B. *March's Advanced Organic Chemistry*, 7th ed. John Wiley & Sons, Hoboken, NJ, ***2013***, pp. 1365–1367.

[29] (a) Michael, A. *J. Prakt. Chem.* ***1887***, *35*, 379; (b) Bergmann, E.D.; Gingberg, D.; Pappo, R. *Org. React.* ***1959***, *10*, 179; (c) Perlmutter, P. *Conjugative Addition Reactions in Organic Synthesis*. Pergamon Press, Oxford, ***1992***; (d) see Smith, M.B. *Organic Synthesis*, 3rd ed. Wavefunction, Inc./Elsevier, Irvine, CA/London, England, ***2010***, pp. 877–888; (e) Smith, M.B. *March's Advanced Organic Chemistry*, 7th ed. John Wiley & Sons, Hoboken, NJ, ***2013***, pp. 943–949.

[30] (a) Connors, T.A.; Ross, W.C.J. *J. Chem. Soc.* ***1960***, 2119; (b) Allan, R.D.; Dickenson, H.W.; Fong, J. *Eur. J. Pharmacol.* ***1986***, *122*, 339. Also see Bernáth, G.; Kovács, K.; Láng, K.L. *Acta Chim. Budapest* ***1970***, *64*, 183 [*Chem. Abstr.* ***1970*** *73*: 34856f].

3-en-1-carboxylate, reacted with *N*-bromosuccinimide to give ***52***.[31] This allylic halogenation enabled subsequent reaction with the amine surrogate potassium phthalimide to give ***53*** in an overall yield of 41%. Catalytic hydrogenation and treatment with methylamine led to *cis*-3-aminocyclohexane-1-carboxylic acid, ***55***.[31]

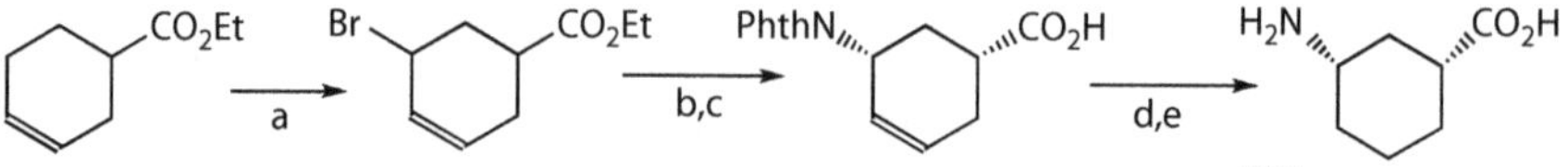

51 52 53 54

(a) NBS (b) K phthalimide (c) aq. H^+ (d) H_2, Pd (e) $MeNH_2$

In another example, allylic acetate ***55*** was prepared and deacetylation gave the corresponding alcohol, which reacted with HBr to give ***56***. Displacement of the allylic bromide moiety with ammonia gave ethyl 4-amino-1-phenyl-2-cyclohexene-carboxylate (***57***),[32] via a S_N2' reaction.[33] A similar reaction with amines gave *N*-alkyl amines.

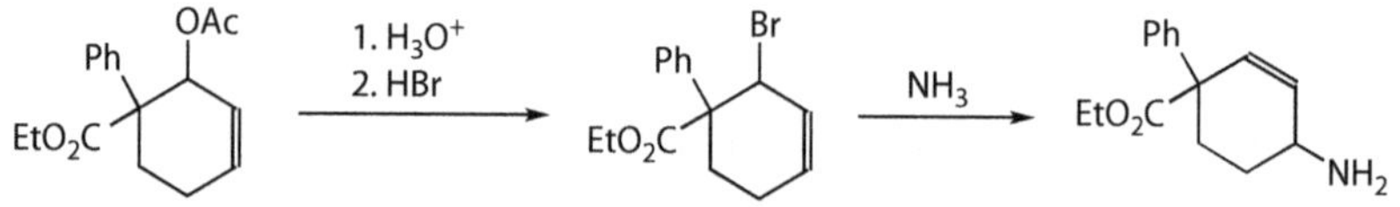

55 56 57

Gabaculine (***61***) is reported to be an inhibitor[34] of GABA-T (GABA transaminase).[35] As first pointed out in Chapter 6, Section 6.3, a transaminase is an enzyme that transfers nitrogenous groups and catalyzes the reaction with 2-oxoglutarate to give succinate semialdehyde.[36] In one synthesis of gabaculines, 3-cyclohexene-1-carboxylic acid (***58***) was converted to allylic acetate ***59***,[37] and subsequent palladium-catalyzed amination gave ***60***.[38] This precursor was subsequently converted to gabaculine, ***61***.

---

[31] Allan, R.D.; Johnston, G.A.R.; Twitchin, B. *Aust. J. Chem.* ***1981***, *34*, 2231. Also see Allan, R.D.; Twitchin, B. *Aust. J. Chem.* ***1980***, *33*, 599.

[32] Satzinger, G.; Herrmann, M. *Ger. Offen.* 2,166,019 [*Chem. Abstr.* ***1973***, *78*: P3822r]. Also see Satzinger, G.; Herrmann, M. *Ger. Offen.* 2,107,871 [*Chem. Abstr.* ***1972***, *77*: P151688w].

[33] (a) Magid, R.M. *Tetrahedron*, ***1980***, *36*, 1901; (b) Bordwell, F.G.; Pagani, G.A. *J. Am. Chem. Soc.* ***1975***, *97*, 118; (c) Bordwell, F.G.; Mecca, T.G. *J. Am. Chem. Soc.* ***1975***, *97*, 123, 127; (d) Bordwell, F.G.; Wiley, P.F.; Mecca, T.G. *J. Am. Chem. Soc.* ***1975***, *97*, 132.

[34] Kobayashi, K.; Miyazawa, S.; Terahara, A.; Mishima, H.; Kurihara, H. *Tetrahedron Lett.* ***1976***, 537.

[35] Roberts, E. *Biochem. Pharm.* ***1974***, *23*, 2637.

[36] (a) Scott, E.M.; Jakoby, W.B. *J. Biol. Chem.* ***1959***, *234*, 932; (b) Schousboe, A.; Wu, J.Y.; Roberts, E. *Biochemistry* ***1973***, *12*, 2868.

[37] (a) Kato, M.; Kageyama, M.; Tanaka, R.; Kuwahara, K.; Yoshikoshi, A. *J. Org. Chem.* ***1975***, *40*, 1932; (b) Trost, B.M.; Timko, J.M.; Stanton, J.L. *J. Chem. Soc. Chem. Commun.* ***1978***, 436.

[38] Trost, B.M.; Keinan, E. *J. Org. Chem.* ***1979***, *44*, 3451.

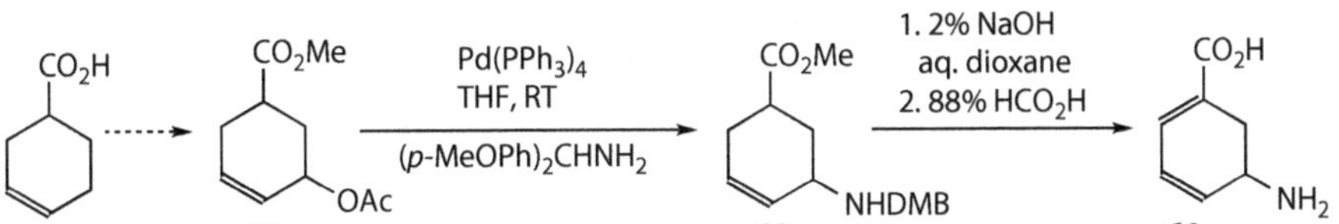

58 59 60 61

The C=C moiety need not be in the ring, but can be exocyclic to the ring as in alkylidene cyclic compounds. Using this modification in the starting material, optically active amino acids were prepared. Reaction of ***62*** with chiral, nonracemic ***63***, for example, gave ***64***.[39] Reduction of the chlorine moiety with zinc/copper and catalytic hydrogenation to reduce the oxazolidine-protecting group gave ***65***.

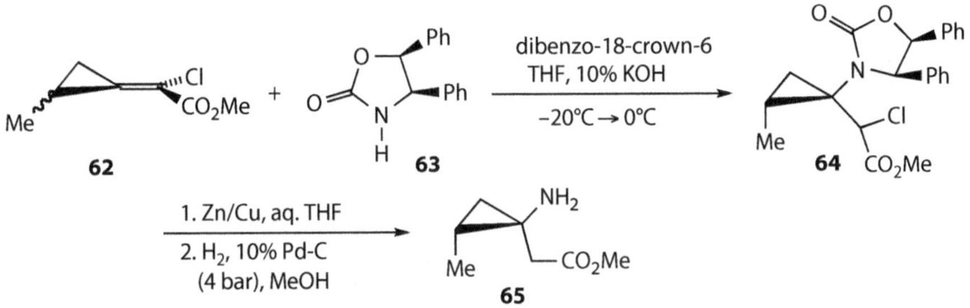

62 63 64 65

Another example that used an alkylidenecarboxylic acid derivative (***66***) incorporated the amine moiety by a different sequence. The reaction of acid chlorides such as ***66*** with ethyl azodicarboxylate (***67***), in the presence of a chiral *Cinchona* alkaloid derivative (TMS-QD HCl), gave cycloadducts ***68*** in good yield.[40] Treatment with trifluoroacetic acid, followed by catalytic hydrogenation, gave the targeted amino acid derivative as amine-amide ***69***. Chiral aminocyclopentane, aminocyclohexane, aminocycloheptane, and aminocyclooctane derivatives containing a pendant carboxyl unit have been prepared using this methodology.

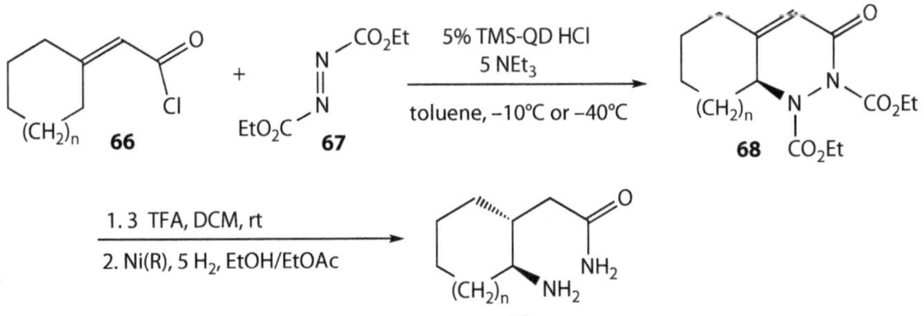

66 67 68 69

Related synthetic strategies may be used. Since cyano is a carboxylic acid by virtue of the fact that acid hydrolysis converts a nitrile to an acid, cyclic nitriles may

[39] Es-Sayed, M.; Gratkowski, C.; Krass, N.; Meyers, A.I.; de Meijere, A. *Tetrahedron Lett.* **1993**, *34*, 289.
[40] Shen, L.-T.; Sun, L.-H.; Ye, S. *J. Am. Chem. Soc.* **2011**, *133*, 15894.

be used to prepare amino acids. However, a "simple" reaction of ammonia and an alkyl halide can sometimes give unusual results. Treatment of ***70*** with ammonia, for example, followed by treatment with acid led to a product with the amine group distal to the carboxyl group. The alkene moiety also "migrated" into conjugation with the acid (derived from hydrolysis of the cyano group). In this reaction, ***70*** was converted to 4-amino-1-cyclohexenecarboxylic acid (***71***)[41] so the cyano group served as the carboxylic acid surrogate.

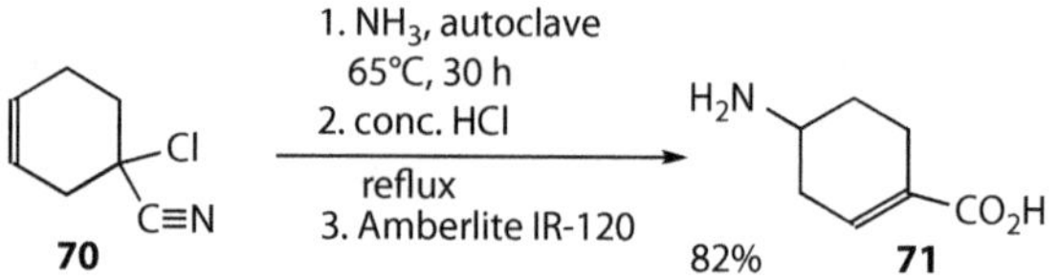

70 **71**

Cyclic nitro compounds are precursors to amino acids due to the fact that reduction of the nitro group gives an amine. Ring-constrained γ-amino acid residues were prepared by the organocatalytic Michael addition of aldehydes to 1-nitro cyclic alkenes to give new aldehydes. Condensation of 1-nitrocyclohexene (***72***) with butanal, for example, gave a 7:1 mixture of ***73***:***74***. Reduction of the aldehyde moiety in ***74***, followed by oxidation to the acid, allowed subsequent reduction of the nitro moiety and protection as *N*-Boc to give the optically active β-substituted δ-nitro alcohol. This alcohol was converted to γ-amino acid residue ***75***.[42] Amino acids such as ***75*** were used to prepare α/γ-peptide foldamers (adopt protein-like, noncovalent folding patterns) that adopted a specific helical conformation in solution and in the solid state.

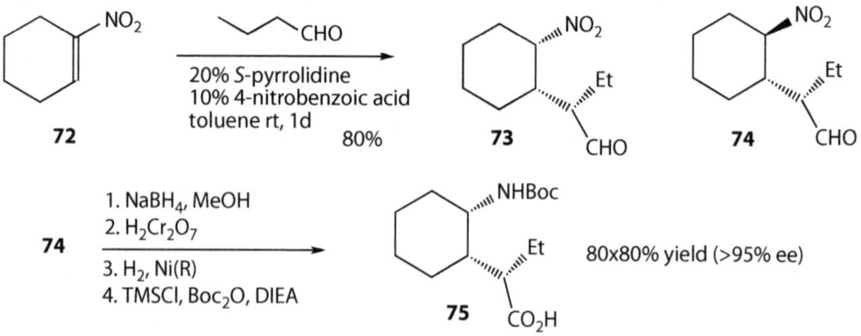

72 **73** **74** **75**

7.2 FROM AROMATIC PRECURSORS

A common synthetic route to aminocyclohexanecarboxylic acids is reduction of nitro- and amino-substituted benzoic acid derivatives. Catalytic hydrogenation of *p*-amino ammonium salt ***76*** gave a 93% yield of 4-aminocyclohexanecarboxylic acid

---

[41] Asahi, Chemical Industry Co. Ltd. *Jpn.* 100 ('67) [*Chem. Abstr.* ***1967***, *66*: P55105f].

[42] Guo, L.; Chi, Y.; Almeida, A.M.; Guzei, I.A.; Parker, B.K.; Gellman, S.H. *J. Am. Chem. Soc.* ***2009***, *131*, 16018.

(*77*), as a 72:28 *cis*:*trans* mixture.[43] Similarly, reduction of the *meta*-derivative (***78***) gave an 88% yield of ***79***.[43] A rhodium–aluminum catalyst was used to reduce the *ortho*-amino-ester (***80***) to ethyl 2-aminocyclohexanecarboxylate (***81***), in 70% yield.[44]

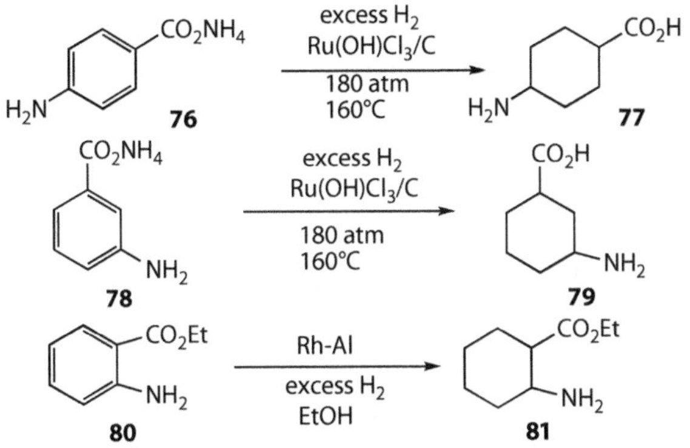

Other aromatic derivatives were reduced to *77* or *79* in a similar manner using a nickel-rhenium catalyst.[45] A ruthenium-on-carbon/ruthenium oxide catalyst was used for the preparation of *77* and *79*[46]; a rhenium–aluminum oxide catalyst was used to prepare *77*[47] or *79*[48]; and a platinum oxide catalyst was used to prepared *77* and *79*.[49] A ruthenium oxide catalyst was used to prepare *77*[50] and also *79*,[51] and a ruthenium catalyst was used to give *77*.[52]

Benzoic acid derivatives provide a route to cyclohexyl derivatives in which the amine unit or the carboxyl unit is not directly attached to the aromatic ring. An example is the catalytic hydrogenation of *82* to give ***83***.[53]

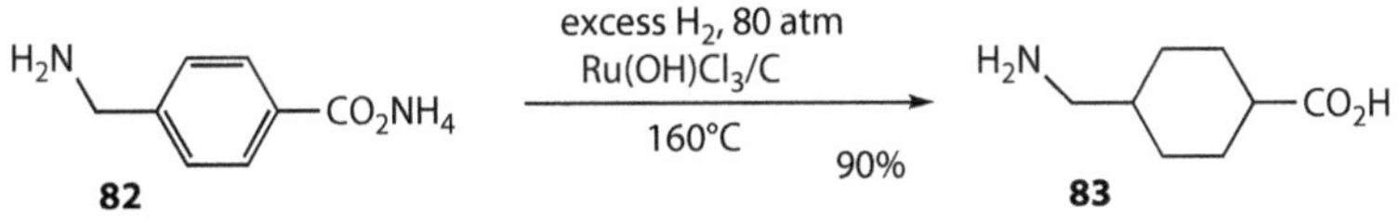

[43] Litvin, E.F.; Freidlin, L.Kh.; Gurskii, R.N.; Istratova, R.V.; Presnov, A.P. *Izv. Akad. Nauk. SSSR Ser. Khim* ***1979***, 2441 (Engl. 2258).

[44] (a) Liska, K.J. *J. Pharm. Sci.* ***1964***, *53*, 1427; (b) Smissman, E.E.; Steinman, M. *J. Med. Chem.* ***1967***, *10*, 1054.

[45] Palaima, A.I.; Poshkene, R.A.; Karpavichyus, K.I.; Kil'disheva, O.V.; Knunyants, I.L. *Izv. Akad. Nauk. SSSR Ser. Khim.* ***1977***, 195 (Engl. 171).

[46] Rhone-Poulenc. SA French Patent 1,473,246 [*Chem. Abstr.* ***1968***, *68*: P12555b].

[47] Vechkanov, G.N.; Zhizdyuk, B.I.; Malukh, V.A.; Chegolya, A.S. *Sin. Volokna* ***1969***, 21 [*Chem. Abstr.* ***1970***, *73*: 66100w].

[48] Schneider, W.; Hoyer, J. *Archiv. Pharm.* (Weinheim) ***1971***, *304*, 637.

[49] Skaric, V.; Kovacevic, M.; Skaric, D. *J. Chem. Soc. Perkin Trans. I* ***1976***, 1199.

[50] Freifelder, M.; Stone, G.R. *J. Org. Chem.* ***1962***, *27*, 3568.

[51] Ponomarev, A.A.; Ryzhenko, L.M.; Smirnova, N.S. *Zh. Org. Khim* ***1969***, *5*, 75 (Engl, 73).

[52] Schneider, W.; Hüttermann, A. *Archiv der Pharm.* (Weinheim) ***1965***, *298*, 226.

[53] Litvin, E.F.; Freidlin, L.Kh.; Gurskii, R.N.; Istratova, R.V.; Presnov, A.P. *Izv. Akad. Nauk. SSSR Ser. Khim* ***1979***, 2441 (Engl. 2258).

An alternative method of reducing the aromatic ring relies on alkali metals (sodium, lithium calcium, etc.) in protic solvents such as ethanol, a so-called dissolving metal reduction. An example is the reaction of ***84*** with sodium in isoamyl alcohol to give ***85***.[54] Sodium in ammonia was used for the same transformation (as early as 1894), but with less *cis/trans* selectivity.[55] Partial reduction of a benzene ring is also possible via a Birch reduction,[56] allowing preparation of a cyclohexenecarboxylic acid. Other reduction methods can be used as well, as in the conversion of ***86*** to ethyl 2-amino-1-cyclohexenecarboxylate (***87***).[57]

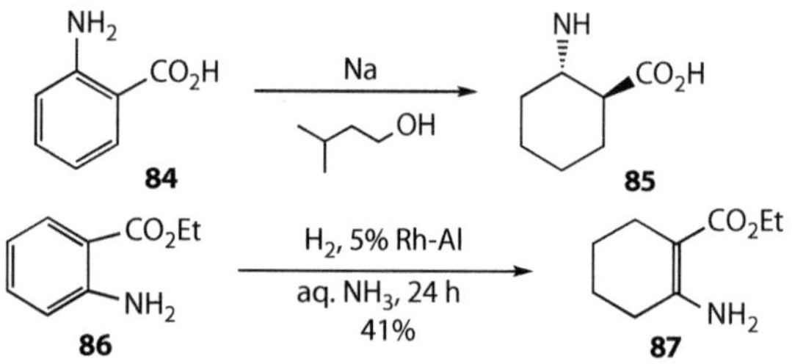

Although they are not highlighted in this section, virtually any aromatic compound with an amine and carboxylic acid unit, or appropriate surrogate precursors, can be converted to the corresponding amino acid. One example that illustrates this statement is the preparation of naphthyl oxazoline derivative ***88***.[58] A subsequent reaction with a lithium amide such as $LiNMe_2$, followed by alkylation with an alkyl halide, gave ***89***. Hydrolysis of the oxazoline required some effort, but the product was amino acid ***90***.[59] In this sequence, the oxazoline is utilized as a carboxyl surrogate, but also activates the aromatic ring to attack by the amide anion.

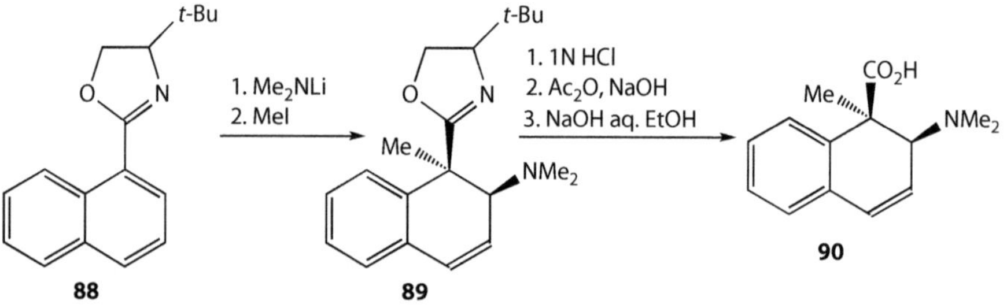

---

[54] Hünig, S.; Kahanek, H. *Chem. Ber.* ***1953***, *86*, 518. Also see Allan, R.D.; Johnston, G.A.R.; Twitchin, B. *Aust. J. Chem.* ***1981***, *34*, 2231.

[55] Einhorn, A.; Meyenberg, A. *Deutsch. Ges. Ber.* ***1894***, *27*, 2466.

[56] For a brief review of the Birch reduction and its application to the synthesis of natural products, see (a) Schultz, A.G. *Chem. Commun.* ***1999***, 1263. Also see (b) Smith, M.B. *Organic Synthesis*, 3rd ed. Wavefunction, Inc./Elsevier, Irvine, CA/London, England, ***2010***, pp. 459–461; (c) Smith, M.B. *March's Advanced Organic Chemistry*, 7th ed. John Wiley & Sons, Hoboken, NJ, ***2013***, pp. 912–916.

[57] Liska, K.J. *J. Pharm. Sci.* ***1964***, *53*, 1427.

[58] Rawson, D.; Meyers, A.I. *J. Org. Chem.* ***1991***, *56*, 2292.

[59] Shimano, M.; Meyers, A.I. *J. Org. Chem.* ***1995***, *60*, 7445.

7.3 FROM CYCLIC KETONES

A pendant ketone moiety was used as an amine surrogate, via formation of an oxime, in Section 7.1.2. There are a variety of ways in which a cyclic ketone can serve as a precursor to an amino acid. Section 7.1.2 discussed methods by which a dicarboxylic acid derivative was converted to a cyclic amino acid. If the focus is placed on the cyclic ketone, methodology is available for the preparation of amino acids. Oxidative cleavage of ketones can lead to α,ω-dicarboxylic acids (also see Section 7.1.2 for other dicarboxylic acids), and if properly functionalized, amino acids. In early work, Noyes[60] converted camphor (***91***) to ***92***. Beckmann rearrangement[61] led to the formation of 3-amino-1,2,2-trimethylcyclopentanecarboxylic acid, ***93***.[62]

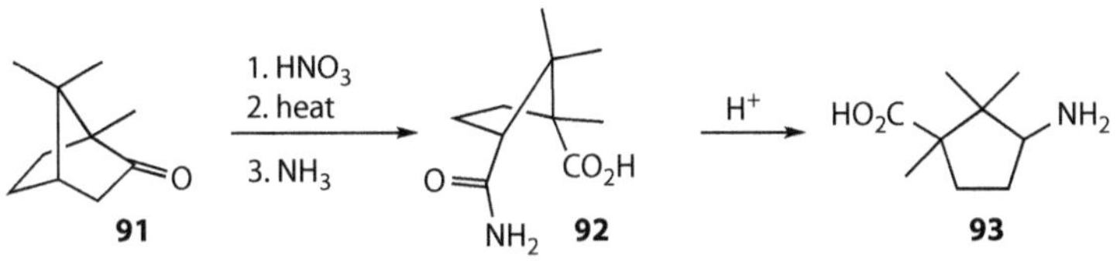

91 92 93

Amines react with ketones to give imines or enamines, depending on whether the reaction is with a primary or a secondary amine, respectively. With these reactions in mind, the reaction of an amine with a cyclic β-keto-ester undergoes elimination of water (dehydration) to give an alkenyl amino acid derivative (an enamino-ester). For example, the reaction of 2-carboethoxycyclopentanone (***94***) with ammonium nitrate is believed to give ***95*** as a transient intermediate, but the final product is ethyl 2-amino-1-cyclopentenecarboxylate (***96***), the product of dehydration.[63] It is noted that β-keto-acid derivatives such as ***94*** are easily prepared by a Dieckmann condensation of an α,ω-diester.[64] Dieckmann postulated the intermediacy of ***95*** in this reaction as early as 1907.[65] This reaction and minor variations have been reported many times.[66] Urethane has been employed as an amine surrogate with cyclic keto-esters,[67] and

---

[60] (a) Noyes, W.A. *J. Am. Chem. Soc.* ***1894***, *16*, 500; (b) Königs, W.; Hörlin, J. *Ber. Deutsch. Chem. Ges.* ***1893***, *26*, 817.

[61] (a) Beckmann, E. *Ber.* ***1886***, *19*, 988; (b) Donaruma, L.G.; Heldt, W.Z. *Org. React.* ***1960***, *11*, 1; (c) see Smith, M.B. *Organic Synthesis*, 3rd ed. Wavefunction, Inc./Elsevier, Irvine, CA/London, England, ***2010***, pp. 188–190; (d) Smith, M.B. *March's Advanced Organic Chemistry*, 7th ed. John Wiley & Sons, Hoboken, NJ, ***2013***, pp. 1365–1367.

[62] (a) Faigle, H.; Karrer, P. *Helv. Chim. Acta.* ***1962***, *45*, 73; (b) Gassman, P.G.; Cryberg, R.L. *J. Am. Chem. Soc.* ***1969***, *91*, 2047.

[63] Prelog, V.; Szpilfogel, S. *Helv. Chim. Acta* ***1945***, *28*, 1684.

[64] (a) Dieckmann, W. *Berichte* ***1894***, *27*, 102, 965; (b) Dieckmann, W. *Berichte* ***1900***, *33*, 2670; (c) Dieckmann, W.; Groenveld, A. *Berichte* ***1900***, *33*, 595; (d) Hauser, C.R.; Hudson, B.E. *Org. React.* ***1942***, *1*, 266 (see p. 274); (e) Schaefer, J.P.; Bloomfield, J.J. *v* ***1967***, *15*, 1; (f) see Smith, M.B. *Organic Synthesis*, 3rd ed. Wavefunction, Inc./Elsevier, Irvine, CA/London, England, ***2010***, pp. 827–829; (g) Smith, M.B. *March's Advanced Organic Chemistry*, 7th ed. John Wiley & Sons, Hoboken, NJ, ***2013***, p. 1235.

[65] Dieckmann, W. *J. L. Ann. Chem.* ***1901***, *317*, 27.

[66] (a) Durbeck, H.W.; Duttka, L.L. *Tetrahedron* ***1973***, *29*, 4285; (b) Takaya, T.; Yoshimoto, H.; Imoto, E. *Bull. Chem. Soc. Jpn.* ***1967***, 40, 2844.

[67] Takaya, T.; Yoshimoto, H.; Imoto, E. *Bull. Chem. Soc. Jpn.* ***1968***, *41*, 2176.

also phosphazines such as $Ph_3P{=}NSiMe_3$.[68,69] This method can also be used for the preparation of cyclohexenecarboxylic acid derivatives.[70]

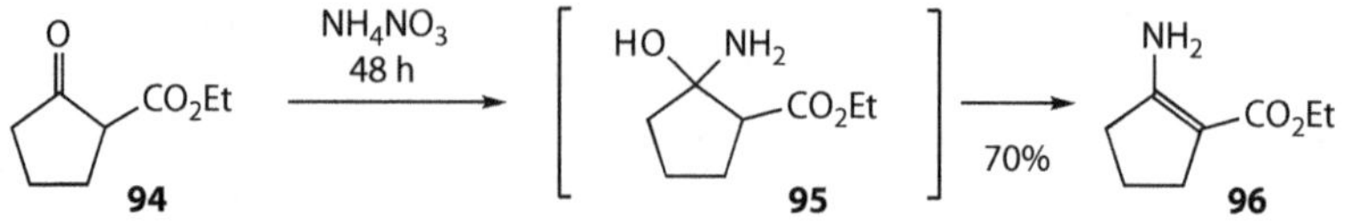

Using ammonia[71a,72] or amines, this synthetic approach is quite common. Amines react with cyclic β-keto-esters such as ***94*** reacted with propanamine, for example, to give ethyl 2-(*N*-propylamino)-1-cyclopentenecarboxylate, ***97***.[73] This work was based on the earlier reported reaction with ***94*** with benzylamine to give the *N*-benzyl derivative of ***97***.[74] Just as with the ammonia reaction, cycloalkenyl amino acids can be reduced to their saturated derivatives. Reaction of ***98*** with methylamine, for example, gave ***99***, which was converted to ***100*** via catalytic hydrogenation.[75]

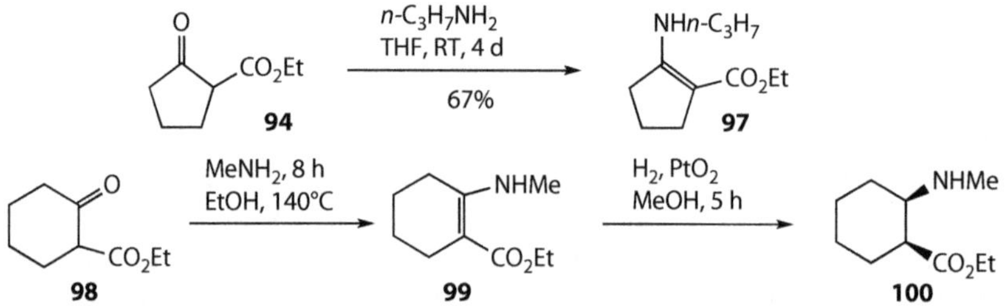

Reductive amination of a cyclic ketone using catalytic hydrogenation in the presence of ammonia gives amino acids. Treatment of ***101*** with ammonia and Raney nickel, for example, led to ***14*** (reductive amination), as a mixture of *cis*- and *trans*-isomers.[76]

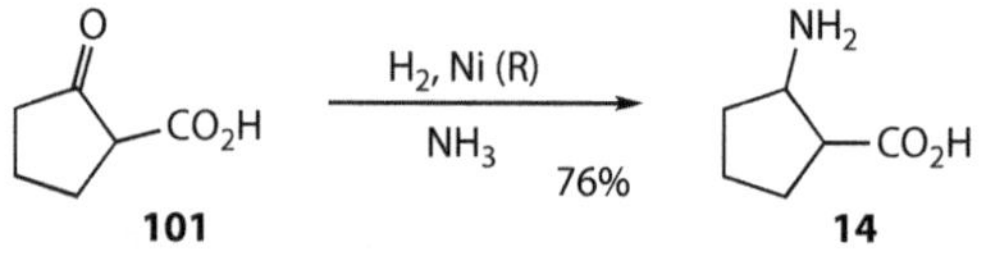

It is possible to refunctionalize other cyclic derivatives to give an amino acid. 1,2-Cyclopentenedicarboxylic acid (***102***) was prepared in several steps from ethyl

[68] Birkofer, L.; Ritter, A.; Richter, P. *Chem. Ber.* ***1963***, *96*, 2750.
[69] Kloek, J.A.; Leschinsky, K.L. *J. Org. Chem.* ***1978***, *43*, 1460.
[70] Hünig, S.; Kahanek, H. *Chem. Ber.* ***1953***, *86*, 518.
[71] (a) Durbeck, H.W.; Duttka, L.L. *Tetrahedron*, ***1973***, *29*, 4285; (b) Takaya, T.; Yoshimoto, H.; Imoto, E. *Bull. Chem. Soc. Jpn.* ***1967***, *40*, 2844.
[72] Buckler, R.T.; Hartzler, H.E *J. Med. Chem.* ***1975***, *18*, 509.
[73] Pennington, F.C.; Kehret, W.D. *J. Org. Chem.* ***1967***, *32*, 2034.
[74] Triebs, W.; Mayer, R.; Madejski, M. *Chem. Ber.* ***1954***, *87*, 356.
[75] Horii, Z.; Watanabe, T.; Ikeda, M.; Tamura, Y. *Yakugaku Zasshi* ***1963***, *83*, 930 [*Chem. Abstr.* ***1964***, *60*: 4106d].
[76] Berger, H.; Paul, H.; Hilgetag, G. *Chem. Ber.* ***1968***, *101*, 1525. See Prelog, V.; Szpilfoge, S. *Helv. Chim. Acta* ***1946***, *28*, 1684.

2-oxocyclopentanecarboxylate (***94***), and subsequent catalytic hydrogenation of the double bond gave ***103***.[77] Dicarboxylic acid ***103*** was converted to an anhydride, and subsequent reaction with ammonium hydroxide gave *N*-hydroxyimide ***104***. Lossen rearrangement[78] led to 2-aminocyclopentanecarboxylic acid (***105***). An alternative preparation of ***103*** from ***94*** has also been reported.[79]

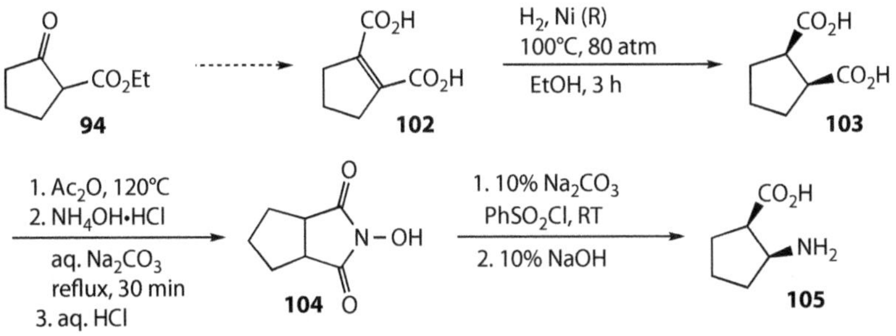

Cyclic keto-esters other than β-keto-esters lead to other amino acids. This structural variation allows other methods to be used for introduction of the amine group. Conversion of *R*-***106*** (resolved from the racemic material with brucine) to oxime ***107*** was accomplished by treatment with hydroxylamine and pyridine.[80] Dissolving metal reduction of the oxime and hydrolysis gave 1*R*,3*R*-3-aminocyclopentane-1-carboxylic acid, ***108***. An identical approach was used to prepare cyclohexane derivatives. The reaction of ***109***[81] with hydroxylamine[82] and subsequent esterification gave ***110***.[83] Reduction of the oxime with Raney nickel gave ethyl 4-aminocyclohexane-1-carboxylate, ***111***.

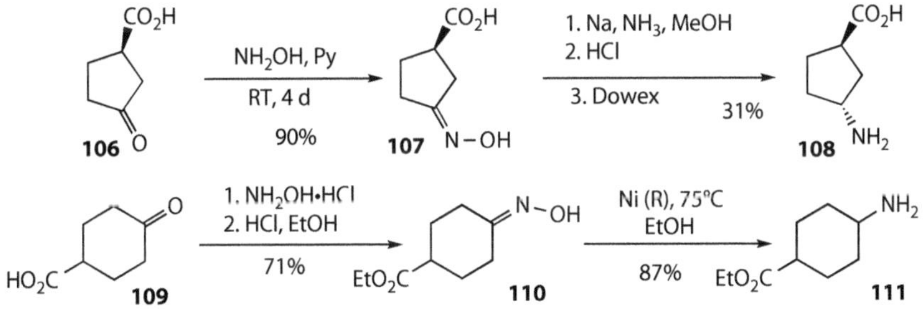

[77] Takaya, T.; Yoshimoto, H.; Imoto, E. *Bull. Chem. Soc. Jpn.* **1967**, *40*, 2844.

[78] (a) Lossen, W. *Ann.* **1872**, *161*, 347; (b) Lossen, W. *Ann.* **1874**, *175*, 271, 313; (c) Popp, F.O.; McEwen, W.E. *Chem. Rev.* **1958**, *58*, 374, and references cited therein. See (d) Smith, M.B. *March's Advanced Organic Chemistry*, 7th ed. John Wiley & Sons, Hoboken, NJ, **2013**, pp.1362–1363.

[79] Nativ, E.; Rona, P. *Isr. J. Chem.* **1972**, *10*, 55.

[80] Also see Allan, R.D.; Johnston, G.A.R.; Twitchin, B. *Aust. J. Chem.* **1979**, *32*, 2517.

[81] (a) Schdeider, W.; Dillmann, R. *Chem. Ber.* **1963**, *96*, 2377; (b) Hardegger, E.; Plattner, Pl.A.; Blank, F. *Helv. Chim. Acta* **1944**, *27*, 793.

[82] Mills, W.H.; Bain, A.M. *J. Chem. Soc.* **1910**, *97*, 1866.

[83] Schneider, W.; Hüttermann, A. *Archiv. der Pharm.* (Weinheim), **1965**, *298*, 226.

The reaction of ***112*** with phenethylamine gave a protected aminocyclopentane derivative. Both ***112a*** and ***112b*** were converted to the corresponding amine (***113a*** or ***113b***). Deprotection by catalytic hydrogenation led to ***114a*** [ethyl 2-(2-amino-1-cyclopentyl)ethanoate; 4% overall yield] or ***114b*** [ethyl 2-(2-amino-1-cyclopentyl) propanoate; 38% overall yield].[84]

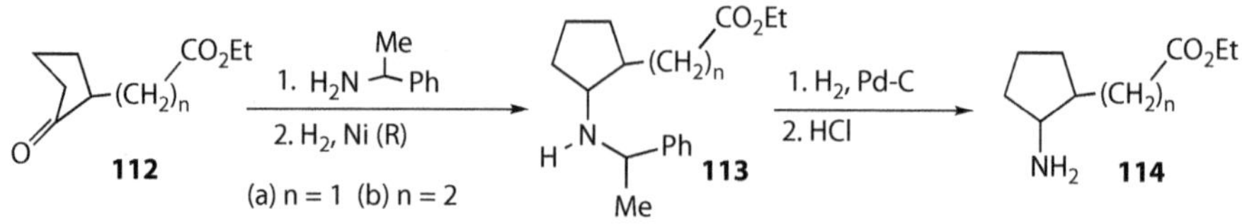

A different approach used keto-ester ***115***[85] to form an enamine, which reacted with 4-methoxybenzyl bromide to give ***116***[86] using the Stork enamine synthesis.[87] Reduction to the alcohol, conversion to the tosylate, and reaction with sodium azide gave ***117***. Subsequent hydrogenation gave ***118***.

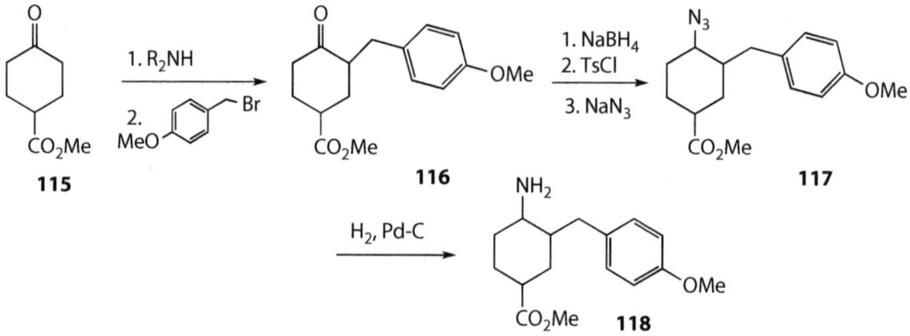

Keto-ester ***94*** was also reacted with ethyl α-cyanoacetate in a Knoevenagel type condensation[88] to give ***119***. Subsequent reaction with ammonium hydroxide led to ***96*** in good yield.[89] Amines also reacted with ***119*** to give *N*-alkyl derivatives of ***96***.

[84] Omar, F.; Frahm, A.W. *Archiv. Pharm.* (Weinheim), ***1989***, *322*, 461.

[85] Stork, G.; Brizzolara, A.; Landesman, H.; Szmuszkovicz, J.; Terrell, R. *J. Am. Chem. Soc.* ***1963***, *85*, 207.

[86] Bégué, J.-P.; Fétizon, M. *C.R. Acad. Sci.* ***1965***, *260*, 3425.

[87] (a) Stork, G.; Terrell, R.; Szmuszkovicz, J. *J. Am. Chem. Soc.* ***1954***, *76*, 2029; see (b) Smith, M.B. *Organic Synthesis*, 3rd ed. Wavefunction, Inc./Elsevier, Irvine, CA/London, England, ***2010***, pp. 873–877; (c) Smith, M.B. *March's Advanced Organic Chemistry*, 7th ed. John Wiley & Sons, Hoboken, NJ, ***2013***, pp. 550–552.

[88] (a) Japp, F.R.; Streatfeild, F.W. *J. Chem. Soc.* ***1883***, *43*, 27; (b) Knoevenagel, F. *Ber.* ***1896***, *29*, 172; (c) Knoevenagel, F. *Ber.* ***1898***, *31*, 730; (d) Jones, G. *Org. React.* ***1967***, *15*, 204; (e) see Smith, M.B. *Organic Synthesis*, 3rd ed. Wavefunction, Inc./Elsevier, Irvine, CA/London, England, ***2010***, p. 829; (f) Smith, M.B. *March's Advanced Organic Chemistry*, 7th ed. John Wiley & Sons, Hoboken, NJ, ***2013***, pp. 1157–1160.

[89] Kasturi, T.R.; Srinivasan, A. *Tetrahedron* ***1966***, *22*, 2575.

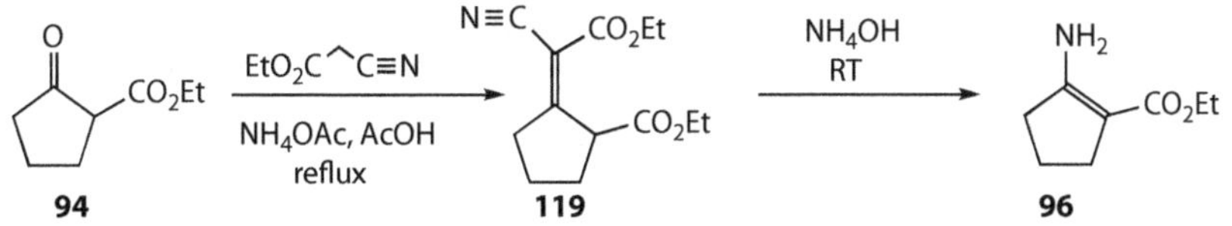

Ketone ***120***[90] also served as an amino acid precursor by reaction with 3-methoxyphenylmagnesium bromide (***121***), to give ***122*** via conjugate addition and elimination of ethoxide.[91] Conjugate addition of potassium cyanide followed by acid hydrolysis gave ***123***. Reaction of the ketone moiety with methylamine to give the imine and catalytic hydrogenation of the imine moiety gave ***124***, but in poor yield. Both the 2-methoxy and the 4-methoxy derivative, as well as the simple phenyl derivative, were prepared by this method. In some cases, amino-esters such as ***125*** cyclized to the bicyclic lactam.[102]

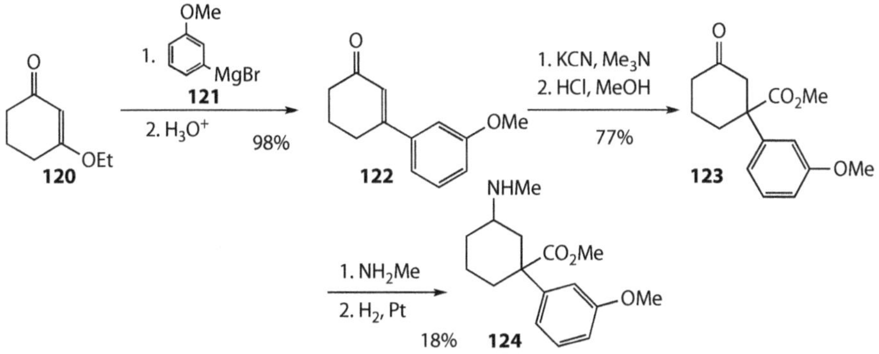

7.4 CYCLOADDITION STRATEGIES

The Diels-Alder reaction is one of the more powerful reactions in all of organic chemistry.[92] There are several variations of a Diels-Alder strategy that can be used to prepare cyclic amino acids.

Dicarboxylic acid derivatives, and also anhydrides, were shown to be useful precursors to amino acids in Section 7.1.2. A good route to bicyclic anhydrides such as ***125*** used a Diels-Alder reaction of 1,3-butadiene and maleic anhydride.[93] This product

[90] Grewe, R.; Nolte, E.; Rotzoll, R.-H. *Chem. Ber.* ***1956***, *89*, 600.

[91] Takeda, M.; Inoue, H.; Noguchi, K.; Honma, Y.; Kawamori, M.; Tsukamoto, G.; Yamawaki, Y.; Saito, S. *Chem. Pharm. Bull.* ***1976***, *24*, 1514.

[92] (a) Desimoni, G.; Tacconi, G.; Barco, A.; Pollini, G.P. *Natural Product Synthesis through Pericyclic Reactions*. American Chemical Society, Washington, DC, ***1983***; (b) see Smith, M.B. *Organic Synthesis*, 3rd ed. Wavefunction, Inc./Elsevier, Irvine, CA/London, England, ***2010***, pp. 1013–1075; (c) Smith, M.B. *March's Advanced Organic Chemistry*, 7th ed. John Wiley & Sons, Hoboken, NJ, ***2013***, pp. 1020–1039.

[93] (a) Kaufmann, St.; Rosenkranz, G.; López, J. *J. Am. Chem. Soc.* ***1946***, *68*, 2733; (b) Wallis, E.S.; Nagel, S.C. *J. Am. Chem. Soc.* ***1931***, *53*, 2787.

was converted to ***126*** by treatment with ammonia and then catalytic hydrogenation.[94] Subsequent reaction with sodium hydroxide and bromine (NaOBr to give a Hofmann rearrangement)[24] gave 2*S*-aminocyclohexane-1*R*-carboxylic acid, ***127***.[94,95]

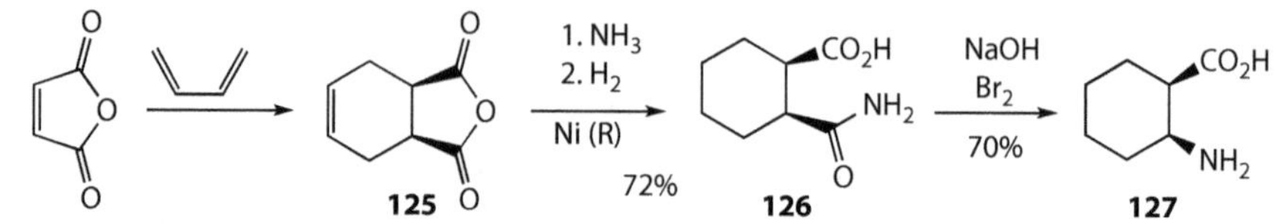

Other anhydrides can be used in the Diels-Alder reaction[92] to ultimately give substituted cyclic amino acids. An example is the reaction of 1,3-butadiene with ***128***[96] to give ***129***.[97] Esterification was followed by conversion to ethyl 6-amino-1-phenyl-3-cyclohexene-1-carboxylate (***130***).

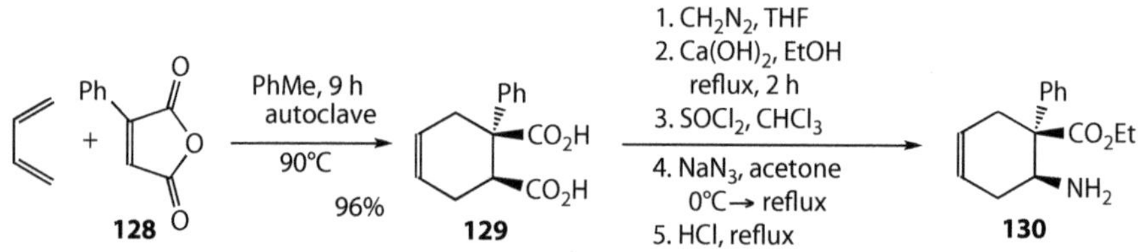

A closely related strategy used citraconic anhydride (***131***) in a reaction with 2-ethoxy-1,3-butadiene. The anhydride moiety in the resultant cycloadduct was hydrolyzed to give ***132***.[98] The ketone moiety was reduced to an alcohol, allowing formation of lactone ***133***. Treatment with acetic anhydride led to a new anhydride (***134***), and the acid moiety was converted to a Cbz amino moiety (in ***135***). Treatment of the lactone with HCl in ethanol led to amino-ester ***136***.

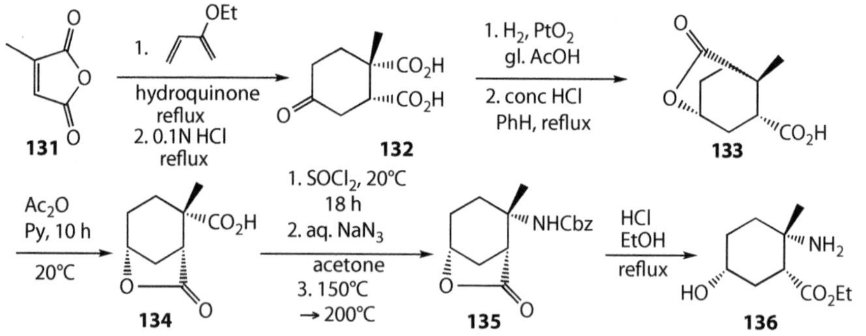

---

[94] Plieninger, H.; Schneider, K. *Chem. Ber.* ***1959***, *92*, 1594.

[95] Also see Hünig, S.; Kahanek, H. *Chem. Ber.* ***1953***, *86*, 518.

[96] (a) Miller, L.E.; Staley, H.B.; Mann, D.J. *J. Am. Chem. Soc.* ***1949***, *71*, 374; (b) Miller, L.E.; Mann, D.J. *J. Am. Chem. Soc.* ***1950***, *72*, 1484; (c) Miller, L.E.; Mann, D.J. *J. Am. Chem. Soc.* ***1951***, *73*, 45.

[97] Satzinger, G. *Liebigs Ann. Chem.* ***1972***, *758*, 43.

[98] (a) Bruck, P.R.; Clark, R.D.; Davidson, R.S.; Günther, W.H.H.; Littlewood, P.S.; Lythgoe, B. *J. Chem. Soc. C* ***1967***, 2529; (b) Davidson, R.S.; Littlewood, P.S.; Medcalfe, T. *Tetrahedron Lett.* ***1963***, 1413.

A variation of this sequence is the reaction of the butadiene surrogate 3-sulfolene with maleic anhydride to give ***125***. Subsequent hydrogenation of the double bond gave anhydride ***137***, and this anhydride treated with trimethylsilyl azide gave ***127***.[99] The transformation of ***137*** to ***127*** probably proceeds via a transient isocyanate,[100] ***138***. A closely related method converted anhydrides such as ***136*** to the corresponding *N*-hydroxy imide, and then to a *trans*-2-aminocyclohexanecarboxylic acid.[101,102]

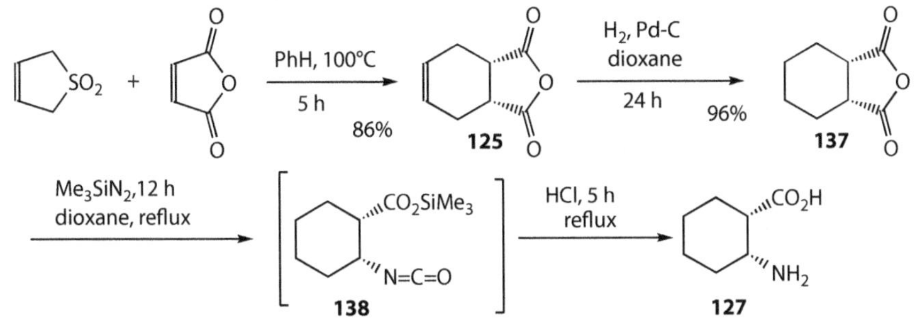

Heteroatom-substituted reactants can be used in Diels-Alder reactions[92] to prepare amino acids. A Diels-Alder reaction of tosyl cyanide and dienes will generate a bicyclic lactam. Cyclopentadiene reacted with *p*-methylbenzonitrile (*N*-tosyl cyanide; TsCN), for example, to give azabicycloheptene derivative, ***139***.[103] Treatment with aqueous acetic acid converted the sulfonyl-imine moiety to lactam ***18***. Subsequent catalytic hydrogenation of the double bond gave ***140***. Acid hydrolysis gave *cis*-3-aminocyclopentane-1-carboxylic acid, ***28***.

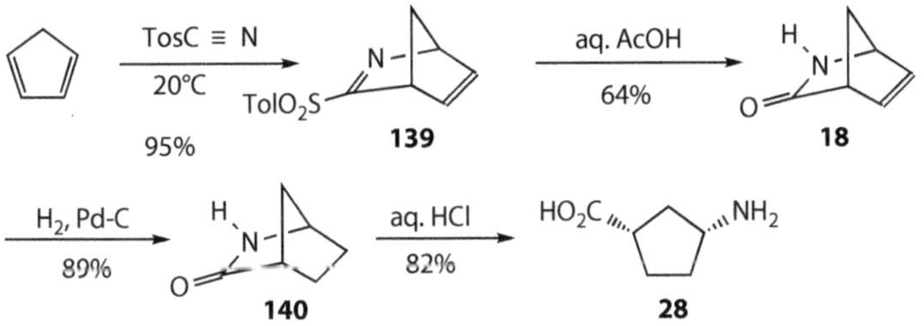

Dienes that have a pendant carboxylic acid moiety are well known, such as penta-2,4-dienoic acid, and dienes that have a pendant nitrogen moiety are also known. These types of diene reactants participate in Diels-Alder reactions.[92]

---

[99] Goodridge, R.J.; Hambley, T.W.; Ridley, D.D. *Aust. J. Chem. 1986, 39*, 591.

[100] Kricheldorf, H.R. *Liebigs Ann. Chem.* ***1975***, 1387.

[101] Orndorff, W.R.; Pratt, D.S. *J. Am. Chem. Soc.* ***1912***, *47*, 89.

[102] Bauer, L.; Miarka, S.V. *J. Org. Chem.* ***1959***, *24*, 1293.

[103] Jagt, J.C.; van Leusen, A.M. *J. Org. Chem.* ***1974***, *39*, 564. Also see Adriaens, P.; Meesschaert, B.; Janssen, G.; Dumon, L.; Eyssen, H. *Recuil. Trav. Chim. Pays-Bas* ***1978***, *97*, 260; Evans, C.T.; Roberts, S.M. *Eur. Pat. Appl.* EP 424,067 [*Chem. Abstr.* ***1991***, *115*: P49382h].

Amide-substituted diene ***141***[104] reacted with ***142***,[105] for example, to give ***143***.[106] Treatment with 1,8-diazabicyclo[5.4.0]undec-7-ene (DBU) gave a 70% yield of ***144***, which was converted to isogabaculine (***145***, 3-amino-1,4-cyclohexadiene-1-carboxylic acid), an irreversible enzyme-activated GABA-transaminase inhibitor.

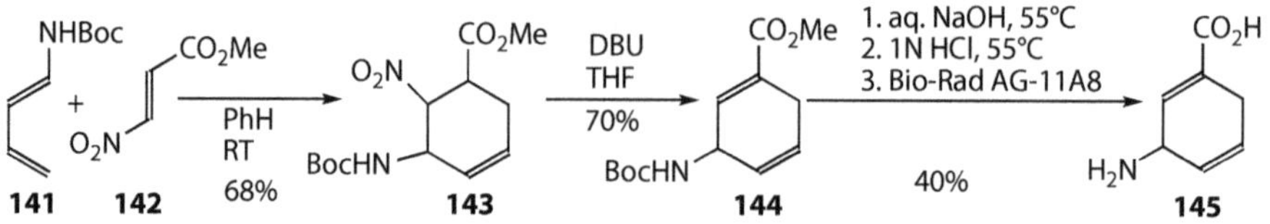

A virtually identical approach involved the reaction of amino butadiene ***146*** and ***147*** to give a 1:3 mixture of ***148:149***.[107] Amino-ester ***148*** is known as tilidine [ethyl 2-(*N*,*N*-diethylamino)-1-phenyl-3-cyclohexenecarboxylate], and it is a clinically used analgesic.[107d] Carbamate ***150*** was also used in a reaction with ***147***, and gave ethyl 2-(*N*-Cbz amino)-1-phenyl-3-cyclohexene-1-carboxylate (***151***), along with 20% of the isomeric ***152***.

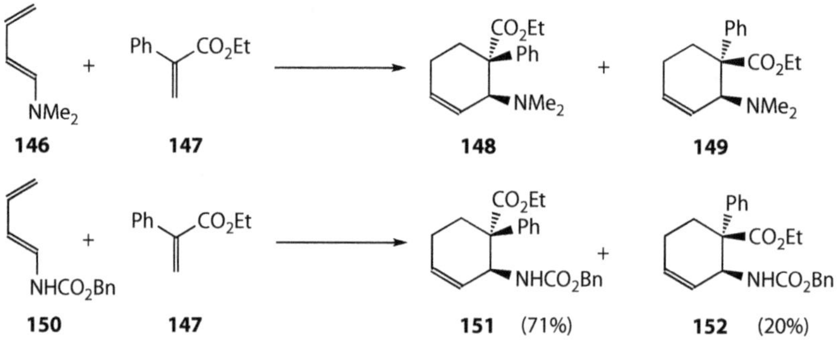

The preparation of four-membered ring amino acids often involves a thermal [2+2]-cycloaddition reaction.[108] The cycloaddition reaction of enamine ***153*** (derived from isobutyraldehyde) and methyl acrylate at 175°C, for example, gave methyl 2-(*N*,*N*-dimethylamino)-3,3-dimethylcyclobutane-1-carboxylate, ***154***.[109] Various

[104] Overman, L.E.; Taylor, G.F.; Petty, C.B.; Jessup., P.J. *J. Org. Chem.* ***1978***, *43*, 2164.

[105] McMurry, J.E.; Musser, J.H. *Org. Synth.* ***1977***, *56*, 65.

[106] Danishefsky, S.; Hershenson, F.M. *J. Org. Chem.* ***1979***, *44*, 1180.

[107] (a) Overman, L.E.; Petty, C.B.; Doedens, R.J. *J.Org. Chem.* ***1979***, *44*, 4183; (b) Warner-Lambert Pharmaceuticals Co. Br. Pat. 1,120,186 [*Chem. Abstr.* ***1968***, *69*: P96062v]; (c) Satzinger, G.; Herrmann, W.; Bovak, R.M. Ger. Offen. 1,923,619 [*Chem. Abstr.* ***1970***, *72*: P78530d]; (d) Satzinger, G. *J. L. Ann. Chem.* ***1969***, *728*, 64.

[108] See (a) Smith, M.B. *Organic Synthesis*, 3rd ed. Wavefunction, Inc./Elsevier, Irvine, CA/London, England, ***2010***, pp. 1076–1097; (b) Smith, M.B. *March's Advanced Organic Chemistry*, 7th ed. John Wiley & Sons, Hoboken, NJ, ***2013***, pp. 1040–1051.

[109] (a) Brannock, K.C.; Bell, A.; Burpitt, R.D.; Kelly, C.A. *J. Org. Chem.* ***1964***, *29*, 810; (b) Brannock, K.C.; Bell, A.; Burpitt, R.D.; Kelly, C.A. *J. Org. Chem.* ***1961***, *26*, 625; (c) Brannock, K.C. (Eastman Kodak Co.). *Ger. Pat.* 1,131,207 [*Chem. Abstr.* ***1962***, *57*: P13642d]; (d) Ficini, J.; Krief, A. *Tetrahedron Lett.* ***1970***, 885.

N,N-dialkylamino derivatives were prepared by using different enamines of isobutyraldehyde.

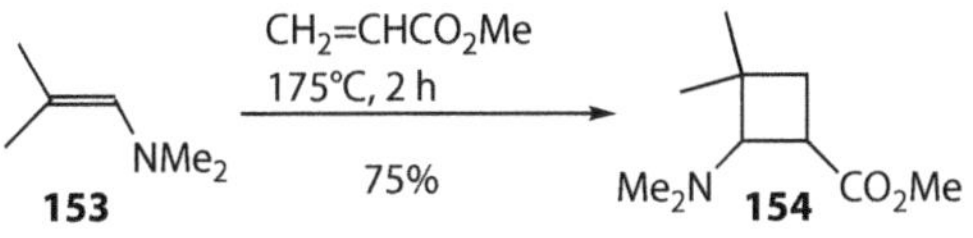

There are cycloaddition processes that do not necessarily involve pericyclic reactions. A three-component cycloaddition strategy was used to prepare functionalized aminocyclohexanecarboxylic acids. The reaction of ***155*** with ethyl 3-oxobutanoate and an amine such as butylamine, in the presence of ceric ammonium nitrate, led to ***156*** in 77% yield.[110] Subsequent reduction with an acyloxyborohydide gave ***157*** in 94% yield. Several derivatives were prepared by this method using different amines in the three-component coupling.

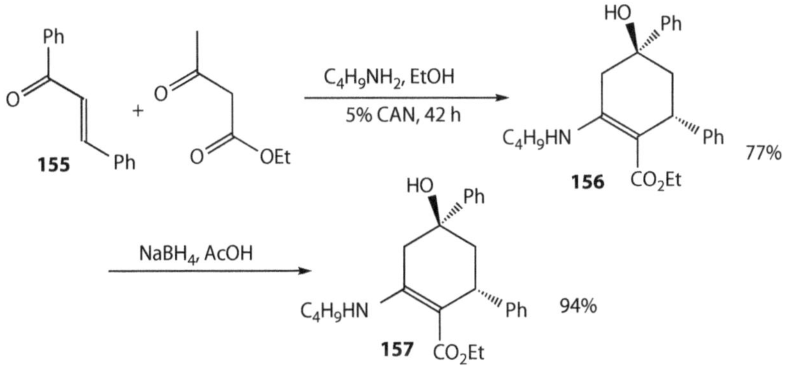

7.5 HETEROCYCLIC AMINO ACIDS

The compounds in this section are similar to the others in this chapter, but as heterocyclic compounds they will incorporate a nitrogen in the ring. Indeed, the main focus of this section involves amino acids with heterocyclic rings that contain nitrogen with a pendant carboxyl group. The final products are substituted aziridines, azetidines, pyrrolidines, piperidines, or hexahydroazepines. Many of the synthetic strategies are unique to this class of amino acids, and for that reason, they have been combined into a separate section. Such compounds are occasionally found in nature. Proline is an obvious example of an α-amino acid that is actually a pyrrolidinecarboxylic acid. Non-α-amino acids are also known. For example, there is a substantial concentration of l-azetidine-2-carboxylic acid in garden or table beets (*Beta vulgaris*).[111] It is also known that this amino acid is fed to farm animals in the form of sugar beet molasses, shredded sugar beet pulp, and pelleted sugar beet pulp. It

[110] Sridharan, V.; Menéndez, J.C. *Org. Lett.* ***2008***, *10*, 4303.

[111] Rubenstein, E.; Zhou, H.; Krasinska, K.M.; Chien, A.; Becker, C.H. *Phytochemistry* ***2006***, *67*, 898.

may be misincorporated into proteins in place of proline, in animals as well as in humans, and causes numerous toxic effects as well as congenital malformations.[112] Azetidine 3-carboxylic acid as well as (*S*)-(–)-4-oxo-2-azetidinecarboxylic acid are also known, and their x-ray crystallography structures have been reported.[113]

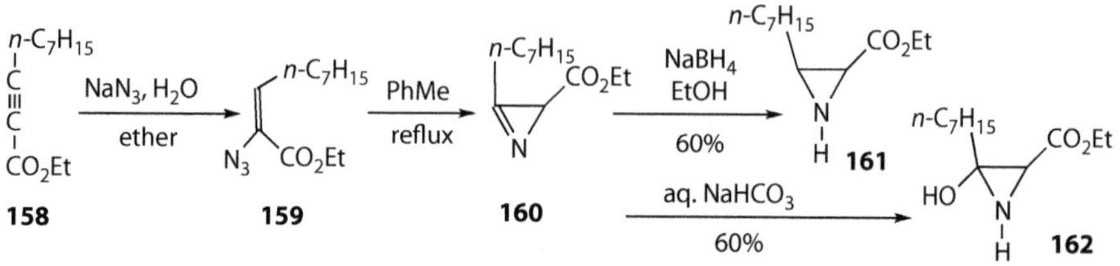

A synthesis of an aziridinecarboxylic acid derivative reacted alkyne ***158*** with sodium azide to give ***159***, and heating the resulting azide in toluene led to formation of ***160***.[114] Reduction of the azirine moiety in ***160*** with sodium borohydride gave ethyl 3-heptylaziridine-2-carboxylate, ***161***. Alternatively, hydrolysis with aqueous sodium bicarbonate led to ethyl 3-heptyl-3-hydroxyaziridine-2-carboxylate, ***162***. A *n*-C_5H_{11} derivative was also prepared from the appropriate alkyne. Note that derivatives such as ***162*** are formally α-amino acids, but they are included to open the discussion on this class of heterocyclic amino acids.

A more elaborate approach produced a chiral, nonracemic amino acid that was used in the synthesis of a statine derivative (see Chapter 6, Section 6.6). Azide ***164*** was prepared from ***163***, and reaction with 1,3-propanedithiol gave ***165***.[115] Conversion of the dithiane to the aldehyde and oxidation with pyridinium dichromate in *N,N'*-dimethylformamide (DMF) to the acid moiety gave ***166***, after esterification. The final sequence reduced the azide and generated the aziridine ring in ***167*** via an internal Mitsunobu reaction.[116,117]

[112] Rubenstein, E.; McLaughlin, T.; Winant, R.C.; Sanchez, A.; Eckart, M.; Krasinska, K.M.; Chien, A. *Phytochemistry* ***2009***, *70*, 100.

[113] Mora, A.J.; Fitch, A.N.; Brunelli, M.; Wright, J.; Báez, M.E.; López-Carrasquero, F. *Acta Crystallogr. Sect. B* ***2006***, *62*, 606.

[114] Haddach, M.; Pastor, R.; Riess, J.G. *Tetrahedron Lett.* ***1990***, *31*, 1989.

[115] Charkraborty, T.K.; Gangkhedkar, K.K. *Tetrahedron Lett.* ***1991***, *32*, 1897.

[116] Mitsunobu, O. *Synthesis* ***1981***, 1.

[117] See (a) Smith, M.B. *Organic Synthesis*, 3rd ed. Wavefunction, Inc./Elsevier, Irvine, CA/London, England, ***2010***, pp. 1213–126; (b) Smith, M.B. *March's Advanced Organic Chemistry*, 7th ed. John Wiley & Sons, Hoboken, NJ, ***2013***, pp. 469–470.

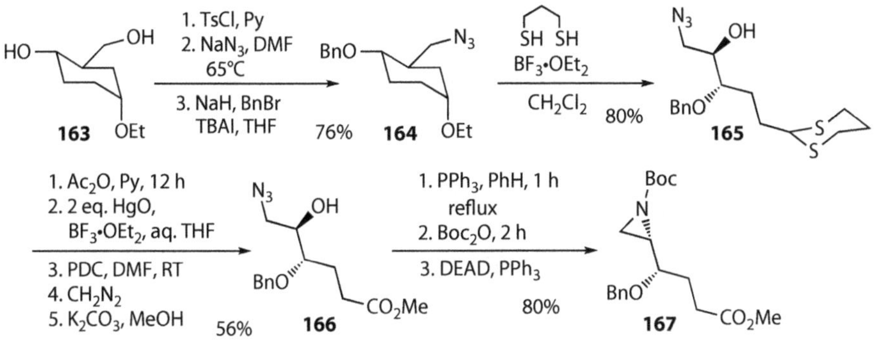

Aziridinecarboxylic acid derivatives are known in which the aziridine moiety is not positioned α- to the carboxylic acid moiety. 2-Alkyl-3,4-iminobutanoic acid derivatives (2-alkylaziridinecarboxylic acids) have been prepared from aspartic acid.[118] The enolate anion of the β-amino-ester of aspartic acid (see ***168***) was alkylated with benzyl bromide and provided ***169*** in 64% yield. Partial hydrolysis was followed by reprotection of the amino unit with Boc. Reduction of the acid moiety allowed conversion of the resulting alcohol to tosylate ***170***. Deprotection of the amino group was followed by cyclization to the aziridine (***171***). Aziridinecarboxylic acids such as this are useful synthons for the preparation of substituted β-amino acids such as ***172*** via catalytic hydrogenation.[118] Modification of this basic approach allowed the synthesis of all four stereoisomers of ***172***.

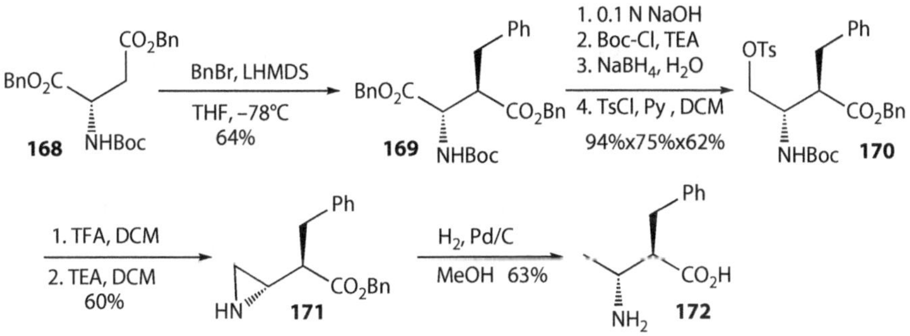

Azetidine amino acids are known. An example is the oxidation of ***173*** to give a ketone, and subsequent treatment with boron trifluoride led to conjugated ketone ***174***.[119] Conjugate addition of azide, reduction to the amine, and cyclization gave diastereomeric methyl 9-(3-hexyl-2-aziridino)nonanoic acid (***175*** and ***176***). The final step in that sequence proceeded in only 37% yield. If the alkenyl moiety was converted to an epoxide moiety, aziridinecarboxylic acids were prepared in good yield, via the azide.[119]

118 Park, J.; Tian, G.R.; Kim, D.H. *J. Org. Chem.* **2001**, *66*, 3696.
119 Lie Ken Jie, M.S.F.; Syed-Rahmatullah, M.S.K. *J. Am. Oil. Chem. Soc.* **1992**, *69*, 359.

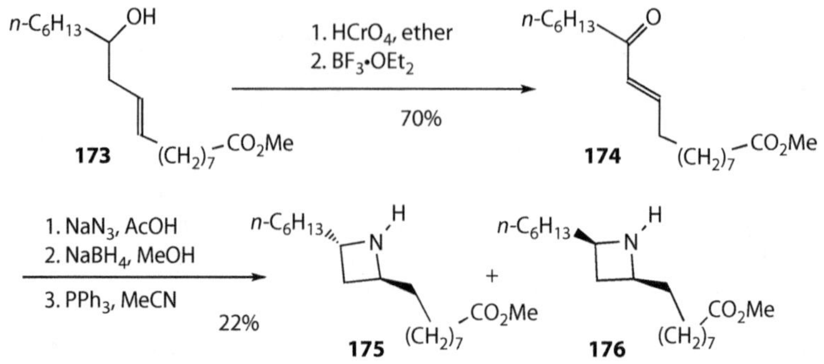

Several syntheses have appeared in which pyrrolidinecarboxylic acid, including proline derivatives, was prepared. One reason, of course, is the availability of proline and its derivatives. Proline was converted to the corresponding aldehyde ***177*** by standard procedures, and subsequent Wittig olefination[120] led to ***178***.[121] In an interesting transformation, the alkynyl moiety in ***178*** was converted to an acid moiety, giving 3-(2-*N*-Boc pyrrolidino)prop-2*E*-enoic acid (***179***).

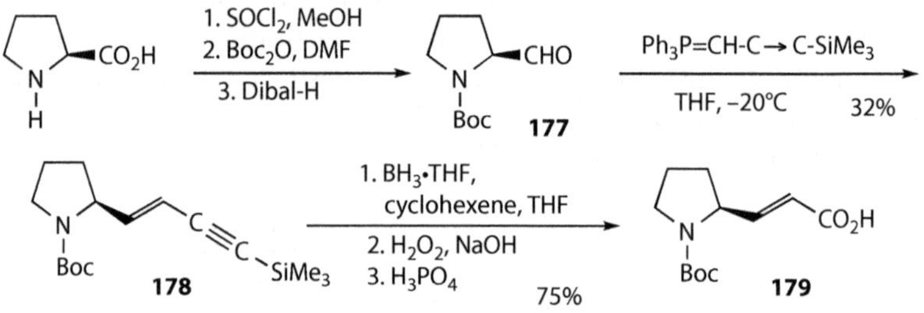

Other functionality can be incorporated by this approach, as in the condensation of ***177*** with ***180*** to give ***181*** (ethyl 2-methyl-3-hydroxy-3-(2-*N*-Boc pyrrolidino) propanoate,[122] and subsequent deprotection gave the pyrrolidine amino acid,[123] which is a component of the anticancer agent, the cyclic peptide dolastatin 10[124] (also see Chapter 6, Section 6.1.4).

---

[120] See (a) Smith, M.B. *Organic Synthesis*, 3rd ed. Wavefunction, Inc./Elsevier, Irvine, CA/London, England, ***2010***, pp. 729–739; (b) Smith, M.B. *March's Advanced Organic Chemistry*, 7th ed. John Wiley & Sons, Hoboken, NJ, ***2013***, pp. 1165–1173.

[121] Miles, N.J.; Sammes, P.G.; Kennewell, P.D.; Westwood, R. *J. Chem. Soc. Perkin Trans. I* ***1985***, 2299.

[122] Tomioka, K.; Kanai, M.; Koga, K. *Tetrahedron Lett.* ***1991***, *32*, 2395.

[123] See (a) Hanson, G.J.; Baran, J.S.; Lindberg, T. *Tetrahedron Lett.* ***1986***, *27*, 3577; (b) Hamada, Y.; Shioiri, T. *Chem. Pharm. Bull.* ***1982***, *30*, 1921.

[124] (a) Pettit, G.R.; Kamano, Y.; Herald, C.L.; Tuinman, A.A.; Boettner, F.E.; Kizu, H.; Schmidt, J.M.; Baczynskyj, L.; Tomer, K.B.; Bontems, R.J. *J. Am. Chem. Soc.* ***1987***, *109*, 6883; (b) Pettit, G.R.; Singh, S.B.; Hogan, F.; Lloyd-Williams, P.; Herald, D.L.; Burkett, D.D.; Clewlow, P.J. *J. Am. Chem. Soc.* ***1989***, *111*, 5463.

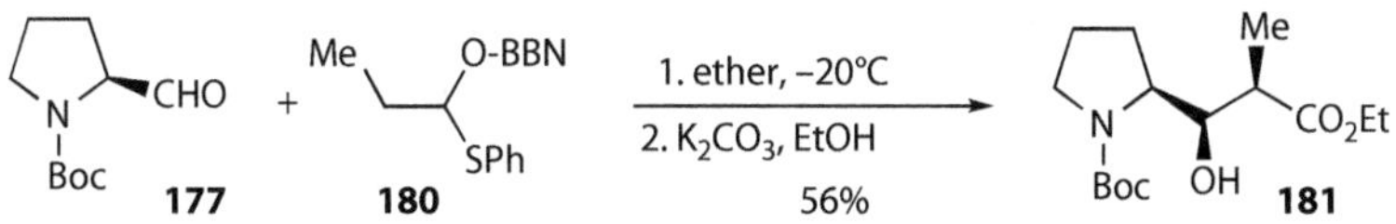

Conjugated alkynyl esters are useful precursors when a five-membered ring is formed via an internal Michael addition.[125] Reaction of ***182*** with methyl 6-aminohexanoate, for example, generated a transient acyclic derivative (***183***), which cyclized to give methyl 6-(2-carbomethoxymethylene-1-pyrrolidino)hexanoate, ***184***.[126] Several other *N*-substituted derivatives were also prepared.

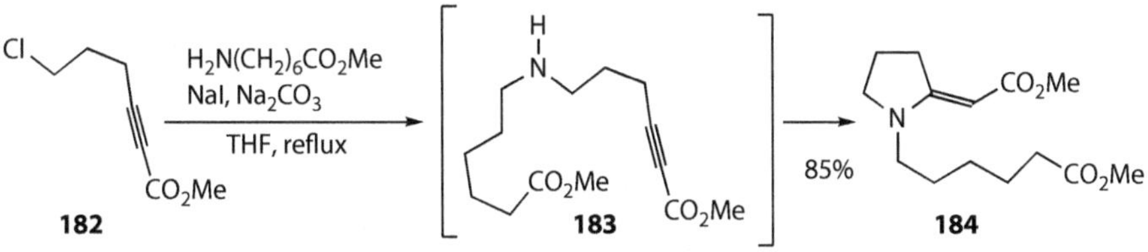

A very different route used an unusual lactone (***185***) in a Mitsunobu[116,117] type transesterification reaction with methanol to give ***186***.[127] Reduction of the ester to an aldehyde and Wittig olefination[120] gave ***187***.

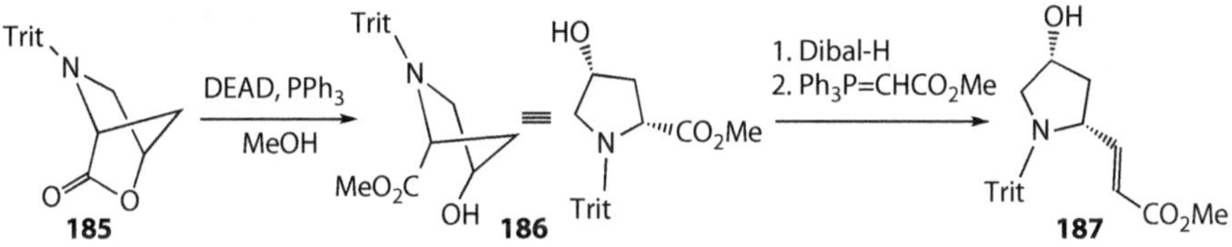

β-Lactams can be used to prepare pyrrolidinone derivatives. The alkenyl group in β-lactam ***188***, for example, was oxidatively cleaved to ***189***, allowing ring expansion[128] to ***190***.[129] Several 2-aryl derivatives and *N*-aryl derivatives were also prepared by this method.

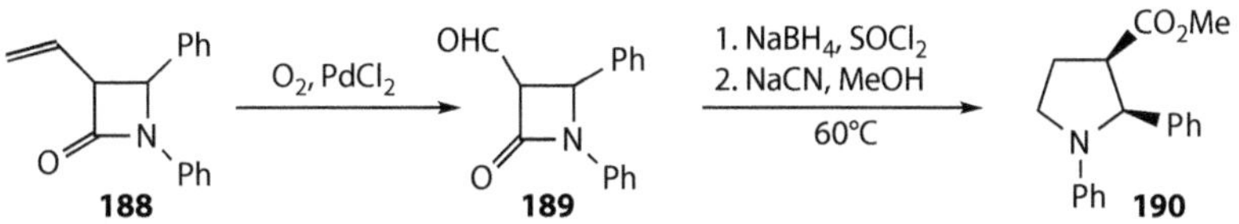

---

125 See Smith, M.B. *March's Advanced Organic Chemistry*, 7th ed. John Wiley & Sons, Hoboken, NJ, ***2013***, pp. 946–947.

126 Wilson, C.A., III; Bryson, T.A. *J. Org. Chem.* ***1975***, *40*, 800.

127 (a) Papaioannou, D.; Staavropoulos, G.; Sivas, M.; Barlos, K.; Francis, G.W.; Aksnes, D.W.; Maartmann-Moe, K. *Acta Chem. Scand.* ***1991***, *45*, 99; (b) Papaioannou, D.; Stavropoulos, G.; Karagiannis, K.; Francis, G.W.; Brekke, T.; Aksnes, D.W. *Acta Chem. Scand.* ***1990***, *44*, 243.

128 See Cavanga, F.; Linkies, A.; Pietsch, H.; Reuschling, D. *Angew. Chem. Int. Ed. Int.* ***1980***, *19*, 129.

129 Bose, A.K.; Krishnan, L.; Wagle, D.R.; Manhas, M.S. *Tetrahedron Lett.* ***1986***, *27*, 5955.

Another synthesis of heterocyclic amino acids used a ring expansion strategy. The six-membered ring in amino-ketone ***191***, prepared from the *N*-benzyl derivative,[130] was expanded by reaction with diazoacetate and rearrangement to give the seven-membered ring amido-ketone, ***192***. Reduction, elimination, and catalytic hydrogenation gave hexahydroazepine-4-carboxylic acid, ***193***. The latter stages of this synthesis were used in a closely related transformation in which a carbamoyl-protected amino-ketone (***194***)[131] was converted to hexahydroazepine-3-carboxylic acid (***195***).

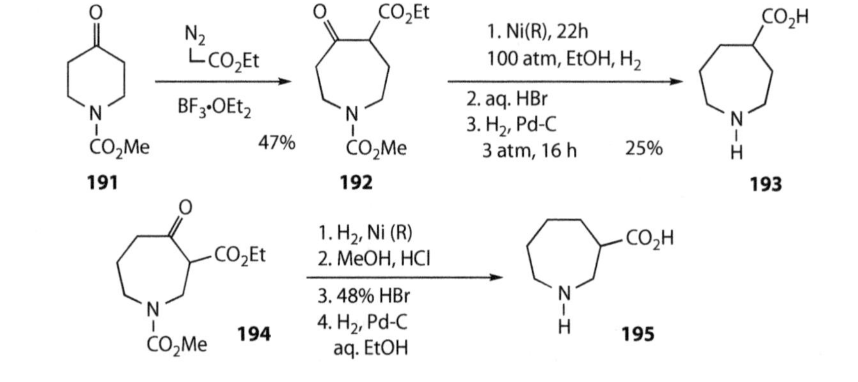

In a different approach that utilized heterocyclic intermediates, the amino alcohol methyl serinol was converted to oxazolidine ***196***, allowing manipulation of the hydroxymethyl group to give the conjugated ester moiety in ***197***.[132] Deprotection and cycloaddition led to ***198*** in good yield.

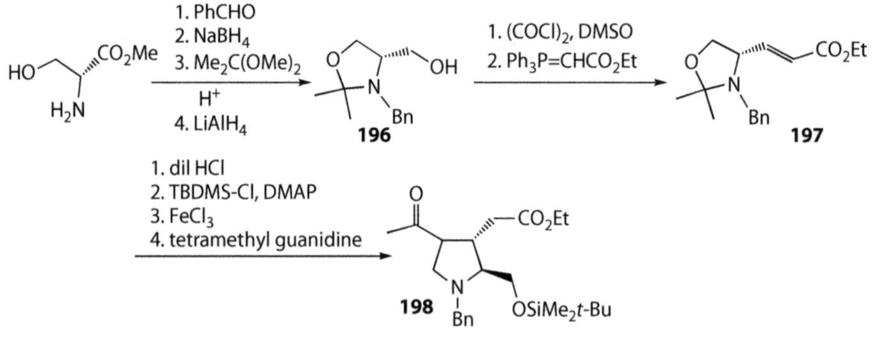

Another synthesis involved the preparation of a thiazole amino acid by an intramolecular acyl addition of the thiol moiety in ***199*** to give give ***200***, which was used

[130] Morosawa, S. *Bull. Soc. Chem. Jpn.* **1958**, *31*, 418.
[131] Krogsgaard-Larsen, P.; Hjeds, H. *Acta Chem. Scand.* **1976**, *B30*, 884.
[132] (a) Barco, A.; Benetti, S.; Spalluto, G.; Casolari, A.; Pollini, G.P.; Zanirato, V. *J. Org. Chem.* **1992**, *57*, 6279; (b) see Garner, P.; Min Park, J. *J. Org. Chem.* **1987**, *52*, 2361.

in Hecht's synthesis of bleomycin A_2.[133,134] This synthesis clearly does not involve a pyrrolidine moiety, which is the main target discussed in this section, but does contain a five-membered thiazole derivative, and so is included here. Specifically, treatment of dipeptide ***199*** with HCl gave thiazoline ***200***.[134] Subsequent oxidation with nickel oxide led to ethyl 2-(2-*N*-acetylaminoethyl)-thiazole-5-carboxylate, ***201***.

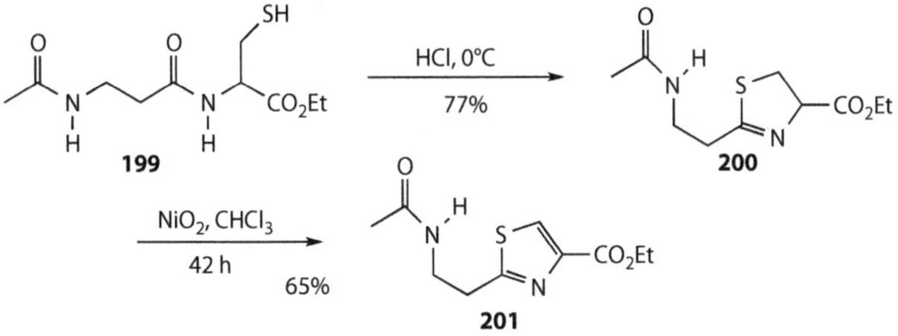

There are many examples of piperidinecarboxylic acids, prepared by a variety of methods. Cyclization reactions from properly functionalized acyclic precursors are an important generic route to cyclic amino acids. An example is the reduction of azide ***202*** to give the corresponding amine, and subsequent internal Michael addition[135] of the amine moiety with the conjugated ester gave methyl 2-(2-piperidino) ethanoate, ***203***.[136]

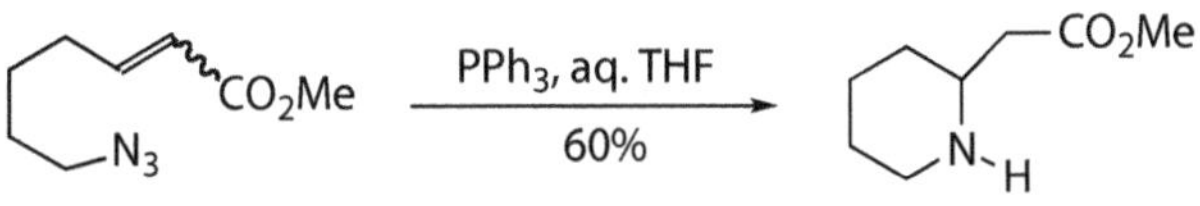

202 **203**

A different cyclization route involved the coupling of two ester moieties via a Dieckmann type condensation.[137] An initial Michael addition of ***204*** to ethyl acrylate was followed by protection of the amine as the *N*-Boc derivative in ***205***. Treatment

---

[133] (a) Ikekawa, T.; Iwami, F.; Hiranaka, H.; Umezawa, H. *J. Antibiot.* (Tokyo) ***1964***, *17A*, 194; (b) Umezawa, H.; Maeda, K.; Takeuchi, T.; Okami, Y. *J. Antibiot.* ***1966***, *19A*, 200; (c) Umezawa, H. Suhara, Y.; Takita, T.; Maeda, K. *J. Antibiot.* ***1966***, *19A*, 210.

[134] (a) McGowan, D.A.; Jordis, U.; Minster, D.K.; Hecht, S.M. *J. Am. Chem. Soc.* ***1977***, *99*, 8078; (b) Arai, H.; Hagman, W.K.; Suguna, H.; Hecht, S.M. *J. Am. Chem. Soc.* ***1980***, *102*, 6631; (c) Levin, M.D.; Subrahamanian, K.; Katz, H.; Smith, M.B.; Burlett, D.J.; Hecht, S.M. *J. Am. Chem. Soc.* ***1980***, *102*, 1452; (d) Hecht, S.M.; Rupprecht, K.M.; Jacobs, P.M. *J. Am. Chem. Soc.* ***1979***, *101*, 3982; (e) Ohgi, T.; Hecht, S.M. *J. Org. Chem.* ***1981***, *46*, 1232; (f) Pozsgay, V.; Ohgi, T.; Hecht, S.M. *J. Org. Chem.* ***1981***, *46*, 3761; (g) Aoyagi, Y.; Suguna, H.; Murugesan, N.; Ehrenfeld, G.M.; Chang, L.-H.; Ohgi, T.; Shekhani, M.S.; Kirkup, M.P.; Hecht, S.M. *J. Am. Chem. Soc.* ***1982***, *104*, 5237; (h) Aoyagi, Y.; Katano, K.; Suguna, H.; Primeau, J.; Chang, L.-H.; Hecht, S.M. *J. Am. Chem. Soc.* ***1982***, *104*, 5537.

[135] See Smith, M.B. *March's Advanced Organic Chemistry*, 7th ed. John Wiley & Sons, Hoboken, NJ, ***2013***, pp. 946–947.

[136] Knouzi, N.; Vaultier, M.; Toupet, L.; Carrie, R. *Tetrahedron Lett.* ***1987***, *28*, 1757.

[137] (a) Smith, M.B. *Organic Synthesis*, 3rd ed. Wavefunction, Inc./Elsevier, Irvine, CA/London, England, ***2010***, pp. 827–829; (b) Smith, M.B. *March's Advanced Organic Chemistry*, 7th ed. John Wiley & Sons, Hoboken, NJ, ***2013***, p. 1235.

with sodium led to Dieckmann cyclization, and subsequent reduction gave ***206***.[138] Other 6-alkyl and 6-aryl derivatives were prepared by the method.

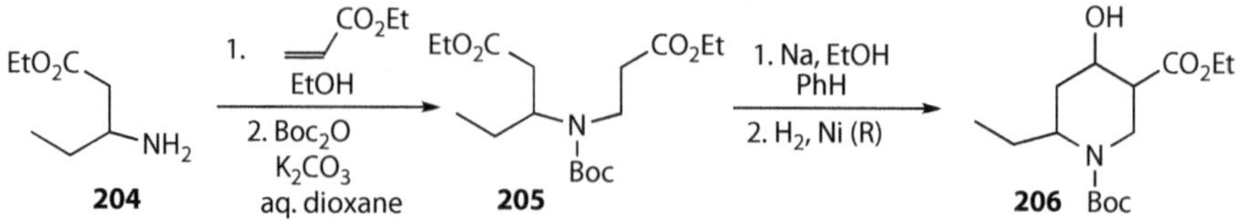

An interesting synthesis proceeded by shrinking a seven-membered ring, specifically α-bromocaprolactam (***207***) to the tetrahydropyridine derivative ***208***.[139] Both the five- and eight-membered ring analogs of ***208*** were prepared from α-bromopiperidone and α-bromoazacyclooctan-2-one, respectively.

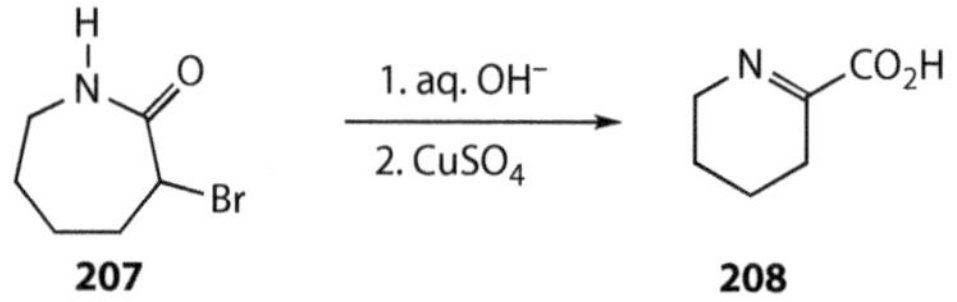

A piperidinecarboxylic acid derivative, ***209*** (known as NNC 05-1869), has been prepared, and shown to be active in animal models of diabetic neuropathy.[140] This work also prepared radiolabeled derivatives in order to study the efficacy of these compounds.

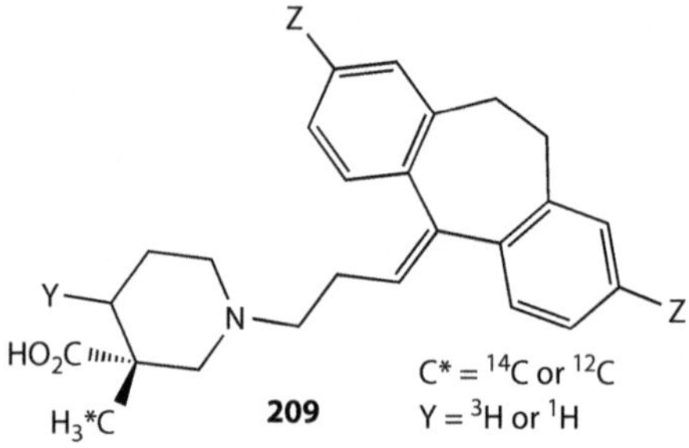

A series of piperidinylacetic acid and piperazinylacetic acid derivatives were prepared and shown to be potent VLA-4 antagonists (Very Late Antigen-4) are expressed by most leukocytes but observed on neutrophils only under special conditions). An example is morpholinyl-4-piperidinylacetic acid derivative ***210***, which showed activity in the *Ascaris* antigen-sensitized murine airway inflammation model by oral administration.[141]

[138] N'Goka, V.; Schlewer, G.; Linget, J.-M.; Chambon, J.-P.; Wermuth, C.-G. *J. Med. Chem.* ***1991***, *34*, 2547.

[139] Botta, A. *Ger. Offen.* 2,359,990 [*Chem. Abstr.* ***1975***, *83*: P97045t].

[140] Valsborg, J.S.; Foged, C. *J. Labelled Comp. Radiopharm.* ***2002***, *45*, 351.

[141] Chiba, J.; Machinaga, N.; Takashi, T.; Ejima, A.; Takayama, G.; Yokoyama, M.; Nakayama, A.; Baldwin, J.J.; McDonald, E.; Saionz, K.W.; Swanson, R.; Hussain, Z.; Wong, A. *Bioorg. Med. Chem. Lett.* ***2005***, *15*, 41.

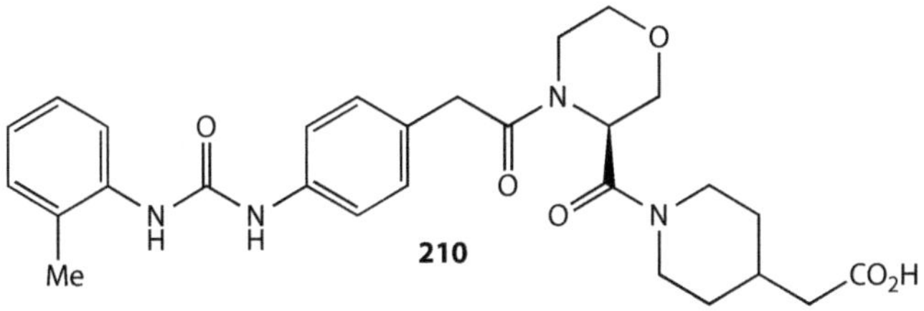

210

Several cyclic peptides have been prepared and shown to stabilize key loops of brain-derived neurotrophic factor, which is a member of the neurotrophin family that includes dimeric factors essential in the development and maintenance of central and peripheral nervous systems.[142] Two examples include ***211***, which contains a 2-amino-cyclohexanecarboxylic acid moiety, and ***212***, which contains an azetidine moiety.

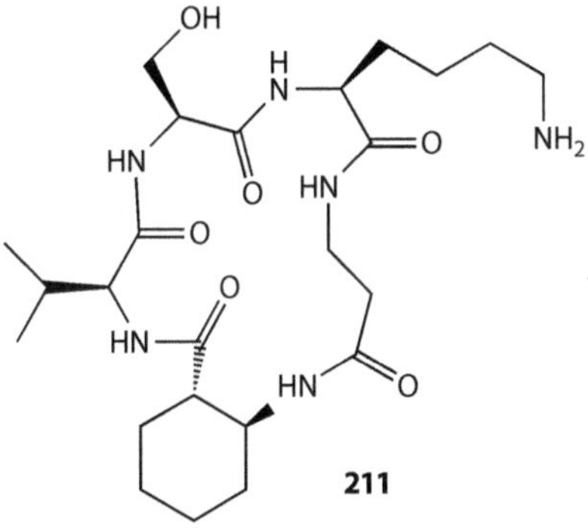

211

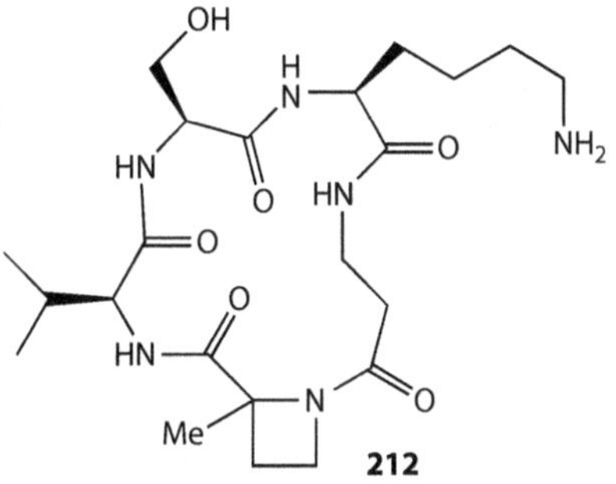

212

Several novel imidazoyl amino acids were prepared, including 1*H*-imidazol-4-ylacetic acid (***213***) from *N*-trityl 2-methylimidazole, ***214***.[143] Generation of the α-lithio derivative by reaction with butyllithium was followed by carboxylation with carbon dioxide and subsequent deprotection. The synthesis of other amino acid analogs involved alkylation of the lithio compound. These compounds were found to be GABA uptake inhibitors. In a different work, [2-{(4-chlorophenyl) (4-iodophenyl)} methoxyethyl]-1-piperidine-3-carboxylic acid, ***215***, was prepared as a radiotracer for GABA uptake sites.[144]

[142] Baeza, J.L.; de la Torre, B.G.; Santiveri, C.M.; Almeida, R.D.; García-López, M.T.; Gerona-Navarro, G.; Jaffrey, S.R.; Jiménez, M.A.; Andreu, D.; González-Muñiz, R.; Martín-Martínez, M. *Biorg. Med. Chem. Lett.* ***2012***, *22*, 444.

[143] Hack, S.; Wörlein, B.; Höfner, G.; Pabel, J.; Wanner, K.T. *Eur. J. Med. Chem.* ***2011***, *46*, 1483.

[144] Van Dort, M.E.; Gildersleeve, D.L.; Wieland, D.M. *J. Labelled Comp. Radiopharm.* ***1995***, *36*, 961.

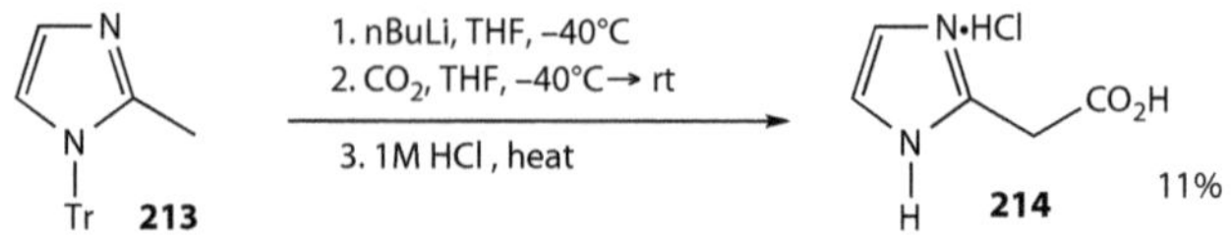

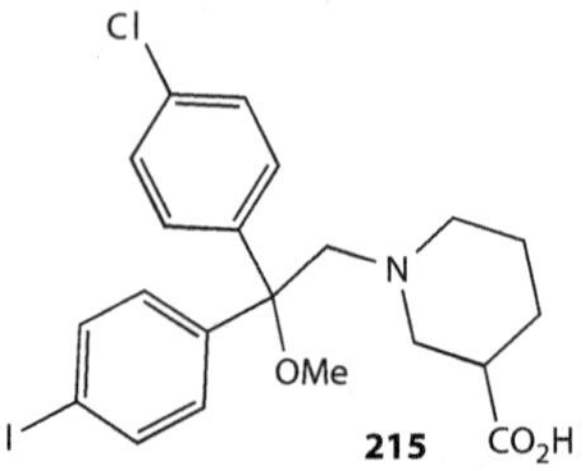

Eprosartan (***216***) selectively inhibits the action of angiotensin II on the AT_1 receptor (it is an antagonist) and is prescribed for treatment of hypertension. ^{11}C-labeled ***217*** was prepared as shown.[145] The structure is an amino acid, but the amine portion is the imidazole moiety. Protection of ***216*** was followed by *N*-benzylation to give ***217***. Oxidation of the alcohol to an aldehyde allowed a Knoevenagel type condensation[146] with ***217*** to give ***219***. Treatment with hydroxide was followed by the Pd-catalyzed carboxylation reaction using ^{11}C-labeled carbon dioxide to give ***220***.

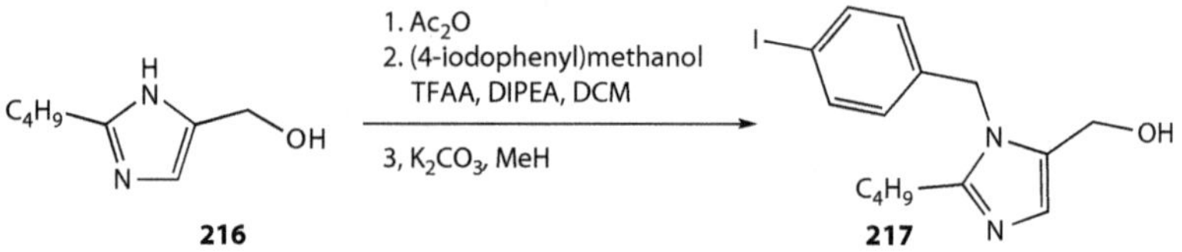

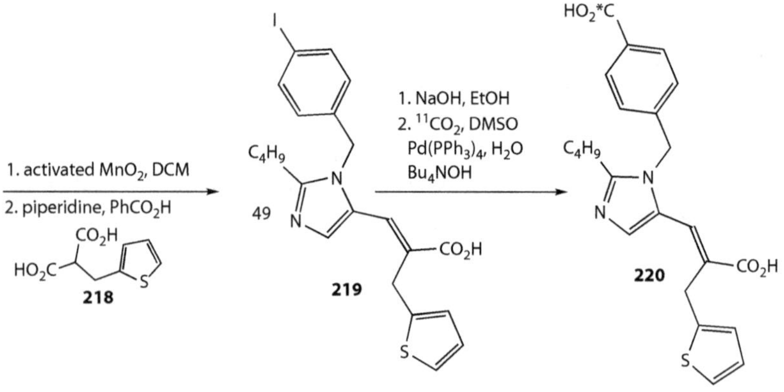

[145] Åberg, O.; Lindhe, Ö.; Hall, H.; Hellman, P.; Kihlberg, T.; Långström, B. *J. Labelled Comp. Radiopharm.* ***2009***, *52*, 295.

[146] (a) Japp, F.R.; Streatfeild, F.W. *J. Chem. Soc.* ***1883***, *43*, 27; (b) Knoevenagel, F. *Ber.* ***1896***, *29*, 172; (c) Knoevenagel, F. *Ber.* ***1898***, *31*, 730; (d) Jones, G. *Org. React.* ***1967***, *15*, 204; (e) see Smith, M.B. *Organic Synthesis*, 3rd ed. Wavefunction, Inc./Elsevier, Irvine, CA/London, England, ***2010***, p. 829; (f) Smith, M.B. *March's Advanced Organic Chemistry*, 7th ed. John Wiley & Sons, Hoboken, NJ, ***2013***, pp. 1157–1160.

Index

(**Named reactions are in boldface type**)

B

D

E

F

G

H

I

J

K

O

T

U

V

W

X

Y

Z

Milton Keynes UK
Ingram Content Group UK Ltd.
UKHW040103071024
449327UK00019B/786

9 780367 379346